Seed Dispersal

Seed Dispersal

Edited by
David R. Murray

Biology Department
The University of Wollongong
New South Wales

ACADEMIC PRESS

(Harcourt Brace Jovanovich, Publishers)
Sydney   Orlando   San Diego   New York   Austin
London   Montreal   Tokyo   Toronto

ACADEMIC PRESS AUSTRALIA
Centrecourt, 25–27 Paul Street North
North Ryde, N.S.W. 2113

United States Edition published by
ACADEMIC PRESS INC.
Orlando, Florida 32887

United Kingdom Edition published by
ACADEMIC PRESS, INC. (LONDON) LTD.
24/28 Oval Road, London NW1 7DX

Printed in Australia

National Library of Australia Cataloguing-in-Publication Data

Seed dispersal.

  Bibliography.
  Includes index.
  ISBN 0 12 511900 3.

  1. Seeds – Dispersal. I. Murray,
  David R. (David Ronald), date– .

582'.01'6

Library of Congress Catalog Card Number: 86-72353

CONTENTS

Contributors   ix

Foreword   xi

Preface   xiii

1. The Aerial Motion of Seeds, Fruits, Spores and Pollen   **1**

F. M. BURROWS

I.	Introduction	2
II.	Aerodynamic Forces and their Effects	4
III.	Trajectory Analysis	13
IV.	Launching and Release Mechanisms	18
V.	Types of Trajectory: Solutions of Equations of Motion	24
VI.	Concluding Remarks	46
	References	46

2. Seed Dispersal by Water   **49**

DAVID R. MURRAY

I.	Introduction	49
II.	Seed Dispersal by Rain, River and Flood	50
III.	Evidence for Seed Dispersal by Ocean Currents	53
IV.	Effective Seed Dispersal by Ocean Currents Compared to Other Vectors	74
V.	The Hazards for Seeding Establishment	78
VI.	Conclusions	80
	Acknowledgements	81
	References	82

3. Seed Dispersal Syndromes in Australian *Acacia*   **87**

DENNIS J. O'DOWD and A. MALCOLM GILL

I.	Introduction	87
II.	Definitions and Limitations	88
III.	Inference of Dispersal Syndromes	89

IV.	Principal Components Analysis	91
V.	Seed Dispersal Syndromes	101
VI.	Characteristics of Seed Dispersal Agents	108
VII.	Comparison of *Acacia* with other Ant- and Bird-dispersed species	111
VIII.	Ecological Consequences of Seed Dispersal	111
IX.	Evolutionary Derivation of Dispersal Syndromes	113
X.	Comparison with American and African *Acacia*	116
XI.	Conclusions	117
	Acknowledgements	118
	References	118

4. Seed Dispersal by Fruit-Eating Birds and Mammals   123

HENRY F. HOWE

I.	Introduction	123
II.	Definitions	126
III.	Foraging for Fruits	128
IV.	Attracting Dispersal Agents	150
V.	Population Ecology of Seed Dispersal	162
VI.	Community Effects	172
VII.	Summary and Conclusions	181
	Acknowledgements	182
	References	183

5. Rodents as Seed Consumers and Dispersers   191

M. V. PRICE and S. H. JENKINS

I.	Introduction	191
II.	Granivorous rodents and the Fates of Seeds	192
III.	Determinants of the Fate Path	197
IV.	Net Effect of Granivorous Rodents	221
V.	Directions for Further Research	230
	Acknowledgements	231
	References	231

6. Seed Dispersal in Relation to Fire   237

R. J. WHELAN

I.	Introduction	237
II.	Fire as a Dispersal Vector	238
III.	Post-Fire Conditions	239
IV.	Long-Distance Dispersal	244
V.	Local Dispersal	245
VI.	General Theories	258
VII.	Summary	264
	Acknowledgements	265
	References	265

7. Evolution of Seed Dispersal Syndromes According to the Fossil Record   **273**

B. H. TIFFNEY

I.	Introduction	274
II.	Materials and Methods	275
III.	The Fossil Record — Data	281
IV.	Discussion	289
V.	Summary	298
	Acknowledgements	300
	Text References	301
	Data References	303

Species Index   **307**

Subject Index   **318**

CONTRIBUTORS

F. M. BURROWS (1) Department of Mathematics, Westminster, Dean's Yard, London SW1P 3PB, England.

A. MALCOLM GILL (87) Division of Plant Industry, Commonwealth Scientific and Industrial Research Organization, P.O. Box 1600, Canberra, A.C.T. 2601, Australia.

HENRY F. HOWE (123) Program in Evolutionary Ecology and Behavior, Department of Biology, University of Iowa, Iowa City, Iowa 52242, U.S.A.

S. H. JENKINS (191) Department of Biology, College of Arts and Science, University of Nevada Reno, Reno, Nevada 89557, U.S.A.

DAVID R. MURRAY (49) Department of Biology, University of Wollongong, New South Wales 2500, Australia.

DENNIS J. O'DOWD (87) Departments of Zoology and Botany, Monash University, Clayton, Victoria 3168, Australia.

M. V. PRICE (191) Department of Biology, University of California, Riverside, California, 92521, U.S.A.

B. H. TIFFNEY (273) Osborn Memorial Laboratories, Yale University, P.O. Box 6666, New Haven, Connecticut, 06511, U.S.A.

R. J. WHELAN (237) Department of Biology, University of Wollongong, New South Wales 2500, Australia.

FOREWORD

The means by which plants are able to disperse provide a fascinating and profitable study. The acquisition of space or territory, remote from parental interference, is often necessary for full development and future propagation of the species. The dispersal of individuals is indeed central to the survival and evolution of plant and animal species alike.

Whilst dispersal may bring benefits for a given plant species, it may irretrievably alter the balance of growth and food supply of the host community. The awe-inspiring spread of pricklypear cactus (*Opuntia fragilis*) in sub-tropical eastern Australia is such an example. The 'escape' of plant species from those organisms that normally consume them is no doubt one good reason for the enhanced performance of many plant species now grown in lands distant from their origin. Dispersal thus allows a change in the ecological standing of the individual, the population, or the species, depending on the scale of the perspective. A knowledge of the means of plant dispersal is therefore vital to the study of ecology, plant geography and the applied pursuits of agriculture, forestry and horticulture.

The ways in which plants have acquired the characteristics necessary for easy travel have been the subjects of much theoretical and empirical research. Although it is tempting to apply causative roles to such features, it is dangerous to ignore the possibilities of chance and the occurrence of the improbable factor. It is certain that local distributions in the short term can be largely accounted for by specific dispersal features and the dispersal vectors available. The cooperation of mistletoe and mistletoe bird in Australia, as described in the book edited by D. M. Calder and P. Bernhardt (Academic Press, Sydney, 1983), is an excellent example. However, it is much less certain that specific dispersal features are important when very long distances are involved. Minuscule chances cannot be ignored if unimaginably long intervals of millions of years are taken into account.

Today, mankind is the supreme mixer of species around the world. The establishment of gardens, urbanization, agriculture and plantation forestry have all ensured mixture of species in a geographic sense. Hybrids have also been developed, and the ruderal elements of the plant kingdom have been encouraged. Weeds are entering Australia at a fairly steady rate of five or six per year, in spite of quarantine. Such introductions may be deliberate or accidental. Fortunately many weed species remain quiescent or local, and do not become rampant invaders.

Indigenous plants are still readjusting their distributions in response to a variety of factors. For example, *Pittosporum undulatum* in Victoria is able to grow and reproduce vigorously 200 km west of its apparent boundary; the introduced European blackbird is all that was necessary for this species to extend its distribution. In southern England, the limits for *Fagus sylvaticus* have not been reached. These are neither climatic nor edaphic, and the eventual distribution of this species is simply a function of time.

The intense interest of 19th century plant geographers led directly to questions of plant dispersal. Hooker's famous introductory essay to the Flora of Tasmania envisaged land bridges around the Southern Hemisphere in congruity with the situation in the Northern Hemisphere. However, since that time, wide disjunctions of whole floral assemblages have been better interpreted in the light of plate tectonics. It is clear that there are many possibilities to account for plant geographic problems: long distance dispersal of species does occur; short distance dispersal continues to modify boundaries and community composition.

In 1930 much information was condensed by Ridley in his book "Dispersal of Plants Throughout the World", but since that time the thrust of botanical research has taken many exciting directions in the wake of technological advances. It is common for research to be cyclic. The problems and concerns about plant dispersal have not gone away, but after 50 years they are able to be tackled with new insights and new technology. The wealth of reference material provided in this sound, multifaceted book should catalyse future research in this important interface of many different and challenging disciplines.

DAVID H. ASHTON
Reader in Botany,
The University of Melbourne.

Angiosperms first appeared in the fossil record about 130 million years ago. The continents that we know today were then *en route* to their present locations, and indeed they are still moving. The evolution of angiosperms during this time has involved interaction with vectors that facilitate the processes of fertilization and seed dispersal (B. H. Tiffney, Chapter 7). As a result, the angiosperms have been supremely effective, frequently displacing gymnosperms and lower plants. Certain groups of gymnosperms have been able to interact with biotic vectors that were supported initially by angiosperms or ancestral gymnosperms, and these are also considered within the scope of this book.

Seed dispersal is one of the central topics in modern biology, with implications for ecology, animal behaviour, plant and animal biogeography, speciation and evolution. The various agents of seed dispersal and the consequences of their action are the subject matter of this book. Wind and water, the first abiotic means of dispersal, are considered first. F. M. Burrows (Chapter 1) describes the movement of spores, fruits and seeds in air currents. His chapter will remain the definitive mathematical account of this topic for many years to come. D. R. Murray (Chapter 2) considers the evidence that viable seeds have been carried long distances by ocean currents. The utility of biochemical methods for establishing phylogenetic affinities among plant taxa from widely separated localities is illustrated, and the potential of such methods is emphasized.

D. J. O'Dowd and A. M. Gill (Chapter 3) present evidence for the development of two major dispersal syndromes among Australian species of *Acacia* — the largest of the four world groups belonging to this important genus. They reach the surprising conclusion that seed dispersal by ants has arisen more recently than seed dispersal by birds. Ant-dispersal is not known in other world groups of *Acacia*, although ant-dispersal is well known for other plant families in Europe and America. Ants are attracted to the oil-rich elaiosome (swollen funicle), which is generally eaten by the ants after the seeds have been transported into the nest. Seeds minus their elaiosomes can be carried out of the nest, or left inside — perhaps in an ideal site for later germination.

H. F. Howe (Chapter 4) provides a comprehensive account of seed dispersal by fruit-eating birds and mammals. As well as identifying and describing the attractive features of fruits and seeds dispersed by each group, he discusses the consequences of seed dispersal for plants whose fruits attract each type of vector, and questions how closely the term 'co-evolution' fits into our current conceptual framework. Similar themes are taken up by M. V. Price and S. Jenkins (Chapter 5), who consider the consequences of seed dispersal by rodents, a highly specialized and numerous group of mammals. If there were no benefit to the plants producing seeds gathered and stored by rodents, it is difficult to imagine that the plants could have survived so effectively up to the present. This problem is considered quantitatively, in terms of a 'fitness' or 'fate-path' diagram.

R. J. Whelan (Chapter 6) describes a situation of major concern in many parts of Australia and other fire-prone areas of the world — the dispersal of seeds in relation to fire. He points out how complex the re-development of vegetation in burned sites is, with seeds already available from the 'banks' of those stored in the soil, from the fruits of plants destroyed by fire, or introduced by any or all of the biotic and abiotic vectors considered elsewhere in this book.

A major agency of seed dispersal has not been considered here in any detail, and this is human activity, particularly in agriculture and transport. To illustrate the latter, I am told that the mud washed from the undersurfaces of motor vehicles and from their wheels is a rich source of weed seeds. This is obviously a problem that those charged with the management of crops, National Parks and wilderness areas must deal with, but plant quarantine and its implementation are outside the scope of this book.

It is a pleasure to record my thanks to the contributing authors, to everyone who supplied information and material for inclusion in the book, and to copyright owners, for permission to reproduce items already published. I also thank Mr J. B. Murray of the University of Wollongong for drawing several figures, and the staff of Academic Press for their diligence and co-operation in publishing this book.

CHAPTER **1**

The Aerial Motion of Seeds, Fruits, Spores and Pollen

F. M. BURROWS

I.	Introduction	2
II.	Aerodynamic forces and their effects	4
	A. The Reynolds number	4
	B. Flows about various shapes: lift, drag, and moments	4
	C. Analytical treatment of air and fluid flows	8
	D. Some geometric shapes and associated flows	10
III.	Trajectory analysis	13
	A. Choice of co-ordinate systems	13
	B. Equations of motion	13
IV.	Launching and release mechanisms	18
	A. Passive release	18
	B. Dynamic release	19
	C. Forced separation	21
	D. Ejection	22
	E. Syphoning	24
V.	Types of trajectory; solutions of equations of motion	24
	A. Aerodynamic classification of seed and fruit groups	24
	B. Aerodynamic coefficients for particles	25
	C. Dust seeds, spores and pollen	27
	D. Variously dense particles	32
	E. Plumed seeds and fruits	35
	F. Plane winged seeds	39
	G. Winged seeds and fruits which rotate when falling	42
VI.	Concluding remarks.	46
	References	46

1

SEED DISPERSAL
ISBN 0 12 511900 3

I. INTRODUCTION

The variety of geometric shapes of seeds, fruits, spores and pollen in nature is probably as extensive as in the multitude of other steady state and evolving objects in the universe. This very diversity makes the prediction and analysis of seed and fruit dispersal by aerial flight difficult, even when the environmental wind motion can be specified accurately. In the aerodynamic sense, seeds and fruits are simply objects of differing mass and geometry and it is possible to classify them under fairly general aerodynamic as distinct from botanical features, but departures of detail in the geometry of individual species are more often than not sufficient to suggest that it may be more appropriate to discuss statistical mean shapes than actual ones. When such shapes are identifiable composites of the kinds of elementary ones normally dealt with in aerodynamics, it is possible to determine analytically at least some of the forces that are developed on them in aerial motion, but when they are not identifiable composites, it becomes necessary to consider alternative representative shapes. Although the degree of geometric simplification required to achieve this aim may be considerable, such a procedure is perhaps the only constructive one that can be followed in some cases.

Each seed or fruit that is separated from its parent plant at some height above the ground describes a trajectory of some sort between release and repose. If the air is quite still, and there is no motion of the release point or velocity of release relative to the ground at the time of release, then the geometry of the trajectory will be of the simplest kind and depend directly and entirely on the aerodynamic characteristics of the species. Different trajectories will occur if the wind is blowing, or if the seed has an initial velocity at the time of release (or both), because aerial flight will then either begin with lateral motion or will acquire it during descent to the ground, irrespective of seed shape.

Although a seed that reaches the ground may sometimes be dragged along the ground or whisked up into the air again, particularly in the case of the plumed varieties, it is the initial or primary trajectory (Burrows, 1973) which is of principal interest. For the primary trajectory the height and type of location of parent plants are of major significance. A plant with a flowering head elevated in isolation on a flat and horizontal plane field can readily take advantage of the wind for dispersal, whereas a similar plant positioned in the midst of a close-packed stand of similar species on a steep hillside is unlikely to be afforded the same opportunity. The likely consequences of such differences are not discussed here, although the distinction is an important one. The reason for not discussing them is the basic difficulty of specifying any natural local wind environment accurately and consistently in the aerodynamic sense, although it may be possible to do this in regions where a reasonable degree of natural or enforced control and uniformity exists for sufficiently long periods. Natural wind motion is rarely steady in time and wind turbulence is normally described in terms of random fluctuations in velocity about an otherwise mean flow state, which is, to a greater or lesser extent, absolutely steady in time at a given point in space. Such air movement leads to fundamental difficulties of observation because the flow stream lines, the paths followed by individual particles of air, and the streak lines delineated by a succession of particles at a particular instant in time following release from a single point in space at earlier times, are all different from one another (Burrows, 1975b; Prandtl and Tietjens, 1957). It is only when the flow is absolutely steady in time that these three flow pattern lines become one and the same, and such as to permit meaningful interpretations to be made of the velocities indicated by marker particles. The relative scales of the flow and the body moving within it are important and turbulence on a micro-scale likely to affect the movement of a pollen grain is hardly likely to affect the movement of a walnut. Conversely, large eddying flows on the scale of the latter may have the proportions of a local environment in comparison to the scale of a pollen grain.

There is no hope of discussing any one of the main components of plant, seed, and wind as a principal dispersal agent in such a way that would guarantee the accuracy of any trajectory shape and dispersal range prediction in other than idealized circumstances. What can be done is to provide, throught the medium of formal analysis, reliable indications of perhaps considerable qualitative, and in some cases quantitative, value of the kind of salient feature pertinent to the flight styles employed and the trajectories followed by different species. By beginning with a basic discussion of some of the principal aerodynamic features involved in such aerial motion, the individual contribution and significance of component terms within the equations of motion can be understood (Section II, III). This in turn defines a need to set out an acceptable and concise breakdown of the enormous range of natural geometric configurations into fairly well defined groups, distinguished on the basis of aerodynamic rather than botanical features. Once suitable characteristic dimensions are selected, the equations of motion can be expressed in a non-dimensional form so as to provide the widest possible range of application. The subsequent quest for solutions leads on naturally to a consideration of the initial conditions for flight (Section IV), and the implications for this of the type of release mechanism employed by particular species. Last, the type of trajectory followed and the kind of range to be expected is discussed (Section V).

II. AERODYNAMIC FORCES AND THEIR EFFECTS

A. The Reynolds number

The kind of aerodynamic force developed on a body such as a seed or fruit having a characteristic length l, moving with speed V_p, relative to the air of kinematic viscosity γ, depends very much on the Reynolds number, $R_e = V_p l / \gamma$. This non-dimensional quantity is directly proportional to the ratio of the dynamic or intertia forces to those arising on account of viscosity. Flows about two geometrically similar bodies of different physical size are identical dynamically if their Reynolds numbers are equal. The flow pattern and the pressures associated with it may change markedly with change in Reynolds number for some geometric shapes and in most cases it is usual to account for these effects by empirical means.

When flight takes place at Reynolds numbers less than unity the viscous forces generated by the flow are the dominant ones and, other than for some perfectly symmetrical body shapes developing only a single force which acts so as to resist the relative motion, it is not easy to evaluate the forces and the stability of motion in free flight, either theoretically or experimentally. Except for smooth spherical particles (Stokes, 1851; Burrows, 1983), there is almost a complete lack of experimental evidence that can be used as a guide to the sort of magnitude and direction of any forces which may be developed on bodies due to viscous flow about them. In contrast to this, a substantial amount of theoretical and empirical information exists for a wide variety of body shapes moving at larger Reynolds numbers. This indicates that, in general terms, the forces developed depend to a greater or lesser extent on the geometry of the wetted surfaces and the pressures induced upon them.

B. Flows about various shapes: lift, drag, and moments

For cases with larger Reynolds numbers, the immediate effects of viscosity are usually confined to a thin boundary-layer extending over the entire surface of a body and, as long as the flow does not separate significantly, those parts of the body profile having convex curvature in the plane of the flow usually tend to induce faster local flow speeds than exist on flatter parts of the surface and on regions having concave curvature. In relation to the speed and pressure of the undisturbed stream well away from the body, the pressures on the body surface are least where the flow speeds are greatest and greatest where the flow speeds are least. The integral effect is to produce a resultant force, which acts at the centre of pressure, together with a moment which tends to change the orientation of the body to some preferred alignment with the flow. Because of the dynamic significance of the contributions they make, it is usual to discuss the resultant force in terms of a lift component, L, which acts along the normal to the undisturbed stream, and a drag component, D, which acts along the direction of the stream. The lift component is developed entirely because of a differential distribution of pressure and does not exist for bodies having complete flow symmetry with respect to the flow direction well away from them. Drag is always present when there is any motion at all relative to the air, for it is caused also by pressure and friction, both of which exist in any case. The combinations of forces and moments produced on a given body may tend to produce rotation or to maximize resistance to motion, as is the case for the elliptical section shown in Figure 1A and the flat plate shown in Figure 1E, F, whilst others, like the symmetrical wing section shown in Figure 1G and the Von Mises wing profile (Milne-Thomson, 1952) shown in Figure 1H, can be self-stabilizing in attitudes associated with the production of a finite lift, provided that in each case these bodies are either suspended at correctly chosen points or have mass distributions that, in given circumstances, are dynamically equivalent to this.

Except for the last two rather special cases, bodies lacking complete spherical symmetry are most unlikely to have attitude stability in free flight, and because of this, it is difficult to determine their motion in a free fall under gravity and in winds. This is so, even in the case of the downward movement of a plane rectangular wing, of moderate to large aspect ratio and of either simple flat or elliptic cross-section (Fig. 1B, D-F) when released from rest. Such a wing is capable of developing and sustaining quite substantial suction pressures near its leading edge, and if the mass of the wing and its moment of inertia about an axis along the span and passing through the centre of mass of the wing as a whole are both relatively small, then very high axial rotational speeds can be developed with the directions indicated in Figure 1 (B, D-F) in a descent under gravity. The two constant total energy trajectories calculated following the methods given by Lamb (1952) for an ellipse moving in frictionless flow and shown as examples in Figure 2 demonstrate the changes of orientation that may occur with position along a flight path. For real fluid flows where friction is present a significant fraction of the total mechanical energy available is required to sustain the rotary motion so that the net forward movement achieved is not as great as that normally possible in a symmetric planing motion in a stable glide. Moreover, because of the almost inevitable presence of some kind of asymmetry in the flow, such rotational motion is usually accompanied by a considerable amount of lateral movement. The transverse

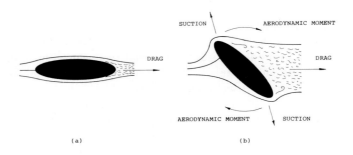

(a)

(b)

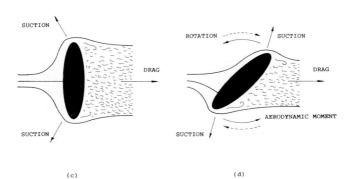

(c)

(d)

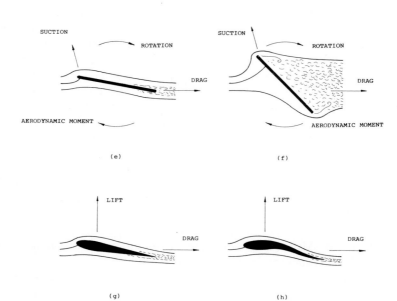

(e)

(f)

(g)

(h)

flows induced by such sideslipping cause the distribution of pressure, and consequently the loading, along the span of any wing-like body to become markedly different from that for symmetric flight and difficult to evaluate. With steady flow symmetry the span loading may be evaluated in principle and for idealized circumstances using any one of a number of available methods (Robinson and Laurmann, 1956).

Transverse flows are always present to a greater or lesser extent near the lateral extremities of bodies possessing anything but perfect and aligned cylindrical or spherical symmetry. When these flows are augmented on one side and diminished on the opposite one by sideslip, either the relatively stable phenomenon of post-stall autorotation occurs (Duncan, 1952; Durand, 1963), or much less predictable kinds of rotation occur, involving toppling and oscillations of appreciable amplitude about a nominal mean attitude of the body. For either case the prediction of free flight motion is difficult because the behaviour of the flow as a whole depends critically on the hysteretic development of the boundary-layer and the amount and varied nature of the flow separation from the otherwise wetted surfaces that is likely to occur.

In contrast to this, the aerial motion of winged bodies resulting from stable pre-stall autorotation (Shapiro, 1955) is one for which the aerodynamic characteristics are, to some extent, quasi-static. For such cases meaningful calculations may be made for a restricted class of trajectories without too much difficulty. The restriction arises on account of the implied stability of motion for, when this is the case, the steady state values of lift, drag, and moment for equilibrium can be derived using a well chosen model. Of the two forces, lift is perhaps the more difficult to determine but the amount of drag developed may be estimated, at least for the higher Reynolds numbers, from the large amount of information assembled by Hoerner (1958). Although it is generally a still more difficult matter to estimate the magnitude of the aerodynamic moments which may be acting, this is of lesser importance for a trajectory calculation for a body moving in a stable configuration, in which the moments must, in any case, be in equilibrium.

In the absence of any form of propulsion, a body moving freely in a gravitational field is always descending relative to the air at every moment of its aerial flight, even though it may appear to be rising away from the ground. Such an illusion is caused when the air moves upwards faster than

◁ Fig. 1. Two-dimensional flows. (A–D) Body of elliptic cross-section in a uniform steady wind. (A) Position of unstable equilibrium. (B) Origin of aerodynamic moment. (C) Preferred position of orientation. (D) Rotation (broken arrows) against the stabilizing aerodynamic moment. (E–F) Flat plate section in a uniform steady wind. (G) Symmetrical wing section in possible stable lifting attitude. (H) Von Mises wing section with reflex camber in possible stable lifting attitude.

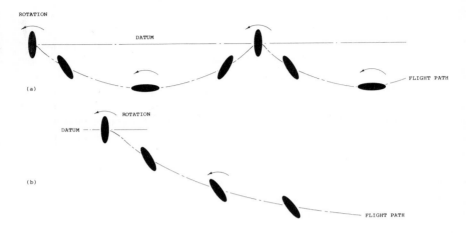

Fig. 2. Constant total energy trajectories for an elliptic section in two-dimensional inviscid flow. (A) Flight path returning periodically to horizontal trajectory datum. (B) Flight path descending ultimately to lower level asymptote.

the body can descend in still air and when sufficient energy required to create the required convective flow is made available by the environment. The relative rate of descent depends on the amount of lift and drag produced, and when complete attitude stability exists in one form or another, the downward acceleration under the action of gravity continues until a steady terminal velocity is attained. In this equilibrium condition the weight of the body is balanced exactly by the upward vertical aerostatic buoyancy and by the aerodynamic force engendered by the motion.

C. Analytical Treatment of Air and Fluid Flows

Although the theory of inviscid fluids is well developed and it is possible, at least in principle, to determine the flow about arbitrary shapes, one of its main defects is expressed in the paradox of D'Alembert (Milne-Thomson, 1949) which is that for bodies moving steadily in such fluids, no drag can exist irrespective of body shape. It is also equally true that no transverse force can be developed in steady motion without imposing the kind of artificial condition devised by Kutta (Milne-Thomson, 1949; Pope, 1951) to ensure the presence of the circulation (Milne-Thomson, 1949) that occurs in the flow of real fluids about aerofoils at moderate angles of attack. For the latter, the difference between the natural flow of an inviscid fluid with and without the Kutta condition, which asserts

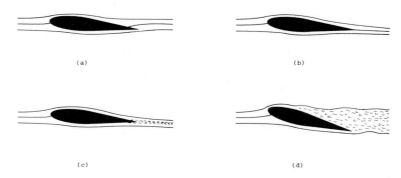

Fig. 3. Aerofoil sections in two-dimensional flow. (A) Inviscid flow without the Kutta condition. (B) Inviscid flow with the Kutta condition. (C) Real fluid flow with boundary-layer and wake. (D) Real fluid flow with aerofoil in stalled condition.

that the flow must leave the trailing edge smoothly and as closely as is possible tangentially, as it can with perfection when the trailing edge is formed by a cusp, is shown in Figure 3A and B, whilst the sort of flow pattern to be expected for real fluids includes the boundary-layer and wake flow indicated in Figure 3C. The existence of circulation is, by the theorem of Kutta and Joukowski for lift (Pope, 1951), sufficient to ensure the presence of a net transverse force in steady flow. When the flow is unsteady, either because of or in spite of the presence of any body immersed in it, forces may be developed even in inviscid fluids, whereas in real fluids the effect of hysteresis is such that it may not be possible to identify specific patterns of flow, except perhaps where there is a clearly defined periodicity.

For many of the flow configurations occurring at Reynolds numbers sufficiently large for inertia forces to dominate, the physical thickness of the boundary-layer is small compared with the characteristic length of the body. The influence this layer has on the effect of the flow as a whole, however, is crucial. In regions of falling pressure, the layer tends to remain thin and vigorous, but boundary-layer flow does not easily penetrate those regions where the pressure is rising. In these circumstances a complete separation of the flow from the body may result, such as that shown for the aerofoil in a stalled condition in Figure 3D. A flow separation in less extreme conditions may or may not be followed by a re-attachment but in either case the effect such a separation has on the forces and moments developed on the body is usually as marked as the changes it causes in the flow pattern. When the boundary-layer remains thin and the flow attached, the differences between corresponding real and inviscid fluid flow patterns are usually quite small for streamlined bodies. In these circum-stances it is permissible to use the relatively simple theories of low-speed inviscid fluid flow for the evaluation of transverse forces and moments, but the only result for drag that can be derived is for that induced in developing lift.

For the very small Reynolds numbers at which viscous forces dominate, the boundary-layer may extend to regions well away from the body surface. It is unfortunate that such flows are unlike those of inviscid fluids because, other than for a few rather special cases involving simple body shapes and restricted enquiry into local detail (Schlichting, 1955), calculations taking account of viscosity are extremely difficult to make. As this is also true for higher Reynolds numbers and thinner boundary-layers, it is necessary to identify the kind of section shapes and the types of flow that can be dealt with effectively by other and less rigorous means.

In determining trajectories for given body shapes, the most productive cases are those for which it is not necessary to examine the local composition of aerodynamic forces and yet which involve real fluid flows that are steady in time and take account of the effect of the boundary-layer without involving the detail within it. The degree of success achieved depends on the accuracy with which the required quantities can be estimated from reliable experimental data, and when the latter are available, results of considerable practical value may often be established without difficulty (see for example, Burrows, 1983). Unsteady flows are an altogether different matter.

The form of the equations used here to describe the motion of seeds, fruits, spores and pollen are purely dynamic ones in the sense that only the Newtonian equilibrium between mass, acceleration, and the overall forces and forces and moments is considered. The kind of breakdown of the aerodynamic forces and moments into Taylor series expansions as proposed originally by Bryan (1911), and which now form the basis of the elegant classical theories of flight stability and control (Duncan, 1952), is not used.

D. Some Geometric Shapes and Associated Flows

Because of the rather drastic simplifications of natural shapes that have to be made in some cases for trajectory calculations, it is convenient to list those that are of the greatest immediate value and to mention significant features of the flow about them. This makes clear the reasons for the later classification of species into groups which are distinguished on the basis of aerodynamic rather than botanical features. The shapes identified are:-

(1) spheres and ellipsoids;
(2) masses attached to either a guide-parachute or a drag-parachute;
(3) winged shapes with either embedded or attached mass concentrations;
(4) porous bundles.

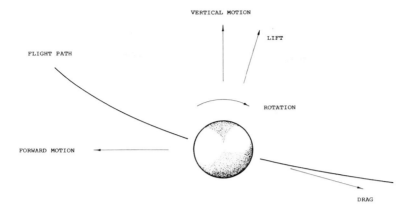

VERTICAL MOTION

LIFT

FLIGHT PATH

ROTATION

FORWARD MOTION

DRAG

Fig. 4. Sphere moving in translation and rotation with lift generated by the Magnus effect.

1. Spheres and Ellipsoids

In the aerodynamic sense the spherical shape is what is known as a 'bluff' body, because it is one for which the dominant aerodynamic force developed is drag. When rotated at significant speeds, cylinders and spheres are capable of developing transverse forces due to the 'Magnus' effect in the flow about them (see for example Prandtl, 1952; Hoerner, 1958) and hence of following flight paths having marked curvature as indicated in Figure 4. When spheres move without rotation the resistance to motion for extremely low flight speeds is given by the classical results of Stokes (1851), together with those of Oseen (1927) and Goldstein (1929). The Stokes solution in particular can be used to very great effect in determining the motion of spherical particles, which in general have very small physical dimensions. To deal with larger sizes the hybrid analytical-empirical derivation of the drag coefficient developed by Burrows (1983) can be used for what amounts to perhaps the entire range of flight conditions likely to be encountered in the flight of seeds, fruits, spores and pollen.

Whether or not an ellipsoid has symmetry about one or more of its axes, aerodynamic forces can be developed both along and normal to the flight path as a consequence of the attitude of presentation to the relative airflow. Since an ellipsoid is not as simple as a sphere in the geometric sense, it is not surprising that exact calculations of drag are restricted to the Stokes flow cases (Lamb, 1952) normally involving very small physical sizes and extremely low flight speeds. Other cases are dealt with on the basis of either approximate theoretical methods, such as slender body theory (Temple, 1958) for the estimation of lift, or by empirical means for estimating the lift, drag, and moments developed in flight.

2. Masses attached to Parachutes

When a parachute is attached to a body there is usually a fairly rapid alignment of the combination with the relative wind so that complications arising as a result of any cross-flows are likely to exist only for short and perhaps insignificant flight times. For steady flows, the dominating force is usually a drag force, whether the parachute is a guide-parachute or a drag-parachute. Under gust loadings there may be sufficiently large deformations of both types of canopy (D. B. O. Savile, pers. comm.) to cause marked alterations in flight behaviour of approximately the duration of the gust, and although somewhat unpredictable, the effect may have significance in the estimation of range.

As long as the orientations of composites of this kind are known, quite reliable estimates of the drag forces which act during flight can be made even when the form geometry is relatively complicated. This is because the range of flight speeds measured relative to the air is quite small for much of the flight time for most cases, and because it is possible to measure, without difficulty, the magnitude of the resistance to motion offered by particular species (Sheldon and Lawrence, 1973).

3. Winged Shapes

For winged shapes that are approximately plane, the methods of classical aerodynamics (Pope, 1951; Milne-Thomson, 1952) offer some scope for analysis even though caution is needed in application. This is because most of the empirical results available for guidance have been established for Reynolds numbers appropriate for flight on the physical scale of aeroplanes, and there is no reason to suppose that the information available for such cases is directly applicable to the much smaller scale of seeds and fruits. Although basically described as plane, the wing of a seed or fruit is likely to be quite crinkled, and this may cause so much difference in flight flow detail that anything but a crude estimate of flight behaviour and performance is impossible. Such estimates may, however, be preferable to the absence of any form of quantitative statement concerning either the sort of ground range achieved or the kind of flight pattern expected from individual species.

Winged shapes that have evolved for rotational descent have certain behavioural similarities to those of rotating-wing aircraft. When due account is taken of the corresponding differences in scale, some of the analytical methods established for the latter may be used as a means of providing estimates of flight behaviour and performance of seeds and fruits that rotate when falling (Norberg, 1973; McCutchen, 1977; Green, 1980), but the extreme difficulty of evaluating trajectory details for other than pseudo-steady flight cases should not be underestimated (Burrows, 1975b).

4. Porous Bundles

'Porous bundle' is a term loosely descriptive of tumble-weeds, and these do not appear to depend on a flight trajectory as a principal element of dispersal. Their movement is mostly confined to the immediate vicinity of the ground and is influenced significantly by the rolling and impact resistance which is an inevitable consequence of this type of motion. Their relatively substantial proportions mean that the driving wind flow is likely to remain effective until motion is arrested in some way. Although meaningful estimates of their aerodynamic resistance may be made as long as the geometric shape of the envelope enclosing the bundle and the distribution of limbs within it can be described, there is little point in doing so because it is the relative wind strengths, below which the bundle remains snagged and above which it becomes freed, which are of principal interest in this kind of dispersal.

III. TRAJECTORY ANALYSIS

A. Choice of Co-ordinate Systems

It is not always possible to choose a co-ordinate system having a universal range of application for a given class of dynamical problems and the case of seed dispersal is no exception to this. In the very simple case of a seed released from rest, the most obvious choice is that of a co-ordinate system with its origin at the point of release in the flowering head of the parent plant and with axis Ox pointing along the downstream horizontal direction of the prevailing mean mainstream wind. The axes Oy and Oz make up the customary right-handed system with Oy aligned with the downward vertical. The corresponding velocities and winds are u, v, w, and U, V, W, respectively (Burrows, 1973, 1975a, 1983). For cases in which seed release occurs from a moving point, it is preferable to choose the origin of co-ordinates at some datum rest position, such as at the base of the plant stem, so as to avoid the complications associated with an accelerated reference frame. Two suitable co-ordinate systems are shown in Figure 5 with the primes denoting quantities defined in the ground frame.

B. Equations of Motion

Figure 6 shows the force and moment vectors acting on a body of mass

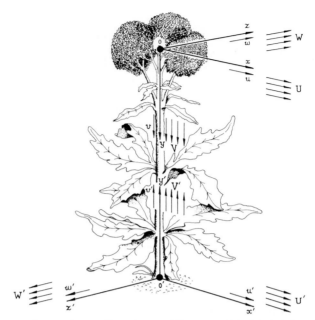

Fig. 5. Release and ground co-ordinate frames. The upper right-handed co-ordinate system Ox, Oy, Oz is the release frame in which the corresponding particle velocities are u, v, w, and the wind velocities U, V, and W. The origin O coincides with the point from which release occurs. The lower co-ordinate system is the right-handed ground frame for which lengths and velocities have been distinguished from those of the release frame by primes. The origin O' of this frame is a fixed reference point.

m and of arbitrary shape in aerial flight with centre of mass at a position $\mathbf{r}$ (x, y, z) in a stationary release co-ordinate frame. The forces consist basically of the weight $\mathbf{w} = m\mathbf{g}$, where $\mathbf{g}$ is the acceleration due to gravity, together with the lift $\mathbf{L}$ (L_1, L_2, L_3) and drag $\mathbf{D}$ (D_1, D_2, D_3) which, by definition, act at the centre of pressure and normal to and along the direction of the relative wind respectively. In general the lift will not necessarily act either in opposed alignment to the gravitational acceleration, or with its axis in coincidence or in axial alignment with the residual aerodynamic moment $\mathbf{M}_0$, which is independent of both lift and drag. The relative wind has speed V_p, where

$$V_p = [(U - u)^2 + (V - v)^2 + (W - w)^2]^{1/2}, \tag{1}$$

and the angles this wind makes with the axes Ox, Oy, Oz are θ, ϕ, ψ respectively. In addition to the aerodynamic and gravitational forces and moments there may be other forces and moments present as is the case

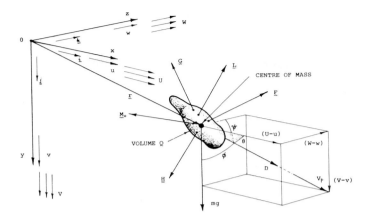

Fig. 6. Stationary release co-ordinate frame with body of arbitrary shape, mass m, and volume Q moving in translation and rotation. The spatial orientation of the frame is defined by the unit vectors **i**, **j**, **k**, and the instantaneous position of the moving body by the vector **r**. The composition of the relative wind V_p, and the angles θ, ϕ, ψ, this wind makes with the co-ordinate directions Ox, Oy, Oz are included in the figure together with the resulting aerodynamic forces of lift **L**, drag **D**, and the residual aerodynamic moment $\mathbf{M_o}$. The vectors **F**, **G**, represent respectively an arbitrary external force and moment whilst **H** is the moment of momentum about the centre of mass. The gravitational force or weight mg acts in the direction **j** and is shown separately from the other force vectors because it is the only one of the forces that always exists.

with the movement of bodies responsive to either electrostatic or magnetic fields, or both. These are represented by the single force vector **F** (F_1, F_2, F_3) and the single moment vector **G** (G_1, G_2, G_3). Motion in the co-ordinate system Ox, Oy, Oz at time t is thus described according to the Newtonian laws by three scalar force equations for translation, and by three scalar moment equations for rotation, and these have the vector form

$$m\frac{d^2\mathbf{r}}{dt^2} = mg\mathbf{j} + \mathbf{L} + \mathbf{D} + \mathbf{F}, \tag{2}$$

$$\frac{d\mathbf{H}}{dt} = \mathbf{G} + \mathbf{M}, \tag{3}$$

where **G** and **M** are respectively the resultant dynamic and aerodynamic moments about the centre of mass. The vector **H** is the moment of momemtum of the body with respect to the centre of mass and includes the six moments and products of inertia which describe the dynamical configuration of the body itself. To proceed further with equation (3) it is usually necessary to introduce a second co-ordinate system with origin at the centre of mass of the body and selected such that the moments and products of inertia are invariants of the motion. Spatial orientation is then defined by the Euler angles (Easthope, 1967).

Except for structurally rigid seeds and fruits of elementary shape, the moments and products of inertia are quantities so difficult to measure or estimate that it is preferable, at least in the first instance, to examine those cases wherein body rotation is not a significant feature of the motion. Although this immediately rules out, for example, any investigation of transients for samaras, it does not preclude the stable auto-rotative states which are quasi-static, nor the more or less stable and purely translational motion exhibited by the wide range of species for which the seeds and fruits have approximate symmetry about one or more axes. For the present purpose, the advantages of retaining equation (3) are outweighed by the disadvantages, so that only equation (2) will be developed and discussed further.

The scalar form of equation (2) is

$$m \frac{d^2x}{dt^2} = L_1 + D(U\text{-}u)/V_p, \tag{4}$$

$$m \frac{d^2y}{dt^2} = L_2 + mg - F_b + D(V\text{-}v)/V_p, \tag{5}$$

$$m \frac{d^2z}{dt^2} = L_3 + D(W\text{-}w)/V_p, \tag{6}$$

where the only force F_b considered at present is the buoyancy force exerted on the body of volume Q and density ρ_p immersed in an environment of density ρ. By making a consistent choice of a characteristic linear dimension l, such as the diameter d for a spherical particle, the dimensional equations (4)-(6) may be expressed in the non-dimensional form

$$\frac{d^2\xi}{d\tau^2} = k \ T_n G \ [4 \ G \ C_{L1}/3 + (G_1 - \xi') \ C_D], \tag{7}$$

$$\frac{d^2\eta}{d\tau^2} = k \ T_n G \ [4 \ G \ C_{L2}/3 + (G_2 - \eta') \ C_D] + 2(1-\sigma), \tag{8}$$

$$\frac{d^2\zeta}{d\tau^2} = k \ T_n G \ [4 \ G \ C_{L3}/3 + (G_3 - \zeta') \ C_D], \tag{9}$$

where ξ, η, ζ are the non-dimensional space co-ordinates

$$\xi = x/h, \quad \eta = y/h, \quad \zeta = z/h, \tag{10}$$

corresponding to a release height h, trajectory number $T_n = \sigma h/1$, $\sigma = \rho/\rho_p$, $k = \pi\, 1^3/6Q$, and τ is the non-dimensional time given by

$$\tau = t\,(g/2h)^{1/2}. \tag{11}$$

In terms of a reference area $S = \pi\, 1^2/4$, the coefficients of lift and drag are

$$C_{L\alpha} = L_\alpha/(\rho V_p^2 S/2), \quad C_D = D/(\rho V_p^2 S/2); \quad \alpha = 1, 2, 3, \tag{12}$$

and the non-dimensional wind speed G is given by

$$G = (3/4)[(G_1-\xi')^2 + (G_2-\eta')^2 + (G_3-\zeta')^2]^{1/2}, \tag{13}$$

where the primes in this case denote differentiation with respect to the non-dimensional time. The components G_1, G_2, G_3 are

$$G_1 = U(2/gh)^{1/2}, G_2 = V(2/gh)^{1/2}, G_3 = W(2/gh)^{1/2} \tag{14}$$

Equations (7-9) show that trajectories are similar geometrically whenever the parameters G_1, G_2, G_3, the trajectory number T_n, the shape factor k, and the coefficients $C_{L\alpha}$ and C_D take the same values for otherwise different cases.

For any case the integration of each one of the equations of motion involves the introduction of two arbitrary constants which can be evaluated from the initial position and velocity of release. Both of these quantities depend on the speed and direction, taken or imparted at the instant of release and, therefore, on the kind of ejection or detachment mechanism employed by the parent and discussed in Section IV. The kind of trajectory subsequently followed depends critically on the effects caused by wind flow and although for well-defined mean wind flow directions, and for steady winds passing over simple ground contours, this flow may sometimes be specified with reliability, it cannot usually be done to a level of accuracy and consistency that is strictly compatible with the mathematical rigour of the equations of motion. This difficulty of specification is likely to remain a fundamental obstacle to any quantitative discussion of airborne seed and fruit movement.

For many cases of practical interest the equations of motion can be reduced substantially to degenerate forms which admit of relatively simple solutions that can usefully be worked with great economy. Rather than discussing these cases and a selection of the other known solutions of the equations of motion collectively it is preferable to deal with them individually under headings based on the preliminary aerodynamic classification of seed and fruit groups suggested elsewhere (Burrows, 1975b). By this means a fairly logical breakdown of an otherwise relatively complicated mathematical system can be made to a reasonable level of simplicity.

IV. LAUNCHING AND RELEASE MECHANISMS

Except for purely vertical projection into still air, two distinct types of start to any flight trajectory exist because a seed has an initial velocity relative to that of the prevailing wind that is greater or less than that of its terminal velocity of fall. The former case occurs when a seed is projected explosively, or by what amounts to a catapulting action, and the latter when separation occurs as a direct result of either passive release or the action of dynamic or aerodynamic force. Since the initial velocity of release is of fundamental importance in defining at least the early geometry of individual trajectories, it is advantageous to classify these into well defined groups which differ from one another on the basis of the release mechanism involved. The groups considered here provide for:-
  (A) passive release;
  (B) dynamic release;
  (C) forced separation;
  (D) ejection;
  (E) syphoning.

A. Passive Release

Passive release is simply a loose start from rest under gravity, and in the mechanical sense, it is not necessary to know the details of the release mechanism. It involves simply an instant of release at which time the parent 'lets go' of the seed in a completely impartial manner and without causing any obstruction to the subsequent motion. For such cases the seed may or may not be at rest relative to the parent plant at the instant of release, but whether or not it is at rest relative to the ground below depends on the presence or absence of movement of the attachment point at the time of release, and on the movement of the seed itself relative to the parent. For dense seeds, a significant if not substantial augmentation in ground range may result from suitable combinations of such movements.

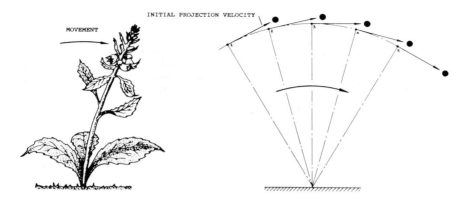

Fig. 7. Seed release from a moving plant showing initial projection velocity vectors corresponding to release positions 1,2,3,4,5.

Plants which are either isolated or in stands experience different amounts of deflection under the action of the wind, and since this is rarely steady in time, a seed may begin its flight with an initial velocity of projection as indicated in Figure 7. The effect of such a given initial velocity on the range achieved is marked for large dense seeds moving at relatively large Reynolds numbers, but minute spherical seeds moving under the Stokes law of resistance, for which the drag coefficient is

$$C_D = 24/Re, \quad Re = V_p d/\gamma, \tag{15}$$

tend to settle quickly into a steady vertical fall at approximately their terminal velocity.

The effect of a favourable wind velocity can be significant if release is made from a plant head moving rapidly upwards and downwind from a deflected position, but it is also possible for wind-induced shock loading of the parent to result in a seed becoming detached at an instant when it has little or no initial velocity in the ground frame. If the gust causing the shock loading and separation is extremely short lived, the seed may simply fall to the ground more or less directly below the parent plant.

B. Dynamic Release

Dynamic release is associated specifically with a forced separation arising through a relative movement between plant and seed. The masses, and consequently the inertias, of plant and seed together constitute a dynamic system with a natural frequency and a characteristic oscillatory mode. Such a system responds to the input of both steady and unsteady forces and

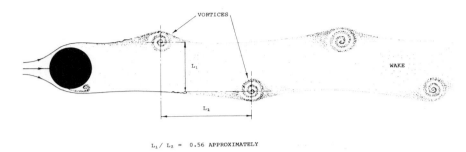

VORTICES

L_1

L_2

WAKE

L_1 / L_2 = 0.56 APPROXIMATELY

Fig. 8. Vortex wake downstream of a circular cylinder for Reynolds numbers between about 100 to 2500.

it is the geometry of each plant and its limbs, together with the structural stiffness of these components, that determine the manner in which vibration takes place. Many plant stems have nominally circular cross-sections and cross-sections of this kind are well known to respond in an unsteady way to a steady streaming flow about them, provided that the Reynolds number is of the right order. The effect arises because of the oscillatory wake produced by the alternate, and more or less regular, shedding of vortices so as to form a vortex wake of the kind shown in Figure 8, and which is not unlike the 'street' considered by von Karman (e.g. Prandtl, 1952). What happens is that just prior to leaving the cylinder, each vortex induces a circulation Γ on the cylinder itself, which, when combined with the free stream moving with speed V and of density ρ, causes a transverse force of magnitude $\rho V \Gamma$ per unit length of cylinder in accordance with the theorem of Kutta and Joukowski (Milne-Thomson, 1949; Pope, 1951). Although not always as ordered as this, flows about many other bodies with bluff cross-sections may similarly develop unsteady transverse forces because of the changing circulations existing on them from instant to instant. Limbs of more streamlined cross-section may experience flutter (Broadbent, 1956) which arises when the combination of streaming flow and aero-elastic distortion is just right for resonance. This phenomenon is particularly common to plants with leaves of thin cross-section like the one shown in Figure 9.

Mechanical vibrations involving amplitudes greater than a critical minimum for any given structure will sooner or later lead to fatigue failure, and seed and fruit attachment points are no exception to this. As such a failure may occur at any part of the oscillatory cycle, it is difficult to specify the corresponding release velocities with the required degree of accuracy for meaningful trajectory calculations.

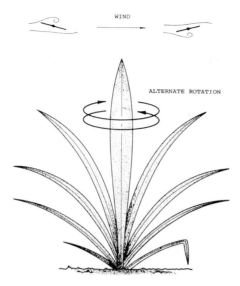

Fig. 9. Leaf flutter in a steady uniform wind. The alternating flow pattern at a leaf cross-section is shown inset.

C. Forced Separation

Forced separation occurs when the aerodynamic loads on a seed or fruit exceed the structural strength of the attachment point. For individual species at a given location it may be possible to identify steady critical mean wind speeds below which, on the average, no separation takes place, and above which, and for equally ripe states, separation is statistically likely. Winds of greater strength and gusts of sufficient strength and duration may cause premature separation of under-ripe seeds and fruits but these cases are less significant because they are ones for which dispersal is unlikely to be effective for germination.

One of the difficulties of estimating the aerodynamic force developed on an attached seed or fruit arises on account of the amount of shelter provided by the parent. A pendant seed that is more or less completely exposed to the stream may, if its shape is a simple one like that shown in Figure 10A, be dealt with in a relatively straightforward way and have separation boundaries defining release loads for ripe and under-ripe states as indicated in Figure 10B. Seeds sheltered from the stream, like the ones shown in Figure 10C, are much less easy to deal with because of the complexity of wake flows which may, in the right circumstances, even act

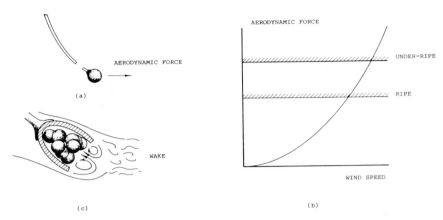

Fig. 10. Forced separation and possible attachment preservation due to wind action. (A) Exposed pendant seed. (B) Separation boundaries. (C) Recirculatory flows tending to preserve attachment.

to preserve attachment rather than to induce separation from the parent. It may be difficult to distinguish the precise source of the separating force and, as with dynamic separation, mechanical vibration may play a significant role in the mechanism of release.

Many different species, from minute spores to fruits of substantial proportions, secure their release in this way and for some of the plumed varieties it may be accompanied by a momentary but marked alteration in parachute geometry. This occurs because of the relatively large aero-dynamic loads that can be developed in a very short time on some canopies and which generally make many species of this kind particularly susceptible to a gust-induced release. When such release occurs in otherwise tranquil conditions the range achieved in the primary trajectory by a plumed seed or fruit may be very short.

D. Ejection

Ejection mechanisms take several different forms but the common purpose is to initiate a ballistic trajectory. Figure 11 shows four examples out of the many possible ones. For the 'splash-cup' (Fig. 11A) of *Mitella* (Scagel *et al.*, 1969) the initial velocity of ejection is acquired from a momentum exchange with an impacting raindrop, whilst the ejection of spores from a 'puff-ball' (*Lycoperdon perlatum*, Fig. 11B) is along a gas jet produced from within and caused by a momentary change in ball volume (Ingold,

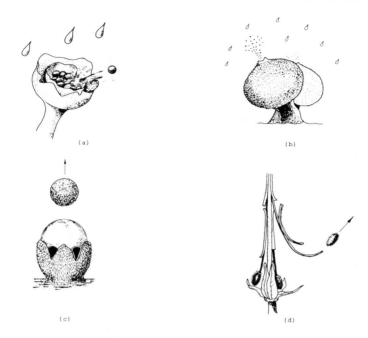

Fig. 11. Ejection mechanisms. (A) 'Splash-cup'; *Mitella.* (B) 'Puff-ball'; *Lycoperdon perlatum.*
(C) Propulsive membrane; *Sphaerobolus stellatus.* (D) 'Spring-limbs'; *Geranium bicknellii.*

1968). Ejection may also be due to impact by rain, but other physical and
mechanical causes can produce the same end result. The ballistospore
Sphaerobolus stellatus makes direct use of a propulsive membrane, shown
at full-stretch in Figure 11C, whilst on a much larger scale, the spring-
like action of the radially distributed limbs of *Geranium bicknellii* (Fig.
11D) is a particularly good example of initial intent with regard to the
prospective azimuthal coverage of local ground by a dispersal mechanism.

Perhaps the most important feature of this type of release is the considerable initial velocity that may be imparted to each seed, fruit or spore
at the start of its aerial motion. This velocity may not always include a
rotational component contributing a 'Magnus' force to augment range,
although some extremely high rotational speeds are apparently achieved
at release in some cases (Swaine and Beer, 1977; Swaine *et al.*, 1979). The
release strategy is not always a good one if the ejection is made either
upwind, or vertically upwards in the absence of wind, but when the wind
is blowing there is a considerable advantage in favour of longer ranges
when the projection velocity is sufficient for carriage to the greatest possible
height.

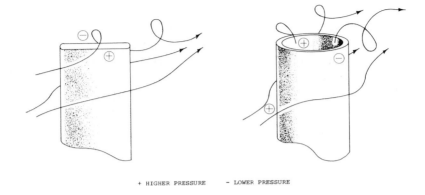

+ HIGHER PRESSURE   − LOWER PRESSURE

Fig. 12. Effects of differential pressures and syphoning in the flow about structures exposed to the wind.

E. Syphoning

In the flow past flattish and streamlined structures having an angle of attack relative to the stream, and in flows past bluff structures, quite substantial suction pressures can become established in regions such as those indicated for the examples shown in Figure 12. These pressures may induce flow from the underside to the upperside of the flattish and streamlined shapes, and from the inside to the outside of tubular ones, and this flow can assist particles to migrate away from initial underside or internal rest positions. Such movement of spores from within 'puff-balls' has been mentioned by Ingold (1953, 1965), but as the possibility of such an aerodynamic syphoning process is a very real one for any tube-like housing in which there is enough room for air to circulate, it is likely to contribute significantly to the initiation of dispersal in a number of other cases.

V. TYPES OF TRAJECTORY: SOLUTIONS OF EQUATIONS OF MOTION

A. Aerodynamic Classification of Seed and Fruit Groups

An aerodynamic classification of seed and fruit groups is a logical consequence of reducing the system of equations (7)-(9) describing general aerial motion to the distinct subsets required in particular cases of practical interest. Since these subsets are based almost entirely on a division of species only according to the similarity of equivalent distributions of mass and geometry, and of flight style and Reynolds number range, this classification is much less complete than the botanical one used by Ridley (1930). The species groups distinguished in this way are (Burrows, 1975b):-

(a) dust seeds, spores and pollen;
(b) plumed seeds and fruits and woolly seeds and fruits;
(c) plane winged seeds and fruits with a central or more-or-less central concentration of mass;
(d) winged seeds and fruits which rotate when falling;
(e) seed-carrying tumble weeds;
(f) aerodynamically unimportant seeds and fruits.

For group (a) the flight Reynolds numbers are usually so small that much of the movement along the trajectory is dominated by viscous forces and the Stokes law of resistance can be used to provide an accurate description of the drag. This is not the case for group (b), because members of this class tend to move at much higher Reynolds numbers and although the drag may be the dominant force developed it arises on account of inertial forces in the flow rather than viscous ones and is less easy to specify accurately. Exceptions to this occur when the amount of drag developed is so large that transients are of short life and it is legitimate to assume a constant value of the drag coefficient for the calculation of a trajectory. Such is the case for substantial and dense parachute canopies or their equivalent. For group (c) usable lift may sometimes be the dominant force developed, whereas for group (d) it is nearly always the dominant one, but the modes of motion occurring in each case are usually quite different. Group (e) is not easy to deal with on account of the changes of geometry likely to occur during movement on the wind and, other than including it for completeness in the above listing, it will not be discussed further. Group (f) is also not discussed because it includes all species for which the weight is relatively large compared with any aerodynamic force which can be developed during release and aerial movement.

B. Aerodynamic Coefficients for Particles

Because of its perfect symmetry the smooth spherical particle takes a prominent place in trajectory studies for, in the absence of rotation, all flight attitudes are similar and the only steady time-averaged aerodynamic force developed in uniform flow is a drag force. It is clear that an accurate specification of the variation of drag coefficient with Reynolds number is the key to a wide range of possible trajectory calculations. For Reynolds numbers less than about unity the analytical result due to Stokes (1851)

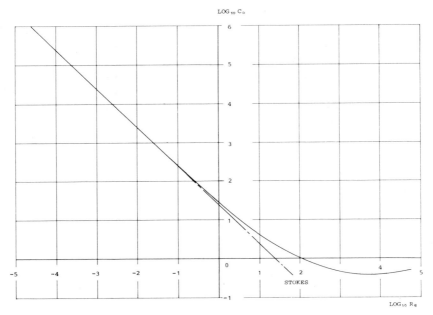

Fig. 13. Drag coefficients for spheres.

is more than adequate for most practical purposes. For Reynolds numbers up to about 60 000 the hybrid analytical-empirical relation derived by Burrows (1983),

Equation 16, p.26 should read:

$$C_D = (24/Re)[1+(3Re/16)]e^{-\phi}+10(1-e^{-\phi})\exp(1+A_1X+A_2X^2+A_3X^3+A_5X^5), \quad (16)$$

$A_1 = -8.819398 \times 10^{-1}$,
$A_2 = +2.842035 \times 10^{-2}$,
$A_3 = +2.219102 \times 10^{-3}$,
$A_5 = -4.300448 \times 10^{-6}$,

and shown in Figure 13 provides comparable accuracy for what amounts to a vast range of particles sizes and speeds. Results like that of Stokes are also available (see for example Lamb, 1952) for a number of other simple geometric shapes such as ellipsoids, circular discs, and circular cylinders and flat blades of infinite length in defined orientation. Bodies of more complicated geometry may not have attitude stability in free flight, for even quite small amounts of initial asymmetry in the flow may be sufficient to induce either wobbling, zooming, plunging, sideslipping, and rotation, or combinations of these movements. When pseudo-static stability exists the aerodynamic drag opposing descent may be either measured or estimated. If the latter, the wealth of information assembled by Hoerner (1958) and Clift *et al.* (1978) is of value.

The production of lift depends on the existence of net cross-flows relative to the flight direction and, for bodies not possessing a clearly identifiable and nominally plane wing of non-negligible aspect ratio presenting itself in a definite attitude to the stream, such flows are not easily specified. A comparable difficulty also exists even when such a wing-like structure is present because the many analytic methods for deriving lift (see for example Milne-Thomson, 1952; Pope, 1951; Robinson and Laurmann, 1956), and the manner in which lift is distributed on a given planform, depend ultimately on the specification for each case of a reference value of the lift coefficient which depends on wing-section geometry. Whilst these values are usually known for man-made wings and other lifting shapes which move at high Reynolds numbers, they are apparently not yet known for botanical ones which move at low Reynolds numbers.

C. Dust Seeds, Spores and Pollen

Because of the tiny size of these particles, and even though they are generally considerably heavier than the volume of air they displace, the flow about them is likely to be well within the range of Reynolds numbers for which viscous forces are dominant. When this is so, the absolute values of the terminal velocities of fall achieved by different species are very small and usually less than about 100 mm.s^{-1} in a majority of recorded cases (Gregory, 1973). As with other and larger species, the variety of geometric shapes is diverse (Moore and Webb, 1978) and a selection of examples is shown in Figure 14.

When the flight Reynolds numbers are less than unity the resistance to motion of these particles is approximately proportional to their velocity relative to the air, the characteristic physical size of each one, and the viscosity of the air itself. For spherical particles of diameter, d, moving without rotation the drag is given by the Stokes law of resistance (Eqn 15) which, in absolute terms, is

$$D = 3\pi\rho\nu d V_p, \tag{17}$$

where V_p is the speed of movement relative to the environment of density ρ and kinematic viscosity ν. There is a similar result available for the flow about ellipsoidal particles having one principal axis aligned with the direction

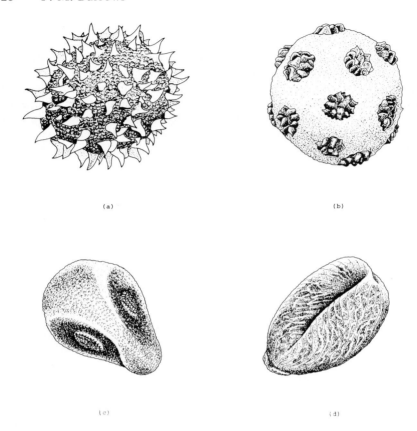

(a)

(b)

(c)

(d)

Fig. 14. Examples of pollen grains. (A) *Tussilago farfara*. (B) *Silene vulgaris*. (D) *Carex pendula*. (E) *Acer pseudoplatanus*.

of the flow, and also for the other cases of perhaps less immediate practical value (Lamb, 1952) mentioned earlier.

Since in a steady free vertical fall made without rotation in a gravitational field, the drag experienced by a spherical particle is equal to the weight, and the weight is proportional to the cube of the particle diameter for uniform distributions of equal particle density, it follows that the relative terminal velocities that can be attained are proportional to the squares of the diameters of particles being compared. For the many other possible shapes the resistance law for Reynolds numbers up to about unity is

$$D = \text{constant} \times V_p, \tag{18}$$

where V_p is the speed of movement relative to the environment and the constant must be determined for each case considered.

An immediate consequence of particle movement dominated by viscous drag is that it is very sensitive to fluctuations in wind speed of an order of magnitude similar to that of the terminal velocities of the particles. In many cases, where a larger physical scale implies the overall existence of a relatively tranquil environment, an included small scale eddy structure of the right sort may result in a considerable amount of either regular or irregular translation of minute particles. In such circumstances many different kinds of movement may occur, from apparent suspension to carriage to great altitude. With diurnal heating and cooling the convection velocities engendered are probably well in excess of the terminal velocities of fall, so that the altitude attained by tiny particles moving under the Stokes law of resistance ensures that vast carriage distances are the rule rather than the exception. Because most natural wind movement is turbulent, the particles may follow trajectories of extremely complicated geometry and it is possible to determine these trajectories only when the wind movement and the aerodynamic forces developed can be specified from instant to instant.

For translation at very small Reynolds numbers the equations of motion (7)-(9) for spherical particles degenerate to the forms

$$\frac{d^2\xi}{d\tau^2} = 36 \ (T_n/R_n)(G_1-\xi'), \tag{19}$$

$$\frac{d^2\eta}{d\tau_2} = 36 \ (T_n/R_n)(G_2-\eta') + 2(1-\sigma), \tag{20}$$

$$\frac{d^2\zeta}{d\tau_2} = 36 \ (T_n/R_n)(G_3-\zeta'); \quad R_n = (d/\gamma)(2gh)^{1/2}, \tag{21}$$

(Burrows, 1983), and the equivalent of these (Burrows, 1975a). Here, the trajectory number $T_n = (\sigma h/d)$. For environments that are time-dependent the only difficulty in proceeding to solutions is that of evaluating the integrals

$$\xi' = 36 \ (T_n/R_n)\int(G_1-\xi') \ d\tau + C_1, \tag{22}$$

$$\eta' = 36 \ (T_n/R_n)\int(G_2-\eta') \ d\tau + 2 \ (1-\sigma)\tau + C_2, \tag{23}$$

$$\zeta' = 36 \ (T_n/R_n)\int(G_3-\zeta')d\tau + C_3, \tag{24}$$

$$\xi = \int\xi' \ d\tau + C_4, \tag{25}$$

$$\eta = \int\eta' \ d\tau + C_5, \tag{26}$$

$$\zeta = \int\zeta' \ d\tau + C_6, \tag{27}$$

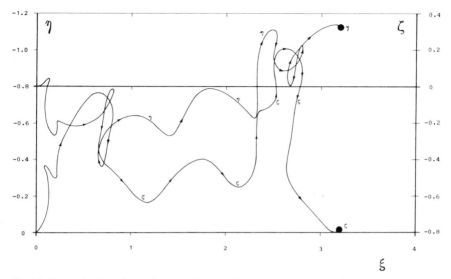

Fig. 15. Example of particle trajectory calculated for movement from rest into pseudo-turbulent flow. The $\xi - \eta$ and $\xi - \zeta$ curves show the trajectory projected onto vertical and horizontal planes respectively. For explanation of symbols, see Equations (25–27).

where C_1, C_2,..., C_6, are arbitrary constants. When the environment has a space dependence the non–linearity of the equations of motion provides difficulties of integration which, except for a few trivial cases, can only be overcome by resorting to numerical techniques. This means that cases have to be discussed in more restricted terms than those possible when analytic means can be used. The selected examples considered by Burrows (1975a) show the differences between the flight paths followed by particles released from rest and those described by individual particles of air in specified environments, and also the complicated trajectory geometry and quite different ultimate positions of migration when particles are released into winds possessing an eddy structure not altogether unlike that expected in turbulent wind–flow. One such trajectory for this pseudo–turbulent flow is shown in Figure 15.

Text on pages 30 and 31 should read:

The discharge of spores of near spherical shape into still air is of special interest because a whole class of plane trajectories can be defined by a universal one-parameter formula (Burrows, 1985) following a suitable re-arrangement of the dimensional result quoted by Buller (1909). For these cases the equations of motion (19)–(20) when solved subject to the initial conditions

$$\tau = 0 \; ; \; \xi' = \xi'_0 \, , \, \eta' = \eta'_0 \, ,$$

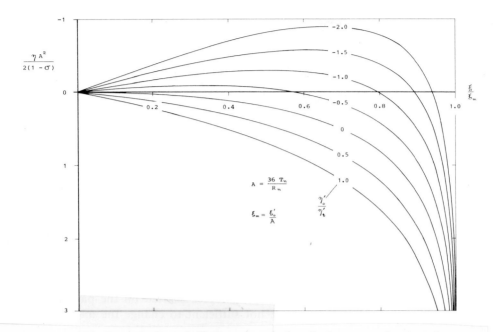

which specify the velocity of discharge and the discharge angle, may be expressed as

$$\eta A^2/\{2(1-\sigma)\} = (\xi/\xi_m)\{(\eta'_0/\eta'_t) - 1\} - \ln\{1 - (\xi/\xi_m)\}, \tag{28}$$

where ξ_m and η'_t are the non-dimensional maximum horizontal distance that can be attained, and the terminal velocity of fall, respectively. Some members of the family of trajectories represented by this result are shown in Figure 16. In absolute terms, the distances moved by these particles in still air are small, but in terms of dimensions relative to those of the particle, they are not. The range ξ_r achieved is specified when η is unity and this can be determined by the somewhat less straightforward procedure of solving the equation

$$A^2/\{2(1-\sigma)\} = (\xi_r/\xi_m)\{(\eta'_0/\eta'_t) - 1\} - \ln\{1 - (\xi_r/\xi_m)\} \tag{29}$$

$$A^2/\{2(1-\sigma)\} = (\xi_r/\xi_m\{(\eta'_0/\eta'_t) - 1\} - \ln\{1 - (\xi/\xi_m)\} \tag{29}$$

numerically for ξ_r. The large class of particles of other than spherical or near spherical shape, such as those shown in Figure 14 (C–D), cannot be dealt with accurately until the constant of proportionality in the drag equation (18) has been established for movement with attitude stability in flight.

When the Stokes and similar laws of resistance define the principal aerodynamic force acting on particles in flight the response to wind movement is not very complicated. As long as the airflow about the particles can be specified in sufficiently fine detail, it is feasible to proceed both to meaningful calculations of trajectories and to equally meaningful interpretations of processes such as impaction and under-surface deposition. Both these processes depend on the amount of momentum possessed by a particle in flight, for it is this quantity which defines the ability of a particle to continue in a specific direction and hence perhaps to cut across the local flow. For a given speed of movement, heavy particles are more reluctant to take up new directions of motion than light ones, as is indicated in the results of the horizontal impaction experiments performed with spores by Gregory (1961). The heavier particles in these experiments showed a marked tendency to cut across the direction of the local flow near to a circular cylinder and in some cases to impact upon it. Although the velocity changes quite rapidly with distance from the surface of such a cylinder, the corresponding changes in the drag forces exerted on the particles in the vicinity of the cylinder were not sufficient to change the momentum of the lighter ones enough to prevent movement with the streaming flow around the cylinder.

Since normal gravitational sedimentation may be wholly prevented whenever the upward momentum acquired by a particle is sufficient to ensure a net upward carriage, deposition may occur on the underside, as well as on the top, of exposed surfaces. This can happen in a turbulent boundary-layer flow as represented schematically in Figure 17A (Burrows, 1975a). Once a surface has been reached, electrostatic and other sufficiently powerful adhesive forces may ensure continued contact and, except perhaps for regions of intense shear in the local airflow, the velocities an attached particle is subjected to there are much less than those experienced in the free-stream, and they are directed along the surface itself. They may, however, still be sufficiently fast to induce a drifting along the surface and, in cases where recirculatory flows are present, a certain amount of deposition concentration of the kind indicated in Figure 17 (B) may result. The extent to which this happens in a given case can be estimated whenever the aerodynamic environment is known in sufficiently fine detail.

D. Variously Dense Particles

Seeds, fruits, spores, and pollen of the more dense and larger kinds than those considered in the previous Section, but again represented by the smooth

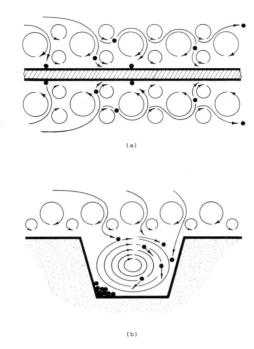

(a)

(b)

Fig. 17. (A) Possible particle movement in a turbulent boundary-layer with upper- and under-surface deposition. (B) Particle deposition concentration due to recirculatory flow in a ditch.

spherical shape, have been dealt with by Burrows (1983). The equations of motion used for this work were similar to equations (7)–(9) describing translation. Rotational motion was not considered and no allowance was made for the Saffman lift due to shear in the flow (Saffman, 1965), so that the cases discussed were those for motion in which the only aerodynamic force developed was that due to drag. The rather wide range of Reynolds numbers considered required the establishment of the hybrid analytical-empirical transcendental relation for the drag coefficient mentioned earlier (Eqn 16) and the calculations provided extensive tables of the ground ranges likely to be achieved in two different kinds of plane trajectories. The cases dealt with were those for release from rest into both steady uniform winds and steady winds in a simplified ground boundary-layer, described by a linear variation of velocity from a prescribed starting value at the release height to zero at ground level. Some calculated trajectories are shown in Figure 18. Consideration was also given to the kinds of trajectory to be expected in a region of pseudo-thermal convection involving a reasonably strong eddying motion in the flow and these showed that some deposition

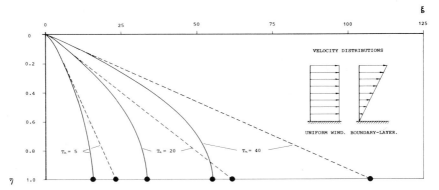

Fig. 18. Some calculated trajectories for dense particles for different values of trajectory number $T_n = \sigma h/d$, the trajectory Reynolds number $R_n = 100$, the non-dimensional horizontal wind speed $G_1 = 10$, and the relative density $\sigma = 0$. The broken lines are for a uniform wind G_1, and the unbroken lines are for a linear boundary-layer velocity distribution.

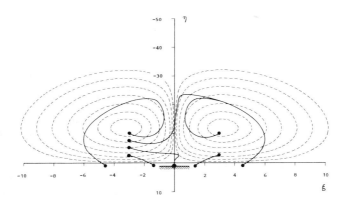

Fig. 19. Particle trajectories in a region of pseudo-thermal convection. The broken lines are the eddy streamlines and the unbroken lines the trajectories. $T_n = 1.0$, $\sigma = 0$, $R_n = 250$.

concentration may result following release from rest from quite widely separated release positions as indicated in Figure 19. For ordinary plane flow the effect on trajectory geometry of an initial projection speed at angles equally spaced at intervals of 45° measured from a horizontal datum, with and without wind flow and a ground boundary–layer, is indicated in Figure 20.

In principle there is little difference between this type of calculation and the ones made for dust seeds, spores, and pollen (Burrows, 1975a),

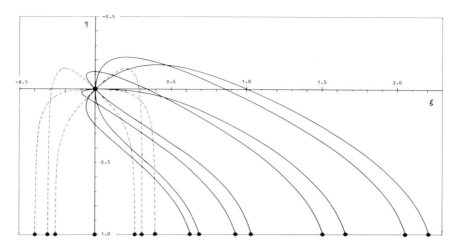

Fig. 20. The effect of an initial projection angle for particles launched into still air (broken lines) and into a linear boundary-layer flow (unbroken lines). The projection angles are equally spaced at intervals of 45°. $T_n = 10$, $\sigma = 0$, $R_n = 500$, $G_1 = 1.0$, non-dimensional projection speed = 2.

but for the dense particles the coupled non-linearity of the equations of motion resulting from the transcendental expression for the drag coefficient precludes the possibility of obtaining analytic solutions for most cases of interest. It is not, however, difficult to proceed to numerical solutions.

For dense non-spherical particles the relative magnitudes of the inertial and aerodynamic forces developed during aerial motion may well be such as to permit a better degree of attitude stability than can be expected for less dense ones. When this is so, the possibility of satisfactorily estimating the various forces and moments developed in flight becomes so improved as to make trajectory calculations a perhaps viable proposition. The same is also true for spherical particles that spin about an axis transverse to the flight path and which develop a substantial amount of lift on account of the Magnus effect. It is not, however, clear what effect spin about axes in alignment, or in approximate alignment, with the flight path has on the flow and the forces it produces and also to what extent transverse forces are developed on non-spherical particles in these circumstances.

E. Plumed Seeds and Fruits

The wide range of species having seeds and fruits incorporating some form of plumage, such as those shown in Figure 21 may, in the aerodynamic

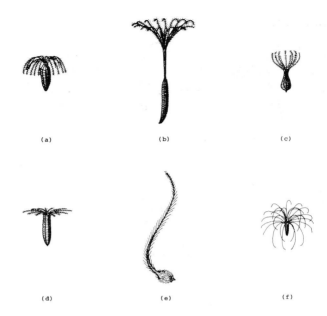

(a)   (b)   (c)

(d)   (e)   (f)

Fig. 21. Types of plumed seeds and fruits. (A) *Sonchus palustris* (Compositae); (B) *Tragopogon porrifolius* (Compositae); (C) *Centranthus ruber* (Valerianaceae); (D) *Centaurea nigra* (Compositae); (E) *Pulsatilla vulgaris* (Ranunculaceae); (F) *Epilobium angustifolium* (Epilobiaceae).

sense, be grouped under one of two headings according to whether the plumage acts as a drag parachute or a guide parachute (Hoerner, 1958). For seeds and fruits having a relatively large pappus of small porosity, aerial motion transients tend to be short so that the terminal velocity in free fall relative to the air is usually attained very rapidly. Such species are very sensitive to fluctuations in wind speed and in particular, and significantly, to variations in convection that may be present throughout the course of the flight path. Many plumed and woolly species of this kind are dynamically and aerodynamically equivalent to the arrangement with concentrations of mass and of drag-producing limbs shown in Figure 22. When the pappus is rather small and very porous its function in some cases is apparently to provide orientation and guidance rather than drag, and dart-like seeds of this type have much larger terminal velocities, which are attained only after a fall through a considerable distance, and without much lateral movement other than in winds of exceptional strength. Seeds and fruits with guide parachutes are, however, quite well equipped for utilizing projection, catapulting or ejection at the moment of release.

When the pappus acts as a drag parachute, the small transients permit the use of the greatly simplified zero–lift forms

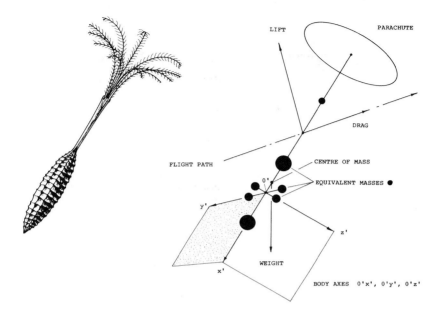

Fig. 22. Plumed seed and equivalent dynamic and aerodynamic system.

$$\xi' = G_1, \tag{30}$$

$$\frac{d^2\eta}{d\tau^2} = kT_n\ C_D|\ G_2\ -\eta|(G_2-\eta') + 2(1-\sigma), \tag{31}$$

(Burrows, 1973) of the equations for plane motion (7)–(8) which amount
to the assumption that the seed moves with the local horizontal wind velocity
at all times following release. Trajectories calculated in this way are wholly
dependent on the vertical convection velocity and a very wide range of
possible flight paths can be derived. When convection currents which vary
rapidly either in time, or in space, or in both, are present, it is no longer
permissible to neglect transients, as markedly different trajectories may result
such as the ones shown in Figure 23 (Burrows, 1973). For these, the series
of convection pulses traversed with the constant velocity of the carrier wind
impart an upward momentum to the seed which is sufficient to provide
for a sustained overall gain in altitude. These results demonstrate a principle
rather than a definitive effect; it is obvious that any flight path geometry
can be constructed in this way by selecting an appropriate distribution
of upward convection and downward sink flows. It is equally obvious that
extremely large carriage distances may result from such convection flows

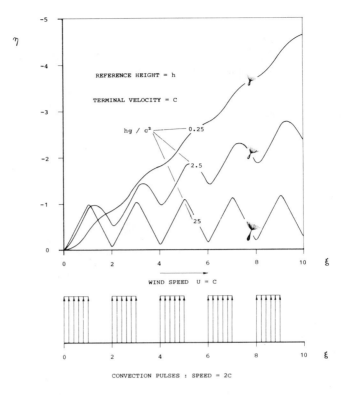

Fig. 23. Trajectories calculated for three different kinds of plumed seed moving with a uniform carrier-wind across a series of convection pulses.

as may often be observed in the field with the movement of thistledown on the wind (Dandeno, 1905; Ridley, 1930). Because plumed seeds and fruits are, in many cases, not particularly small, the possibility of secondary and subsequent trajectories may also be significant factors in dispersal in a suitable habitat on account of the relative ease with which such species may be either whisked up into the air again or dragged along the ground after alighting there.

The sensitivity of pappus geometry to atmospheric humidity is such that between totally moist and totally dry conditions the corresponding opening of the pappus causes, in some cases, a substantial increase in drag coefficient (Sheldon and Burrows, 1973). The benefits are not totally clear, because the apparent disadvantage of the shorter range achieved by transport in a given wind when moist, may be more than offset by the greater suitability for germination of the different angles of repose of Compositae after alighting on the ground. For a maximum rate of spread of a population downwind in a given habitat it is almost axiomatic that an optimum combination of wind speed and humidity should therefore exist, but the link between these two factors is not known for any one case.

F. Plane Winged Seeds

Whenever what amounts to a wing membrane forms part of the physical geometry of a seed there is the possibility of an extensive linear trajectory in regular flight, even in still air. Such a trajectory may exist when lift is developed and when all the aerodynamic forces produced act in such a way that there is equilibrium in an attitude of stable trim. The ratio of lift to drag is a measure of the gliding performance that can be attained, and when this quantity is large, the rates of sink in the glide are small. The requirements for stable trim may be met with perfectly plane and symmetrical wing membranes having a suitably shaped and located mass concentration and also in much less obvious ways by asymmetrical and crinkled ones. For the former, somewhat artificial, configuration it is not too difficult to estimate the aerodynamic forces and moments acting and to deduce trajectories, both for still air and for movement in winds that are totally steady in time and space. When there is geometric asymmetry defining departures from a plane configuration and planform, such as is very likely to occur because of uneven development and ripening of the seed as a whole, the specification of the local flow over it becomes a matter of extreme difficulty. In some circumstances in which stable flight is known to occur it may be possible to identify actual airflows with equivalent ones providing the same end result and in this sense a statistical assessment of the aerodynamic properties of such seeds may be of value. Many asymmetric configurations are much more likely to induce flow patterns, so complicated that the almost inevitable result is a curved flight path, at best having some degree of regularity, such as that shown in Figure 24A, or a very erratic flight path as indicated in Figure 24B.

Because seeds and fruits of this type consist largely of an identifiable wing they are, to some extent, not unlike aeroplanes so that some guidance as to flight path may be derived from the well established results for these machines in unpowered flight (see Durand, 1963; Pope, 1951), provided that allowance is made for the very different range of flight Reynolds numbers involved. The Reynolds numbers may affect significantly, not only the amount of lift and drag which can be developed during flight, but also the positions at which these forces, and changes in them from instant to instant, act. Combinations in fore and aft curvature of the wing membrane,

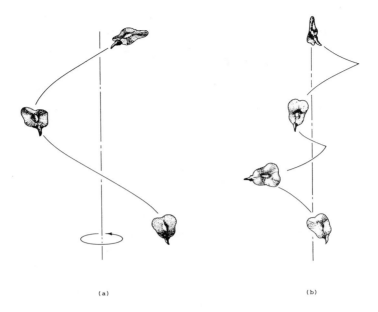

(a)
(b)

Fig. 24. Possible flight paths in still air for a plane winged seed.

reflex camber of the kind shown in Figure 1(H), dihedral, twist, warp and sweep may each contribute separately, or in combination, so as to produce flight stability, or the lack of it. Even when flight is basically stable because pseudo-static equilibrium exists there remains the likelihood that some of the self-excited modes of motion known to occur in the flight of aeroplanes will develop during the course of a trajectory. These modes include not only the oscillatory ones characteristic of flight along a path that lies wholly in the plane of symmetry, or the equivalent of this plane, but also the more complicated and coupled lateral directional combinations, which may be either oscillatory with various combinations of amplitude and phase, or simply divergent. The Phugoid oscillation (Duncan, 1952), which occurs when the flight path followed lies wholly in the plane of symmetry, is shown in Figure 25 and amounts, more or less, to a simple exchange of potential and kinetic energy about a mean glide path. It has a periodic time at a mean forward speed V given approximately by $\pi\sqrt{2}V/g$, which is clearly very short when the flight speeds are small. Figure 26 shows the spirally divergent mode that usually exists in the uncontrolled motion of most winged lifting bodies.

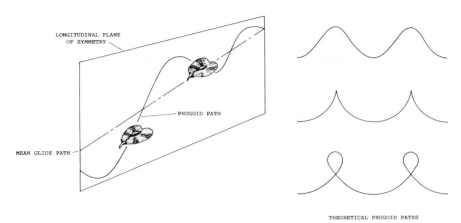

Fig. 25. Flight paths for a plane winged seed moving with a Phugoid oscillation.

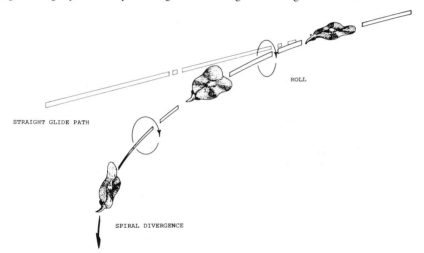

Fig. 26. Plane winged seed following a flight path with a spiral divergence.

Dynamic modes of this kind can be dealt with using the classical theories of stability and control (Duncan, 1952). However, as this involves a breakdown of the acting aerodynamic forces into components, describing in fine detail more or less exactly what is happening due to the flow over each and every portion of the body surface, there is probably little point in pursuing such analysis for seed movement if only because of the differences of geometry between and within species.

G. Winged Seeds and Fruits which Rotate when Falling

Lift may sometimes be so distributed on a winged body as to permit the existence of one or other of two states of auto-rotation. Such a distribution can be achieved with the right combination of local angles of attack and of sideslip, and may provide the means of producing a significant reduction in the terminal velocity of fall of relatively heavy bodies. It may thus enable species to achieve significant dispersal ranges in both the presence and absence of winds. Because the local speeds developed at the extremities of a rotating wing can be quite high, the amount of aerodynamic force developed there, together with the corresponding moments, may be large enough and in the right combination to produce dynamic stability.

For angles of attack below that at which stalling occurs the concentration of lift near the tip of a wing with dihedral, or the equivalent of this, becomes marked on the advancing tip with moderate increase in sideslip and, when the stall is reached, the well known state of post-stall auto-rotation (Duncan, 1952) may occur. Once rotation has begun, and this can be induced by a sudden increase in the amount of sideslip present, the down-going portion of the wing becomes progressively more stalled and the up-going portion less stalled, or even un-stalled. Post-stall auto-rotation is the state utilized for the spinning manoeuvre which is permissible with many types of aeroplanes but, whilst there is little doubt that some seeds and fruits experience this type of aerial motion, it is not clear that it is one deliberately employed and it is certainly not one affording a very predictable result, at least for the incipient phase of the motion.

For the kind of rotational motion usually employed by samaras it is only in the early stages of a flight, and perhaps during transients due to wind gusts, that difficulties emerge in the specification of flight conditions. The initial presentation of seeds and fruits of this type relative to the airflow may markedly influence the kind of flight path eventually established, for even though their geometry is such that pre-stall auto-rotation (Shapiro, 1955) is inherently possible, it does not always follow that this motion will take place in all circumstances. For example, a perfectly plane winged seed can fall in a perfectly straight line without rotation, and without any inclination to the vertical, when released from rest into absolutely still air, and yet the same seed, when released with an initial angle of inclination of the wing plane to the vertical, may very quickly proceed to an incipient rotation and progress to the more stable auto-rotative state combined with the more or less steady vertical descent, which is perhaps the most obvious feature of the motion of seeds of this type. Quite small amounts of geometric asymmetry in wing planform, camber, and mass distribution for otherwise similar species may cause differences in the pattern of motion and, in some cases, lead to preferred attitudes and directions of rotation.

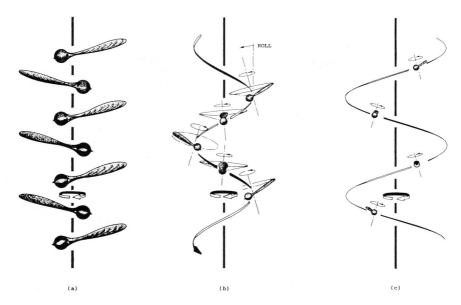

Fig. 27. Trajectories for rotating winged seeds descending in still air. (A) Simple helical trajectory. (B) Compound helical trajectory. (C) Compound helical trajectory with large lateral dimensions.

Some aspects of the motion of winged seeds which rotate when falling can be discussed using results derived for helicopter rotors (Shapiro, 1955; Norberg, 1973; Burrows, 1975b) and it is evident that for falls in still air trajectories of the kind shown in Figure 27 are possible. Figure 27A shows what is perhaps the most familiar case in which the centre of mass of the seed descends roughly in a straight vertical line whilst the wing tip describes a single, and fairly regular, helical path. For the trajectory shown in Figure 27B the deviation of the path followed by the centre of mass from the vertical straight line (Fig. 28A) occurs if the plane of the wing tip path during rotation is inclined to the horizontal and flight equilibrium is established in a sideslipping motion with stable auto-rotation. The amount of lift developed and the way in which it is distributed depends then on the sideslip and it is locally increased or decreased according to whether the wing is advancing or retreating along the direction of the sideslip. The rolling this produces, together with the gyroscopic effects of the rotating mass, result in the compound helical trajectory shown in Figure 27B. As is usual with inverted gyroscopic precession (Easthope, 1967) the direction of the rotation of the helical path described by the centre of mass is opposite to that of the rotation about the centre of mass. Compound helical trajectories of considerably greater lateral dimensions become possible when

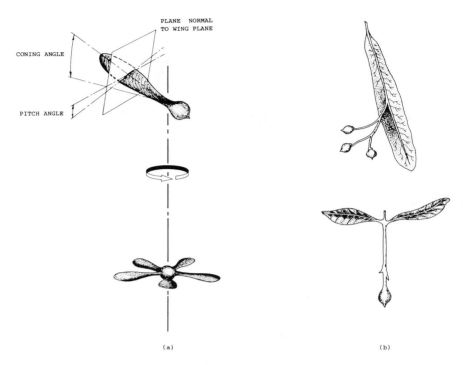

(a)   (b)

Fig. 28. Elementary in-flight attitudes for rotating winged seeds. (A) Pitch and coning angles for a single winged seed and for a nominally plane multi-winged seed. For the multi-winged seed the wing plane is normal to the axis of rotation. (B) Winged seeds with offset and suspended masses.

seeds of this type are projected with sufficient speed in downward directions inclined to the vertical (Fig. 27C).

Although these trajectories are relatively stable, it is possible for seeds of this type to become toppled from their flight paths by sudden cross-winds of sufficient strength. If this happens then falling, and possibly carriage in an unpredictable manner, may occur and stable auto-rotation will only become re-established if more steady regions of wind flow are reached.

Unlike the single winged varieties, multi-winged seeds of the type shown in Figure 28A have no choice of wing pitch and coning angles, and if their centre of mass is concentrated in the plane of the wings their fall is made either right way up or upside down. When there is a concentration of mass in suspension below the plane of the wings (Fig. 28B) the ultimate attitude of steady fall is well defined.

Certain winged species, such as *Liriodendron tulipifera*, have partial concentrations of mass away from the short-time averaged positions of

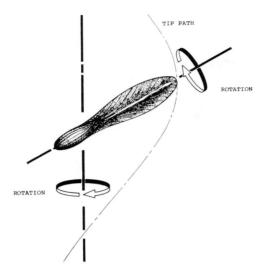

Fig. 29. Trajectory for a rolling winged seed descending in still air.

the axes along which the aerodynamic forces act, and there are others, such as *Fraxinus pennsylvanica*, with a wing planform arranged more or less symmetrically about a central spine. The effect of this in the former case is to produce, and in the latter to permit, a rotation in pitch about an axis which is roughly aligned with the span of the wing, in addition to the components of the downward spiral (McCutchen, 1977) as is indicated in Figure 29. During such a motion the precise effect on the production and magnitude of the aerodynamic forces developed is not at all clear, mainly because so little is known of the quantitative significance of hysteresis in the flow. What is clear is that, as the wing goes through the full 360° range of angles of attack, the work done by the relatively powerful suction forces developed near the leading edge of the wing as the stalling angle is reached, is sufficient to supply enough rotational kinetic energy to the mechanical system in this phase of the motion for there to be a surplus. This surplus energy can be stored in the rotating mass and utilized to maintain the rotation against the aerodynamic resistance developed in the less favourable portions of the cycle. For a trajectory initiated from rest in still air, the total amount of energy available to the system is the potential energy due to the height of the release point above the ground. Since part of this energy is required to maintain the rotation in pitch there is less available for producing lift, so that the terminal velocities of fall for this kind of samara are greater than for those that do not rotate in pitch (Green, 1980).

To determine the trajectories followed by rotating winged seeds moving from release in variable winds the complete set of equations (2) and (3) are required. This in turn means that the aerodynamic force and moment coefficients in these equations, together with the inertial ones defined by the mass distribution, must be specified for the wide range of different flight attitudes employed by the species in dispersal. The difficulty of doing this satisfactorily is enormous. For steady rotational flight with descent at the terminal velocity, approximate calculations of trajectories may be made in a similar way to those for plumed seeds (Burrows, 1973, 1975b).

VI. CONCLUDING REMARKS

Whilst it is clear from the foregoing that the aerial movement of seeds, fruits, spores and pollen can be discussed on a theoretical basis for circumstances and environments that are selected models of reality, it is equally clear that, because of the diversity of natural pattern exhibited in real species geometry and in real wind flows, the ultimate value of the theory is perhaps in quantifying effects of interest, rather than in providing definitive results for specific problems. An accurate specification of any real wind flow is at best likely to be descriptive only of local atmospheric conditions, although local on the atmospheric scale may be more than sufficient on the environmental scale for dealing with trajectories emanating from species in a particular habitat. In some matters of detail the theory is particularly useful, and is especially so in indicating the sort of differences to be expected in the trajectories followed by different species in the same environment, in the very different positions of ultimate migration that may be expected when the winds are turbulent, and in the circumstances in which deposition concentrations may occur. Such deductions can be made as long as the aerodynamic coefficients are known for individual species.

REFERENCES

Broadbent, E. G. (1956). 'The Elementary Theory of Aero-Elasticity. An Aircraft Engineering Monograph'. Bunhill Publications Limited, London.
Bryan, G. H. (1911). 'Stability in Aviation'. Macmillan, London.
Buller, A. H. R. (1909). 'Researches on Fungii'. Vol.1. Longmans Green London.
Burrows, F. M. (1973). *New Phytololologist* **72**, 647–664.
Burrows, F. M. (1975a). *New Phytologist* **75**, 389–403.
Burrows, F. M. (1975b). *New Phytologist* **75**, 405–418.
Burrows, F. M. (1983). *Proc. Roy. Soc. Lond., Series A* **139**, 15–66.
Burrows, F. M. (1985). Calculation of the primary trajectories of ejected spores. (Unpublished work).

Clift, R., Grace, J. R., and Weber, M. E. (1978). 'Bubbles, Drops, and Particles'. Academic Press, New York.

Dandeno, J. B. (1905). *Science* **22**, 568.

Duncan, W. J. (1952). 'Control and Stability of Aircraft'. Cambridge University Press, London.

Durand, W. F. (Editor) (1963). 'Aerodynamic Theory. Volume V'. Dover Publications, New York.

Easthope, C. E. (1967). 'Three-dimensional Dynamics'. Butterworth, London.

Goldstein, S. (1929). *Proc. Roy. Soc. Lond., Series A* **123**, 225.

Green, D. S. (1980). *Am. J. Bot.* **67**(8), 1218–1224.

Gregory, P. H. (1961). *Ann. Appl. Biol.* **38**, 357–376.

Gregory, P. H. (1973). 'The Microbiology of the Atmosphere'. Leonard Hill. International Textbook Company, Aylesbury, UK.

Hoerner, S. F. (1958). 'Fluid-Dynamic Drag'. Hoerner, New York.

Ingold, C. T. (1953). 'Dispersal in Fungi'. Clarendon Press, Oxford.

Ingold, C. T. (1965). 'Spore Liberation'. Clarendon Press, Oxford.

Ingold, C. T. (1971). 'Fungal Spores'. Clarendon Press, Oxford.

Lamb, H. (1952). 'Hydrodynamics'. Cambridge University Press.

McCutchen, C. W. (1977). *Science* **197**, 691–692.

Milne-Thomson, L. M. (1949). 'Theoretical Hydrodynamics'. Macmillan, London.

Milne-Thomson, L. M. (1952). 'Theoretical Aerodynamics'. Macmillan, London.

Moore, P. D., and Webb, J. A. (1978). 'An Illustrated Guide to Pollen Analysis'. Hodder and Stoughton, London.

Norberg, R. A. (1973). *Biol. Rev.* **48**, 561–596.

Oseen, C. W. (1927). 'Hydrodynamik'. Akademische Verlagsgesellschaft, Leipzig.

Pope, A. (1951). 'Basic Wing and Aerofoil Theory'. McGraw-Hill, New York.

Prandtl, L. (1952). 'Fluid Dynamics'. Blackie, London and Glasgow.

Prandtl, L., and Tietjens, O. G. (1957). 'Fundamentals of Hydro- and Aero-mechanics'. Dover Publications, New York.

Ridley, H. N. (1930). 'The Dispersal of Plants Throughout the World'. Reeve, Ashford, UK.

Robinson, A., and Laurmann, J. A. (1956). 'Wing Theory'. Cambridge University Press, London.

Saffman, P. G. (1965). *J. Fluid Mechanics* **22**, 385.

Scagel, R. F., Bandon, R. J., Rouse, G. E., Schofield, W. B., Stein, J. R., and Taylor, T. M. C. (1969). 'Plant Diversity'. Wadsworth Publishing Company, Belmont, California.

Schlichting, H. (1955). 'Boundary-layer Theory'. Pergamon, London.

Shapiro, J. (1955). 'Principles of Helicopter Engineering'. Temple, London.

Sheldon, J. C., and Burrows, F. M. (1973). *New Phytologist* **72**, 665–675.

Sheldon, J. C., and Lawrence, J. T. (1973). *New Phytologist* **72**, 677–680.

Stokes, G. G. (1851). Trans. Cambridge Phil. Soc. **9**, 51.

Swaine, M. D., and Beer, T. (1977). *New Phytologist* **78**, 695–708.

Swaine, M. D., Dakubu, T., and Beer, T. (1979). *New Phytologist* **82**, 777–781.

Temple, G. (1958). 'Fluid Dynamics'. Clarendon Press, Oxford.

CHAPTER 2

Seed Dispersal by Water

DAVID R. MURRAY

I.	Introduction	49
II.	Seed Dispersal by Rain, River and Flood	50
	A. Arid Zones	50
	B. The Amazon River in Brazil	51
	C. *Xanthium occidentale* in Australia	52
III.	Evidence for Seed Dispersal by Ocean Currents	53
	A. The Distribution of *Cycas* Species	55
	B. Palms	60
	C. Legumes	62
	D. Mangroves	72
IV.	Effective Seed Dispersal by Ocean Currents Compared to	
	Other Vectors	74
	A. Krakatau	74
	B. The Hawaiian Islands	76
	C. The Galapagos Islands	77
V.	The Hazards for Seedling Establishment	78
VI.	Conclusions	80
	Acknowledgements	
	References	

I. INTRODUCTION

The rise of the angiosperms to dominance in many habitats has been an intricate and complex process, involving the development of superior means of propagation. The effectiveness of long-distance methods of seed dispersal has been well illustrated by those who have studied the vegetation patterns of oceanic islands, such as Carlquist (1970, 1974, 1981, 1983), Porter (1976, 1979, 1983), Raven (1973) and Sykes (1977). A newly emergent solitary island can acquire vegetation only as a result of long-distance diaspore

SEED DISPERSAL
ISBN 0 12 511900 3

dispersal. The same principle may also apply to continents: certain habitats, such as alpine zones or rainforests, can be regarded as islands surrounded by other different habitats. A chance arrival might need to be successful only once for that species to gain a foothold, and perhaps a new centre for active radiation and speciation. There can be no doubt that modern inter-continental distributions of angiosperms have been profoundly influenced by long-distance seed dispersal.

In this chapter, examples are described of the ways in which seeds of both angiosperms and gymnosperms are dispersed by water. In terrestrial environments, seeds may be carried by running water for as long as they may float, and sometimes this action of water is indiscriminate. However, certain species of angiosperms are adapted for seed dispersal by flowing water, either as a solitary mechanism, or as one possible means of dispersal among several. This theme is considered further in Section II.

The balance of this chapter is concerned with the more spectacular prospect of seed dispersal by ocean currents. As a school student I enjoyed reading Thor Heyerdahl's books about his attempts to drift to Pacific Islands from South America on balsa rafts, to demonstrate the feasibility of an idea that was contrary to the conventional wisdom of the time (Heyerdahl, 1950, 1958). What man-made rafts can do, seeds and fruits have been doing for millions of years.

Nelson (1978) has reviewed the history of observations on tropical (West Indian) seeds that were deposited naturally on the shores of the British Isles (Section V), and notes that the earliest such reports were made by Pena and de L'Obel (1570) and Sloane (1696, 1725). Oceanic drift of fruits and seeds has occasionally been a fashionable subject for research. Interest in the subject of island revegetation was aroused following the massive eruption of Krakatau in August 1883, when the surviving portion of this island was completely denuded (Ernst, 1908; Section IV,A). In several books, Guppy (1887, 1890, 1906, 1917) documented many instances of successful flotation of tropical fruits and seeds in sea-water. The most recent book on the subject is that by Gunn and Dennis (1976), which is an excellent aid to the identification of drift fruits and seeds.

II. SEED DISPERSAL BY RAIN, RIVER AND FLOOD

A. Arid Zones

Attenborough (1984) has described the dispersal of seeds from dried plant remains by the action of rain. Tensions that develop from the selective uptake of water by the dead outer tissues of the seed-heads can shoot seeds several feet into the air (p. 156 of Attenborough, 1984). On landing, such seeds may germinate, or be carried still further by running water. The life cycles of many of these ephemeral species are completed rapidly. Germination in response to rain is risky if insufficient rain falls for the transpiration needs of the growing plants. Adaptations which encourage germination only after substantial rain has fallen include the presence of salt-containing bracts around the seeds. Inhibitory amounts of sodium chloride (up to about 0.9 M in water-saturated tissues; Beadle, 1952) occur in the bracts from several Australian species of salt-bush (*Atriplex*). Seeds of these species have been shown to germinate only when substantial winter rain has fallen (between one and three inches), sufficient to dilute the salt below concentrations that completely inhibit germination, and allow the salt-bush seedlings to complete a growth cycle, or their entire life-cycle in the case of annuals (Beadle, 1952).

B. The Amazon River in Brazil

Plant species that prefer to grow along the banks of streams and rivers probably shed many of their seeds directly into running water. For example, Pedley (1978) indicates that the seeds of *Acacia aulacocarpa, A. salicina* and *A. stenophylla* are normally dispersed by streams in northern Australia.

A far more complex system of water-dependent seed dispersal occurs in the flood plain of the Amazon River in Brazil. When the river overflows its banks and floods the surrounding forests, numerous species of fish swim among the trees, eating the fruits of figs and Brazil nuts (*Bertholletia excelsa*) as they fall. Thus the fruits are not simply stranded in new locations when the flood-waters recede. Fish closely related to pirhanas, but which are not carnivorous, and cichlids depend substantially on their annual intake of fruits and seeds. The soft fruits of figs (*Ficus* spp.) are eaten readily, and the small seeds are passed by the fish some distance away from the parent tree. The fruits of the Brazil nut can weigh up to four pounds (1.8 kg) and contain from 12–24 seeds (for illustration, see p. 137 of Rix, 1981). Embryos of the Brazil nut can be excreted undamaged by species of characin fish that have crushing molars and jaw muscles powerful enough to crack the shells (p. 216 of Attenborough, 1984). The Brazil nut embryos that escape crushing are deposited ready to germinate.

This complicated system of seed dispersal involves mutual adaptations of fish and plant species, and presents a very interesting example of co-evolution.

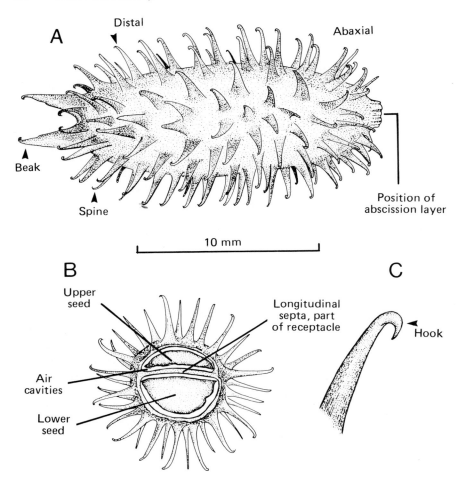

Fig. 1. The fruit of *Xanthium occidentale*: (A) the burr; (B), transverse section of the burr showing unequal seeds in chambers with air cavities; (C) hooked spine (from Liddle and Elgar, 1984).

C. *Xanthium occidentale* in Australia

Xanthium occidentale is an annual weed of tropical or subtropical origin, a component of the widespread *Xanthium strumarium* complex (Löve and Dansereau, 1959). In Australia it is known as the Noogoora burr, after the estate in Brisbane (Queensland) where its first rapid spread was recorded between 1871 and 1896 (Maiden, 1920). Between 1897 and 1920, this weed also spread rapidly into northern New South Wales. Maiden attributed

Table I. Physical characteristics of burrs of *Xanthium occidentale*.

Fresh weight on day of collection	0.29 g
Dry weight	0.22 g
Waterlogged weight	0.596 g
Volume	0.57 mL
Density (dry)	0.39 g cm^{-3}
Density (saturated)	1.05 g cm^{-3}
Sinking time	69.9 h
Sinking rate	0.083 m s^{-1}
Sinking rate with spines removed	0.091 m s^{-1}

Data of Liddle and Elgar (1984)

its capacity for spreading to two main means of dispersal: attachment of burrs to stock animals, or flotation of burrs in flood waters (p. 101 of Maiden, 1920). The most recent studies of Noogoora burr dispersal are those of Wapshere (1974a,b) and Liddle and Elgar (1984).

The burr structure is shown in Figure 1. The outer spines permit attachment to the wool of sheep, or less readily to the hair of cattle or horses. Air cavities, which surround each seed (Fig. 1), assist the burrs to float for about 70 h (Table 1). This would enable burrs to travel a distance up to 420 km in a current moving with average speed 6 km h^{-1}.

Burrs are released at maturity following the development of an abscission layer at the base of the fruit (Fig. 1). This release may be passive, or modified by the action of animals, wind or water. Liddle and Elgar (1984) summarized these multiple pathways for dispersal as shown in Figure 2. They considered that dispersal by water would probably deposit burrs in 'safe' habitats downstream from parent plants, whereas animal dispersal, although potentially more wasteful of seeds, would permit more rapid movement of plant populations upstream.

Liddle and Elgar (1984) arrived at a conclusion that may be generally true: the range of dispersal pathways open to any diaspore is probably much wider than usually considered.

III. EVIDENCE FOR SEED DISPERSAL BY OCEAN CURRENTS

The present ocean basins have been developing since the breakup of the former super-continents from about 225 million years ago (for reviews see Raven and Axelrod, 1974; Sullivan, 1977; Fleming, 1979). Major fragments of Gondwanaland, including India, Australia, Antarctica, South Africa and South America, once shared a common vegetation characterized by *Glossopteris* and *Gangamopteris*. These plants were once considered to be ferns or cycads (p. 124 of Laseron, 1955) but are now considered to be seed

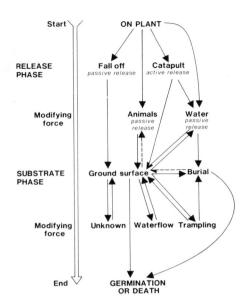

Fig. 2. The multiple pathways available for dispersal of burrs of *Xanthium occidentale* (from Liddle and Elgar, 1984).

ferns (7.III,D). From early Jurassic times to early Cretaceous, Gymnosperms of the Pentoxylales became important elements of the vegetation of India (Rao, 1974, 1981), New Zealand (Blaschke and Grant-Mackie, 1976) and Australia (Walkom, 1918, 1919; Drinnan and Chambers, 1985). Various possible scenarios for the separation of components of Gondwanaland suggest a separation of India from Australia in the early Cretaceous, then a later separation of New Zealand, from about 80 million years ago (Raven and Axelrod, 1974; Fleming, 1979).

With exceptions such as New Zealand, Madagascar, the Seychelles, Sri Lanka, New Caledonia and the Falkland Islands, most oceanic islands have resulted from volcanic action associated with sea-floor spreading, and their ages are often correlated with the distances they have been carried from mid-oceanic ridges (Wilson, 1963a,b). The system of ocean currents that prevails today is also of recent origin. A new land-bridge connecting North and South America was completed only about four million years ago (p. 159 of Sullivan, 1977), and this closure forced currents to circulate in the Atlantic and Pacific Oceans. Sea levels have fluctuated with intermittent ice-ages, and many islands have at times been totally submerged, resulting in the complete loss of existing plant cover (Renvoize, 1979). Such islands must be revegetated *de novo* on re-emergence.

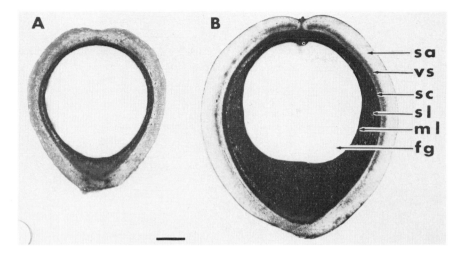

Fig. 3. Transverse sections of mature seeds of *Cycas revoluta* (A) and *Cycas circinalis* (B). Sarcotesta (sa), vascular bundle in sarcotesta (vs), sclerotesta (sc), spongy layer (sl), membranous layer (ml) and female gametophyte tissue (fg) are all present in *C. circinalis* (B), but note the absence of the spongy layer in *C. revoluta* (A). The sclerotesta is so hard that it was necessary to use a jeweller's saw to obtain the sections. Bar = 1 cm (from Dehgan and Yuen, 1983).

A. The Distribution of *Cycas* Species

1. Structural Features of Seeds Permitting Flotation
Buoyancy of the fruits and seeds of angiosperms may result from one or more adaptive modifications to structure. The retention of air in chambers in the fruits of *Xanthium* has already been described (Fig. 1). This occurs also in the false fruits of *Atriplex inflata* and *Atriplex spongiosa* (p. 50 of Beadle, 1952), and in the seeds of orchids such as *Cypripedium parviflorum* (p. 372 of Eames and MacDaniels 1947). Specialized cell layers that contain air spaces at maturity may also develop within the seed coats, as in *Menyanthes trifoliata* and *Calla palustris* (Johri, 1984).

Recent anatomical studies by Dehgan and Yuen (1983) have revealed that similar modifications may occur within the seeds of certain species of Gymnosperms, namely the cycads *Cycas circinalis*, *C. rumphii* and *C. thouarsii*. In the seeds of *C. revoluta*, representing the majority of modern *Cycas* species, the seed coats consist of an external sarcotesta, a sclerotesta, and a third membranous layer enclosing the central female gametophyte (Fig. 3A). The seed coats of *C. circinalis* comprise the same layers, and an additional layer between the sclerotesta and the innermost membranous layer (Fig. 3B). This extra parenchymatous tissue is described as spongy,

Fig. 4. External surface of the sarcotesta. Bar = 50 μm.*

and is derived from an inner fleshy layer which receives a distinct branch of the vasculature, as do the sarcotesta and the membranous layer.

Further anatomical details of the four seed coat layers of C. circinalis are shown in Figures 4 to 9. The outer surface of the sarcotesta (Fig. 4) is thickly covered with wax extruded from the epidermal and adjacent cell layers (Fig. 5). Many of these cells are filled with wax or resin (Fig. 5). The remaining sarcotesta is composed mainly of storage parenchyma filled with starch grains (Fig. 6). Resin canals (Fig. 6) and vascular bundles (Fig. 7) are also prominent in the sarcotesta, and the tracheids of the xylem are scalariformly pitted (Fig. 7).

The arrangement of cells within the sclerotesta is difficult to discern, as the cells are thickened and apparently interlocked (Fig. 8), resulting in a very hard layer. In contrast, the extra tissue inside the sclerotesta consists of large, thin-walled parenchyma, with intercellular air spaces (Fig. 9).

Dehgan and Yuen (1983) interpret their observations to indicate that the seed coats of C. revoluta and most species of Cycas are modified by the loss of the spongy parenchyma layer at maturity, whereas this layer develops and is retained in C. circinalis, C. rumphii and C. thouarsii.

2. Seed Flotation and the Distribution of Cycas

The seeds of Cycas species that lack the extra parenchymatous layer in the seed coats will float in water only if the seed is non-viable, as an air space is created by the shrinkage of the female gametophyte away from

* Figs 4 to 9 are all from Dehgan and Yuen (1983).

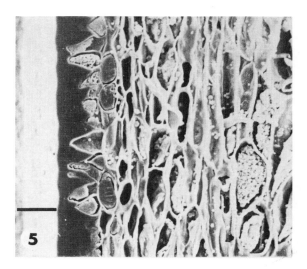

Fig. 5. Section of sarcotesta showing epidermal cells with exterior cuticle. Bar = 100 μm.

the sclerotesta (Giddy, 1974). Sound seeds of these species will sink, and this is often used as a guide to viability. However, the seeds of the few species of *Cycas* with the fourth spongy layer in their seed coats are able to float in sea-water for at least five weeks (Dehgan and Yuen, 1983), possibly for three months or longer (Ridley, 1930; Gunn and Dennis, 1976).

The distributions of taxa with floating or non-floating viable seeds are shown in Figure 10. *C. circinalis* and *C. rumphii* are more widely distributed in the zone shown for taxa with floating seeds than *C. thouarsii*, which is confined to the western extremity. Numerous reports of the discovery of seeds of *C. circinalis* and *C. rumphii* on sandy beaches are cited by Dehgan and Yuen (1983), and these two taxa may form 'luxuriant littoral stands'. The majority of *Cycas* species with non-buoyant seeds are of narrow individual distribution, and often occur inland.

The dispersal of *Cycas* seeds by flotation across the Indian Ocean is more likely to be successful in the direction east to west during the southern winter (Fig. 11). Westward intensification of currents in the oceans is believed to be a consequence of the Earth's rotation (see Robinson, 1963). On reaching the west coast of Madagascar, the south equatorial current divides: the northern branch continues around northern Madagascar, the Comoro Islands and part of the East African coast; the southern branch moves towards the Cape of Good Hope (Fig. 11; see also p.123 of Renvoize, 1979). Dehgan and Yuen (1983) have drawn attention to the failure of *C. thouarsii* to have become widely dispersed, correlating this failure with the lack of a west-east current at the time of the year when mature seeds are shed. Thus *C. thouarsii* is apparently derived from *C. circinalis*.

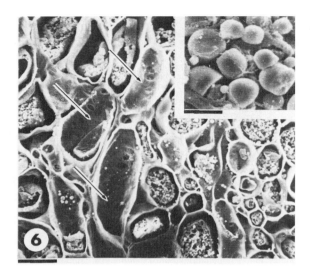

Fig. 6. Starch-filled parenchyma cells and resin canals (arrows) in the sarcotesta. Bar = 100 µm. Inset: starch grains (bar = 33µm).

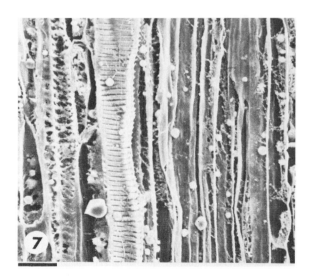

Fig. 7. Vascular bundle. Bar = 50 µm.

Fig. 8. Sclerotesta. Bar = 100 μm.

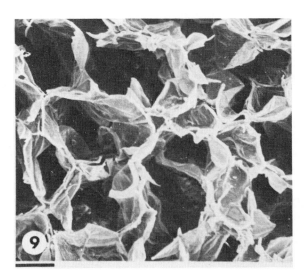

Fig. 9. Spongy parenchyma of the spongy layer. Bar = 100 μm.

60   *David R. Murray*

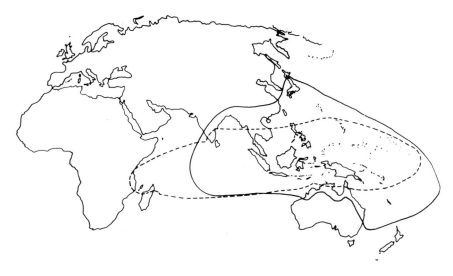

Fig. 10. Distribution of the genus *Cycas*. Species with mature seeds capable of floating are distributed within the dashed line; other species are found within the solid line (from Dehgan and Yuen, 1983).

It will be many years before phylogenetic relationships within the genus *Cycas* are fully elucidated, but Dehgan and Yuen (1983) have concluded that the many geographically isolated island populations referred to *C. circinalis*, *C. rumphii* and *C. thouarsii* should not be distinguished at the level of species, since there are no cytological barriers to breeding within this group, and the morphological differences are so minor.

A point made earlier concerning multiple pathways of seed dispersal (Section II,C) is pertinent also to cycads. The bright orange seeds of *Cycas*, as well as of other cycad genera (Bauman and Yokoyama, 1976) are frequently taken and carried away by a variety of animals that eat the starchy sarcotesta (see Burbidge and Whelan, 1982; Dehgan and Yuen, 1983), then discard the rest of the seed. *Cycas rumphii*, for example, is dispersed both by ocean currents and by bats (van der Pijl, 1957). It would appear that the edible sarcotesta (Fig. 6) is free of cycasin and related glycosides of methylazoxymethanol, first discovered in seeds of *C. circinalis* (for review, see Bell, 1984).

B. Palms

A sprouting coconut (*Cocos nucifera*) on the sandy shore of a tropical beach is displayed on the cover of 'The Living Planet' (Attenborough, 1984).

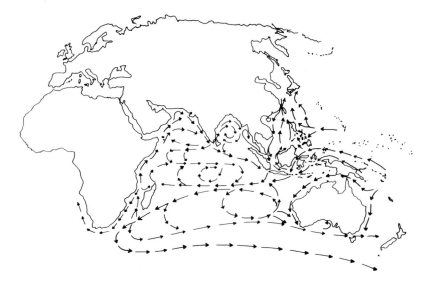

Fig. 11. Major currents in the Indian Ocean (from Dehgan and Yuen, 1983).

No better symbol could be chosen to illustrate the success of flotation as a means of dispersal. Coconut palms grow close to and often overhang the shore, and the shed fruits have ready access to the sea. Their original pattern of spread throughout the tropics is now concealed by the overlying patterns of human migration, so desirable is the fruit for food, drink and materials (Gilmore, 1978).

There is evidence, nevertheless, that the coconut is one of the earliest colonizing species whenever a new littoral habitat becomes available. Renvoize (1979) has described the vegetation of Aldabra, an island of intermediate elevation near Madagascar. Aldabra emerged about 80 000 years ago (Braithwaite *et al.*, 1973), and apparently the coconut became established there before recorded human visits to the island. The coconut was reported to be the largest of the drift fruits arriving on the newly formed shores of Krakatau after the eruption of 1883 (p. 27 of Ernst, 1908). The large number of sprouting nuts on the ground and young plants already 1 m high in 1906 indicated that the parent population had become established very soon after the eruption (p. 30 of Ernst, 1908).

The progenitor of the loulu palm *Pritchardia remota* probably arrived in the Hawaiian Islands by means of floating fruits. *P. remota* prefers a littoral habitat, and its dry fruits will float. Within the Hawaiian Islands, *P. remota* is today found mainly on the low islands Laysan and Nihoa, but it is also found in the Tuamotu Islands in the South Pacific (p. 169 of Carlquist, 1970). The Hawaiian Islands are presumably the source of fruits floating to other islands, as fossil *Pritchardia* stems about 40 000 years old have been found in volcanic tuff from the Salt Lake Crater of Oahu (p. 61 of Carlquist, 1970). The Hawaiian endemic loulu palms *P. lowreyana* and *P. martii* are adapted to wet forest areas, and probably evolved from the same progenitor as *P. remota*. Both produce larger fruits than *P. remota* (p. 170 of Carlquist, 1970), and these no longer have the opportunity for dispersal by flotation.

Another palm that has lost the ability to distribute viable seeds by ocean currents is the coco-de-mer (*Lodoicea seychellarum*). This palm is endemic to the Seychelles, and found only on the islands of Praslin and Curieuse (p. 186 of Gunn and Dennis, 1976). The spectacular double fruits take six years to complete their development, and the seed is reported to be the largest produced by any plant, finally weighing up to 90 kg (Good, 1951). Although denser than a coconut, the fruits will float, and indeed were washed ashore in India and neighbouring countries for centuries before their source was discovered by Mahé de la Bourdonnais in 1743 (Gunn and Dennis, 1976). The dispersal mechanism that must have brought this palm's progenitor to the Seychelles has become ineffective, because the embryo inside the seed can no longer survive prolonged immersion in sea-water.

C. Legumes

Legumes from all three (sub)families have been considered capable of effective seed dispersal by ocean currents (Table II). *Caesalpinia bonduc* (*Guilandina Bonducella*) and *Abrus precatorius* are typical of such species, with pantropical distributions. Both were recorded by Banks in Brazil and the coast of New Guinea (p. 144 of Beaglehole, 1962), and both were among the colonists of Krakatau and Verlaten Island after 1883 (Ernst, 1908). The seeds of *Caesalpinia bonduc* have long been recognized among the drift seeds of Irish beaches (Sloane, 1725; Nelson, 1978), and this species is established in the Galapagos Islands (Porter, 1983) and Lord Howe Island (Pickard, 1983; Rodd and Pickard, 1983).

The identification of these species has never posed any problem, as populations in different locations differ little in morphology. The legumes that have provided a challenge so far as definitive identification is concerned are now considered in some detail, as they illustrate the diagnostic potential of biochemical methods of assessing identity and affinity.

1. Acacia

The seeds of most species of *Acacia* are robust, with thick impermeable seed-coats. These account for a high proportion of seed mass: from 33%

Table II. Examples of legumes whose seeds are capable of long-distance dispersal by ocean currents.

Species and affiliation	References
Caesalpiniaceae	
Caesalpinia bonduc	Ernst (1908) and see text
Parkinsonia aculeata	Porter (1983)
Mimosaceae	
Acacia	Section III,C,*1*
Entada gigas	Sloane (1696), Nelson (1978)
Prosopis juliflora	Porter (1983)
Papilionaceae	
Abrus precatorius	Ernst (1908)
Canavalia maritima	Ernst (1908), Porter (1983)
Castanospermum australe	Green (1979)
Desmodium umbellatum	Ernst (1908)
Dioclea reflexa	Nelson (1978)
Erythrina variegata (indica)	Ernst (1908), Carlquist (1970)
Mucuna sloanei	Nelson (1978)
Sophora tomentosa	Ernst (1908), Section III,C,*2*
Stylosanthes sympodialis	Porter (1983)

to 43% in the few species for which this has been measured (Murray *et al.*, 1978; Ashcroft and Murray, 1979). The structure of the seed coats allows the seed to remain impermeable to water as long as the lens is undisturbed (Tran and Cavanagh, 1980, 1984). The longevity of *Acacia* seeds under ambient conditions is excellent. The seeds can remain in soil for many years (50 or more; see Tran and Cavanagh, 1984), eventually germinating when the lens is shocked by the heat of a passing fire (Chapter 6), or by the scouring action of water and soil particles.

The impermeability of *Acacia* seed coats extends to sea-water. Cambage (1919, 1923) conducted experiments with seeds of several Australian species. These seeds germinated (in response to a boiling water treatment) after periods in contact with sea-water as long as six years (Table III). Given the opportunity, it is clearly feasible for seeds of *Acacia* to survive lengthy periods afloat in rivers and oceans.

Most Australian species of *Acacia* are characterized by the ability to replace pinnate and bipinnate leaves with phyllodes, which are modified petioles. This modification is possibly an adaptation to water stress (Tunstall and Connor, 1975; Walters and Bartholomew, 1984), or may originally have been so. Species with phyllodes are uncommon outside Australia, and many of these occur on islands (Table IV). Are some or all of these species derived from Australian ancestors, carried as seeds by ocean currents?

A way of testing this idea has been provided by Evans *et al.* (1977). By analysing the free amino acid content of seeds of more than 100 species

Table III. Germination records for single seeds of *Acacia* after periods of immersion in sea-water.

A. farnesiana	3¾ years	A. falciformis	6 years
A. melanoxylon	5 years	A. podalyriifolia	4 years

Data of Cambage (1919, 1923).

Table IV. Phyllodinous species of *Acacia* occurring outside Australia.

Species	Location
A. heterophylla	Mascarene Islands (Réunion, Maurice)
A. confusa	Taiwan, Philippines
A. koa	Hawaiian Islands
A. koaia	Hawaiian Islands
A. kauaiensis	Hawaiian Island (Kauai)
A. richei	Fiji
A. mathuataensis	Fiji
A. simplicifolia (syn. A. laurifolia)	Fiji, New Caledonia
A. spirorbis	New Caledonia
A. mangium	Amboine, Oma, Harocha
A. willardiana	Sonora (Mexico), Southern California.

After Vassal (1972).

of *Acacia*, these authors determined that four distinct world groups could be recognized. Seeds from all of the Australian species (with bipinnate foliage or phyllodes alike) accumulate an assortment of non-protein amino acids typified by albizziine, S-carboxyethylcysteine, S-carboxyisopropyl-cysteine, Djenkolic acid, α-amino-β-acetylaminopropionic acid, pipecolic acid, 4-hydroxypipecolic acid and 5-hydroxypipecolic acid.

Another group, comprising Bentham's series Gummiferae, possesses seeds which accumulate a much narrower set of compounds: Djenkolic acid, several derivatives of Djenkolic acid including N-acetyldjenkolic acid (invariably present), pipecolic acid and 4-hydroxypipecolic acid. The seeds of a third group were shown to contain a set of non-protein amino acids similar to that of the Australian group, but with α, β-diaminopropionic acid and one or both of the neurotoxins α-amino-β-oxalylaminopropionic acid and α-amino-γ-oxalyaminobutyric acid in addition. This group consists mainly of African and Asian species, which Bentham had placed in his series Vulgares.

Finally, there are species whose seeds do not accumulate high concentrations of free amino acids. These species are American, with some African, the 'pennata' group of series Vulgares.

Of the island species of *Acacia* with phyllodes (Table IV), three were included in the analyses reported by Evans *et al.* (1977). Only one species exhibited a pattern of seed non-protein amino acids typical of Australian species: *A. heterophylla* from the Mascarene Islands. Both of the Pacific Island species tested, *A. kauaiensis* and *A. confusa*, displayed profiles of seed non-protein amino acid that included α-amino-β-oxalylaminopropionic acid, indicating an Asian origin.

The genus *Acacia* is relatively young. The earliest fossil pollen grains in Australia are about 25 million years old (H. A. Martin, pers. comm. 1985), and in Africa they are about 45 million years old (Guinet, 1981). Therefore it is not possible to account for the present distribution of *Acacia* in terms of ancestral species present in Gondwanaland, then dislocated by 'ancient continental drifts' (Vassal, 1972). The most plausible explanation of the present distribution of island species with phyllodes is that the progenitors of at least some were carried as seeds by ocean currents: across the Indian Ocean from Australia for *A. heterophylla*, and into the Pacific Ocean from Asia for *A. confusa* and *A. kauaiensis*. *A. mangium*, which occurs on some Pacific islands (Table IV), is otherwise confined to high rainfall areas of north-eastern Queensland (p. 96 of Pedley, 1978). It is possible that seeds of this species could have been dispersed from Australia by ocean currents, but this is not certain, as the seeds are typical of those dispersed by birds (Table V of Chapter 3).

Mildenhall (1972, 1975) described fossil pollen grains of *Acacia* present in Pliocene sediments in the North Island of New Zealand. The grains formed aggregates (polyads) structurally similar to those of Australian species (Cookson, 1954; Knox and Kenrick, 1983), comprising either 12 or 16 pollen grains. Considering the distance of New Zealand from Australia at the time, it is improbable that the quantities of *Acacia* pollen found in New Zealand could have accumulated directly from Australian sources. The heavy polyads are not usually carried far by winds, and Mildenhall concluded that *Acacia* had migrated from Australia to New Zealand by transoceanic seed dispersal (p. 221 of Mildenhall, 1975). Unfortunately, *Acacia* became extinct in New Zealand in the Pleistocene, probably as a result of unfavourable climatic changes (Raven, 1973).

Of the remaining species of *Acacia* from the Hawaiian Islands (Table IV), *A. koaia* is rare (p 287 of Carlquist, 1970; St. John, 1980), but *A. koa* is plentiful and appears to have adapted well to its present circumstances. It predominates in dry forests at altitudes around 1600 m (Carlquist, 1970; Walters and Bartholomew, 1984). Both *A. koaia* and *A. koa* are considered to be closely related to *A. kauaiensis*, but it would be useful to confirm the suspected relationships by analysing seed non-protein amino acids. Electrophoretic patterns and activities of seed proteinase inhibitors might also be helpful in assessing these relationships (Weder, 1978; Weder and Murray 1981; Murray and Weder, 1983).

Drift of *Acacia* seeds from North and South America is also possible. Porter (1983) considers this the probable explanation for the origin of three indigenous species in the Galapagos Islands: *A. insulae-iacobi, A. macracantha* and *A. rorudiana*. Profiles of seed non-protein amino acids have not yet been obtained, but would prove diagnostic.

2. Sophoreae

The nomenclature of members of this tribe is unfortunately confusing, but the taxonomic treatment is being revised (Polhill, 1981; Baretta-Kuipers, 1981). Species of interest in the present context belong to *Sophora (sensu stricto)* and *Edwardsia*. Members of *Sophora* (7 to 10 species) are generally regarded as 'seaborne' (Polhill, 1981). *Sophora tomentosa*, now the lectotype for *Sophora* (Yakovka, 1972), is distributed widely from tropical east Africa and Asia to Indonesia and northern Australia. The specimens of *S. tomentosa* presently growing in the Royal Botanic Gardens in Sydney arrived as seeds washed up on Woolgoolga Beach (N.S.W.) in 1961 (L. A. S. Johnson, pers. comm. 1980).

Members of *Edwardsia* (about 10 'species') are distributed on islands representing every ocean, and in mainland Chile (Green, 1970, 1979; Donoso-Zegers and Landrum, 1975; Bean, 1980; Polhill, 1981). Evidence has accumulated that many of these widespread taxa occupy their present locations because of drift dispersal of seeds originating from *Sophora microphylla (Edwardsia microphylla*; Table V) in New Zealand. Most support for this idea, which has been discussed for more than a century, has come from the studies of E. J. Godley and colleagues (Sykes and Godley, 1968; Markham and Godley, 1972; Godley, 1975).

The New Zealand species distinguished as *S. tetraptera* and *S. prostrata* (Table V) have distributions which overlap those of *S. microphylla*, and both are known to form hybrids with *S. microphylla* (Markham and Godley, 1972; Godley, 1975). *S. tetraptera* is found on the eastern side of the North Island, from near East Cape in the north, to the east of Carterton in the south. *S. prostrata* is found in the east of the South Island, from near Blenheim to just south of the Waitaki River. *S. microphylla* has several local variants, and may be found inland at altitudes up to 700 m, as well as near the coast. Here it frequently has the opportunity to shed pods and seeds into rivers, such as the Waitangiroto River depicted by Godley (1975), and estuaries.

Of all the New Zealand species and varieties, only *S. microphylla* produces yellow seeds with a high proportion buoyant in both fresh and salt water (Sykes and Godley, 1968; Godley, 1975). Buoyancy has been attributed to the lower density of the embryo (Sykes and Godley, 1968). The seed coats may account for almost 40% of seed mass, but occupy a low proportion

Table V. Synonyms for Species of *Sophora (Edwardsia)* from New Zealand, Lord Howe Island and Chile.

Sophora microphylla group :
   S. microphylla Ait. = *Edwardsia microphylla* (Ait.) Salisb.
   S. microphylla [Chile] = *Edwardsia macnabiana* (Graham)
   S. macrocarpa Sm. = *Edwardsia chilensis* Miers ex Lindl.
Remainder :
   S. howinsula (Oliver) P. S. Green
   S. masafuerana (Phil.) Skottsb.
   S. prostrata Buchan.
   S. tetraptera J. S. Miller = *Edwardsia grandiflora* Salisb.
   S. toromiro (Phil.) Skottsb.

Sources: Green (1970, 1979); Godley (1975); Bean (1980); C. Donoso-Zegers (pers. comm., 1985).

of seed volume (Murray, 1979; Fig. 12). Recently I estimated the following values for specific gravity (g cm^{-3}): seed coats, 1.38; embryo, 0.83; whole seed, 1.05.

The seeds of *S. microphylla* have been shown to be capable of floating in salt water for at least eight and a half years, and a sample tested (chipped) after five and a half years showed almost complete germination (Godley, 1975). Viable seeds identified as belonging to *S. microphylla* have been recovered from beaches of islands where the plant does not grow, such as Raoul Island in the Kermadec group 960 km from northern New Zealand, as well as from ocean beaches in Chile (Guppy, 1906; Sykes and Godley, 1968). Furthermore, Markham and Godley (1972) have obtained biochemical evidence which firmly establishes a relationship between *S. microphylla* from New Zealand and the taxa resembling *S. microphylla* from Chile and Gough Island in the south Atlantic Ocean. Comparative studies of the phenolic constituents of the leaves confirm that these geographically disparate populations are all *S. microphylla*, which is readily distinguished from both *S. prostrata* and *S. tetraptera*. Most of the phenolics separated from the leaves were identified as glycosides of the flavonol pigments apigenin and luteolin (Fig. 13). The diagnostic features of *S. microphylla* leaves are the presence of apigenin-6-C-rhamnosylglucoside and (or) apigenin-8-C-rhamnosylglucoside, and the absence of apigenin-7-glucoside, luteolin-7-glucoside and apigenin-7-rhamnosylglucoside-4'-glucoside. *S. microphylla* may also be distinguished from *S. prostrata* by its seed coat phenolic constituents (Markham and Godley, 1972).

A thorough comparison was made of variation within and between leaf phenolic constituents for all the samples in the *S. microphylla* populations. The nearby Chatham Island population (p 22 of Given and Williams, 1984) was more like the New Zealand population (84.5% agreement) than like

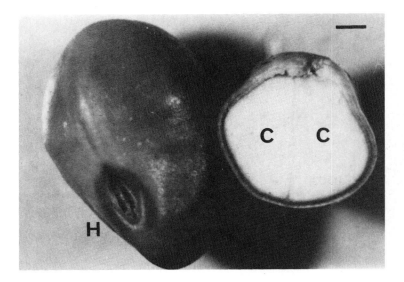

Fig. 12. Intact seed of *Sophora microphylla (Edwardsia microphylla)* together with a seed sectioned transversely through the hilum (H). Note the two storage cotyledons (C) of the embryo inside the seed coats (generally < 200 μm thick). Bar = 1 mm.

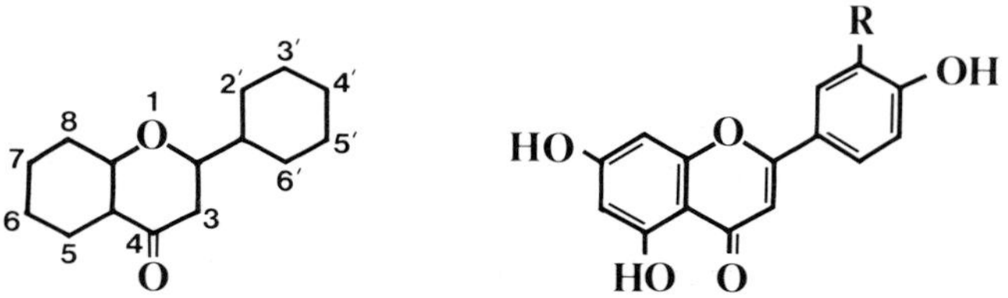

Fig. 13. The convention for numbering atoms of the flavonol skeleton (left) and the formula for apigenin (where R = H) and luteolin (where R = OH).

either the Chilean or Gough Island populations (79.3% and 76.7% agreement respectively). The Chilean samples were more like the New Zealand population (87.5%) than the Chatham Island population (79.3%), and were very similar to the Gough Island population (92.0% correspondence). Markham and Godley (1972) concluded that the populations of *S. microphylla* in the Chatham Islands and in Chile were derived independently from New Zealand seed sources, although it was not possible to precisely define the locality in New Zealand from which each came. Secondly, they concluded that the Gough Island population, localized near a single stream, had been established from a Chilean seed source.

Seed alkaloids have also been extracted and identified. Seeds from the Chilean population of *S. microphylla* have a higher content of cytisine, but less methylcytisine than seeds from New Zealand plants (Urzúa and Cassels, 1970). Analysis of alkaloids could well be employed to confirm the Chilean origin of the Gough Island population proposed by Markham and Godley (1972).

Comparative studies of seed protein content and composition were commenced in 1978 with a sample of *S. microphylla* seeds from a tree growing at the University of Melbourne (Murray, 1979; Murray and Porter, 1980). These studies have since confirmed that *S. microphylla* from New Zealand should be regarded as the progenitor of *S. microphylla* in Chile, and have been extended to include *S. macrocarpa (E. chilensis,* Table V). This second Chilean species is found inland away from the coastal margins and river edges, and at higher altitudes.

The distinguishing morphological features of these two Chilean taxa have been described by Donoso-Zegers and Landrum (1975) and Donoso-Zegers (1978). The imparipinnate leaves are divided into fewer pairs of leaflets in *S. macrocarpa,* and the leaflets are longer (Fig. 14). The leaflets length: width ratios are higher in *S. macrocarpa* than in *S. microphylla*: 3–4:1 compared to 1.5–2:1. The texture of the leaves is leathery (coriaceous) and the mid-veins are more prominent in *S. macrocarpa.* The fruit of *S. macrocarpa* (Fig. 14B) is pubescent, not as dark in colour, and lacks the four marginal flanges (wings) typical of *S. microphylla* pods (Fig. 14). The two are regarded as distinct species in Chile. They are interfertile, and produce a hybrid with intermediate leaf and pod characteristics (Donoso-Zegers and Landrum, 1975).

Nevertheless, the seeds of *S. macrocarpa* cannot be distinguished from those of *S. microphylla* in protein content or polypeptide composition (Fig. 15). The phenolic compounds from seed coats and embryos have also been compared, by extracting these parts of individual seeds from the same samples used to obtain the polypeptide profiles shown in Figure 15. All the seed coats displayed almost identical chromatographic patterns (Fig. 16A). The embryos from Chilean *S. microphylla* displayed three areas (a, b and c; Fig. 16B), which were all different from the seed coat phenolics (Fig. 16A). By comparison, most embryos of *S. macrocarpa* displayed four areas, three identical with those of *S. microphylla* embryos, plus one distinguishing area (d, Fig. 16B).

The conservation of important seed characteristics (Figs. 15, 16) is evidence of a very close relationship, and consistent with the recent derivation of *S. macrocarpa* from *S. microphylla.* It seems likely that other representatives of *Edwardsia* not yet studied biochemically will also prove to be derived from *S. microphylla,* for example, *S. masafuerana* from the Juan Fernandez Islands, *S. toromiro* from Easter Island, and *S. howinsula* from Lord Howe Island (Table V).

Fig. 14. (A), fruits and leaves of *Sophora microphylla (Edwardsia microphylla)* from Chile; (B), fruits, flowers and leaves of *Sophora macrocarpa (Edwardsia chilensis)* from Chile. Bar = 2 cm. (courtesy of C. Donoso-Zegers).

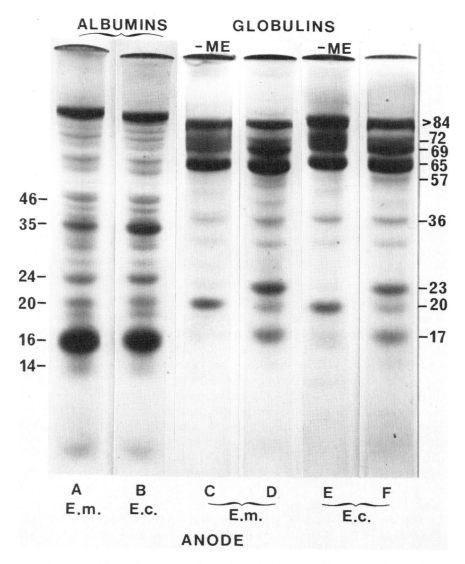

Fig. 15. Polypeptide profiles of albumin (A, B) and globulin (C-F) fractions from embryos of *Sophora (Edwardsia) microphylla* (E.m.) from Valdivia, Chile, and *Sophora macrocarpa (Edwardsia chilensis,* E.c.) also from Chile. Electrophoresis was performed with disc gels of 10% polyacrylamide, with 50µg of protein per gel, and sodium dodecyl sulphate in the presence or absence (C, E) of 2-mercaptoethanol (ME) to reduce disulphide bridges between polypeptides. Gels were stained with Coomassie Brilliant Blue. Estimated molecular weights x 10⁻³ daltons. These values are revised from Murray (1979) using internal reference standards according to Murray *et al.* (1983).

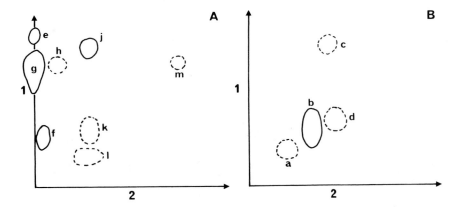

Fig. 16. Two-dimensional chromatography of flavonoids from seed coats (A) and embryos (B) of Chilean samples of *Sophora microphylla (Edwardsia microphylla)* and *Sophora macrocarpa (Edwardsia chilensis).* For methods, see Murray et al. (1978).

D. Mangroves

A mangrove is 'any woody plant that grows between the tidal limits' (p 81 of Dowling and McDonald, 1982). By their tolerance of saline conditions, they are well suited to a pioneer role on the margins of lagoons or estuaries. Contradictory views have sometimes been expressed about the abilities of mangroves to undertake long-distance dispersal by flotation of propagules (fruits, seeds or seedlings). This confusion is no doubt due to lack of information, the large number of species regarded as mangroves, and differences in the nature of the propagule (Table VI). The term 'viviparous' is used to indicate precocious germination and growth of the embryo, such that the seedling itself is shed after considerable extension of the hypocotyl-radicle (part 3 of Table VI). This is clearly an adaptation that would permit the new seedling to spear into soft mud, facilitating its establishment while subject to tidal inundation.

Saenger (1982) states that the propagules of all mangroves are buoyant. The period of buoyancy may be as brief as four days, as in *Avicennia marina* (Clarke and Hannon, 1969), but longer periods are more usual. Even the elongated seedlings of members of the Rhizophoraceae float, first in a horizontal position, but later in a vertical position (p. 677 of Gore, 1977). In *Rhizophora mangle*, new roots develop within 10 to 17 days, and the leaves begin to expand after 40 to 50 days, irrespective of whether the seedling is stranded or still floating (Banus and Kolehmainen, 1975). This is sufficient time to travel hundreds of kilometres, according to the speed of prevailing currents.

Table VI. Propagule types and times taken for development for mangroves from north-eastern Australia.

Species	Mature propagule	Months for development
1. Seeds not viviparous		
Barringtonia racemosa	4-cornered drupe	3–5
Camptostemon schultzii	ovoid capsule	2–3(?)
Cynometra iripa	flat rugose capsule	6–9
Diospyros ferrea	ovoid berry	
Dolichandrone spathacea	linear capsule	
Excoecaria agallocha	trilobed capsule	2
Heritiera littoralis	keeled ovoid 'nut'	7–9
Hibiscus tiliaceus	globose capsule	
Lumnitzera littorea	ovoid fibrous drupe	3
Lumnitzera racemosa	ovoid fibrous drupe	2–3
Osbornia octodonta	globose capsule	1–2
Pemphis acidula	globose capsule	
Scyphiphora hydrophyllacea	cylindrical capsule	4–5
Sonneratia alba	globular berry	3–4
Sonneratia caseolaris	globular berry	5–6
Sonneratia X gulngai	globular berry	3–5
Sonneratia lanceolata	globular berry	
Xylocarpus australasicus	globose capsule	3–4
Xylocarpus granatum	large globose capsule	6–8
2. Seeds cryptoviviparous		
Acanthus ilicifolius	ovoid capsule	2–4
Aegialitis annulata	elongate terete capsule	2
Aegiceras corniculatum	elongate terete cpasule	6–7
Nypa fruticans	fibrous drupe cluster	3–4
3. Seeds viviparous		
Avicennia marina	ovoid fleshy capsule	3
Bruguiera cylindrica	extended embryo	5(?)
Bruguiera exaristata	extended embryo	5
Bruguiera gymnorhiza	extended embryo	4–9
Bruguiera parviflora	extended embryo	4–5
Bruguiera sexangula	extended embryo	
Ceriops decandra	extended embryo	0–3 (+ 12)
Ceriops tagal var. *australis*	extended embryo	0–2 (+ 12)
Ceriops tagal var. *tagal*	extended embryo	1–3 (+ 12)
Rhizophora apiculata	extended embryo	11
Rhizophora X lamarckii	extended embryo	2,6,8 (?)
Rhizophora mucronata	extended embryo	12
Rhizophora stylosa	extended embryo	0–2 (+ 12)

Source: N. C. Duke (pers. comm. 1985); Duke *et al.* (1984)

The distributions of mangroves around Australia have been described by Gill (1975), Bridgewater (1982), Dowling and McDonald (1982), Galloway (1982), Kenneally (1982) and Wells (1982). Many species are distributed precisely according to their physiological requirements, reflecting different capacities for coping with leaf temperature stress (Clough *et al.*, 1982) and salinity tolerance (Winter *et al.*, 1981).

Several examples are sufficient to confirm the capacity for widespread dispersal of mangroves by flotation of propagules. *Ceriops tagal*, a viviparous species (Table VI), is distributed widely around the Indian Ocean, Micronesia and Taiwan; it was collected from the northern coast of Western Australia as early as 1821, by Allan Cunningham (Kenneally, 1982). *Bruguiera gymnorhiza*, *Nypa fruticans*, *Pemphis acidula* and *Sonneratia alba*, pantropical species referred to by Ernst (1908) as colonists of Krakatau (Section IV, A), are also distributed around the Australian coastline (Table VI). *Pemphis acidula* has reached the New Hebrides (Green, 1979), and *Aegiceras corniculatum*, *Avicennia marina* and *Hibiscus tiliaceus* are present on Lord Howe Island (Rodd and Pickard, 1983).

Despite their obvious successes, mangroves, like other species dispersed by ocean currents, are subject as seedlings to numerous hazards (Section V). Mangroves are absent from some islands, such as the Chagos Archipelago south of the Maldive Islands (p.115 of Renvoize, 1979). On other islands, they may be few in species number. Only three pantropical species have become established in the Galapagos Islands: *Avicennia germinans*, *Hibiscus tiliaceus* and *Rhizophora mangle* (Porter, 1983).

IV. EFFECTIVE SEED DISPERSAL BY OCEAN CURRENTS COMPARED TO OTHER VECTORS

It is important to support general statements about the different agencies of seed transport contributing to the vegetation of an oceanic island or an 'island habitat' with observations on both the flora and fauna. The different modes of immigration can be identified readily for most original plant inhabitants, and their quantitative contributions ranked, as shown in the following instances.

A. Krakatau

Several references have already been made to the eruption of Krakatau in 1883, and to the series of visits organized by M. Treub and his colleagues to follow the progress of natural revegetation (Sections I, III). On the first expedition to Krakatau in June 1886, almost three years after the eruption, Treub found the seedlings of nine species of seed-plants established near the shore, all of which had grown from seeds washed up by the sea. Further inland, the number of seed-plants totalled eight, and six of these were different from the species near the shore. The seeds of these plants, two grasses and four Compositae, had clearly been brought by the wind, as had the spores of eleven species of fern found throughout the Malay Archipelago.

In describing this first visit, Ernst (1908) comments particularly on how Treub had expected to find the vegetation of the shore-line re-established from drift seeds well in advance of the revegetation of the interior of the island. However, Treub was surprised to find that revegetation of the shore-line and the interior, including the slopes of the volcanic peak Rakata, had occurred simultaneously: "The flora of the interior of the island had developed not only independently of the strand-flora, but also with much greater rapidity." (p. 7 of Ernst, 1908).

A follow-up expedition did not return until March 1897. Of the 53 seed-plants recorded as established on Krakatau, Verlaten and Lang Islands (the last-named still devoid of vegetation in 1886), 32 species (60%) had arrived by oceanic drift, 17 species (32%) by wind, and only four species (7.5%) had been introduced by fruit-eating animals or by humans (estimates of O. Penzig, cited p. 8 of Ernst, 1908). As three species noted by Treub for Verlaten Island in 1886 were omitted from this calculation by Penzig, the above estimates should be modified slightly by the addition of *Senecio* sp., *Conyza angustifolia* and *Conyza indica* (all Compositae), the last two species being confined to the strand flora (p. 45 of Ernst, 1908). This brings to 56 the actual number of seed-plant species re-established by 1897.

By the time of the third expedition in April 1906, the total number of seed-plant species had increased to 92: one cycad, represented by a solitary female plant of *C. circinalis* (Section III, A) 17 monocotyledons, and 72 dicotyledons. Out of this total of 92 species, 67 species belonged mainly or entirely to the strand-flora; however, the increase in species number that took place between 1897 and 1906 was distributed evenly between strand and inland vegetations (Ernst, 1908). The vegetation of Lang Island was very poorly developed compared to that of Verlaten Island and Krakatau. Ernst attributed the relatively short time taken for the revegetation of Krakatau and Verlaten Island to their proximity to the coast of Java, Sumatra and neighbouring islands with vegetation, and to favourable currents, such that even seeds that would not survive a long period in seawater might survive for several days, sufficient to travel the short distances involved. Another contributing factor could have been the absence of crabs, known to consume seedlings of coconut and other species (p. 58 of Ernst, 1908).

Allowing some uncertainty as to the precise contribution of each vector, Ernst finally estimated that the 92 species of seed-plants established by 1906 were derived 39–72% by sea currents (the upper limit being all 67 strand species), 19–10% birds, and 30–16% by wind (p. 67 of Ernst, 1908).

B. The Hawaiian Islands

The origins of Hawaiian plants and animals have been considered in detail by Carlquist, in two books (1970, 1974) and a more recent article (Carlquist, 1981). About 95% of Hawaiian plant species are endemic, having evolved in the Hawaiian Islands following the successful establishment of their progenitors. A high degree of endemism in an island flora is exactly what impressed Charles Darwin (1839, 1845, 1859) in regard to the Galapagos Islands (Section IV, C).

The various mechanisms known for long-distance dispersal have been allocated to the hypothetical minimum number of progenitors necessary to account for the present indigenous Hawaiian flora. This number has been refined from 255 (Carlquist, 1970) to 272 (Carlquist, 1981). The following estimates for dispersal mechanism are based on the earlier number.

More than half the original flowering plant immigrants were carried as seeds by birds, either internally (39.0%) or by external attachment (12.8%). Only 1.4% came as seeds carried by wind, and almost 23% could have arrived by flotation of seeds or other plant parts. This proportion is further subdivided into 14.3% for species whose seeds are clearly adapted for this means of dispersal, and 8.5% for species that may have arrived by "rare or freak flotation events" (p. 102 of Carlquist, 1970). Since the Hawaiian Islands are not now in the direct path of major currents, and indeed are cut off from the south Pacific by equatorial currents, this accumulation of species from floating seeds is remarkably high.

Progenitors arriving and becoming established in strand habitats did not necessarily remain confined to these regions, but they were also able to generate derivative species that could occupy wetter regions inland, or relatively dry habitats at higher altitude as the islands gained height from continued volcanic activity. Examples already referred to are the loulu palms (*Pritchardia* spp., Section III), and the Hawaiian species of *Acacia* (Section III, C, *1*). Other examples described by Carlquist involve the loss of fruit or seed flotation aids. Thus the Hawaiian cotton (*Gossypium sandvicense*) produces seeds tolerant of salt water, but the seeds do not float, as they are no longer covered by cotton fibres. The wiliwili (*Erythrina sandwicensis*) produces seeds that do not float (pp. 104 and 169 of Carlquist, 1970), in contrast to those of *E. variegata* (*indica*) (Table II, Section III, C). This latter species is widespread, but not found naturally in Hawaii. Its possible relationship as a progenitor of *E. sandwicensis* has not been tested biochemically.

C. The Galapagos Islands

The indigenous vegetation of this famous archipelago comprises 543 species of vascular plants: 107 pteridophytes, 85 monocotyledons and 351 dicotyledons (Porter, 1983). The overall degree of endemism is estimated to be at least 37% (Porter, 1979) and probably 43% (Porter, 1983): eight species of pteridophytes, 18 monocotyledons and 205 dicotyledons. The minimum number of immigrant progenitors necessary to account for the indigenous Angiosperms is 306 (Porter, 1983).

J. D. Hooker (1847a,b) displayed considerable insight into the relationships of the distinct elements of the vegetation and their dispersal mechanisms. He described 253 species of flowering plants and ferns, mostly from specimens collected by Darwin. Both Darwin (1839, 1845) and Hooker correctly recognized the affinities of the 'peculiar' floristic element with plants of South America. This is the element Porter now defines as 'Andean': from western South America, between Venezuela and Chile (p. 34 of Porter, 1983). Hooker's 'non-peculiar' element corresponds to both tropical American and pantropical contributions (Porter, 1983).

Porter (1976) estimated that the immigrant flora had been brought by the agencies birds, wind and oceanic drift in an approximate ratio 6:3:1. The plants in the wind-dispersed category are mainly ferns; only 28 species of angiosperms have been successfully introduced by wind-borne seeds (Porter, 1983). Evidently 35 species have been introduced as seeds brought by ocean currents: two monocotyledons representing one family (Poaceae), and 33 dicotyledons representing 15 families (distinguishing three families of legumes, Table II). Most of these species are plants of the littoral zones and salt flats, such as the mangroves referred to previously (Section III, D). With two exceptions, these species have remained in the habitats to which they are already so well adapted.

The exceptions are two endemic species of *Gossypium*, which have now lost the tomentose seed covering (cotton) that would have enabled these seeds to float: *G. klotschianum*, similar to the Mexican *G. davidsonii*, and

G. barbadense var. *darwinii*, with Andean affinities (Porter, 1983). A similar modification was noted for *G. sandvicense* from the Hawaiian Islands (Section IV, B).

A lower proportion of total indigenous species has been introduced to the Galapagos Islands by seed flotation compared to the Hawaiian Islands (Section IV, B). This may partly be due to the relative youth of the Galapagos Islands, for which a maximum age of three million years has been estimated by potassium-argon dating (Bailey, 1976), compared to 70 million years for the earliest Hawaiian islands (Dalrymple *et al.*, 1973). Paradoxically, the lower proportion of species introductions attributed to oceanic drift may stem from the closeness of the Galapagos Islands to the mainland, which at its nearest point is 800 km away (Porter, 1976). This relative proximity has facilitated seed dispersal by birds (Porter, 1983).

V. THE HAZARDS FOR SEEDLING ESTABLISMENT

"Adverse ecological conditions may be a greater obstable to the establishment of a species in a new location than transport itself." (Carlquist, 1981).

Many authors have referred to the discovery of stranded and viable seeds of species not established in the locality. The seeds of *Sophora microphylla* (Section III, C, 2), for example, are found on the shores of the Kermadec Islands, but this species is not established there (Sykes and Godley, 1968; Sykes, 1977). Renvoize (1979) quotes the number of such species for Aldabra at about 50.

The arrival in Ireland of water-borne seeds from the West Indies has been discussed by Nelson (1978), and is illustrated in Figure 17. None of the species represented by these drift seeds has established itself in the British Isles.

What reasons can be advanced to explain the failure of seedlings to become established, even when the supply of germinable seeds has extended through many centuries? Darwin was well aware of the arrival of West Indian seeds in the British Isles (see Chapter XII of Darwin, 1859), and of the failure of these species to become established. He identified two factors that might be generally important for the establishment of seedlings: the absence of 'destructive insects or birds' that might eat the seedlings, and a suitable climate. In the case of the British Isles, failure of drift seeds to become established is likely to be the result of an unsuitable climate. Often the temperature would be below that required to ensure both germination (Labouriau, 1978) and a reasonable rate of seedling growth after emergence. As well as insects, crabs have also been identified as a major

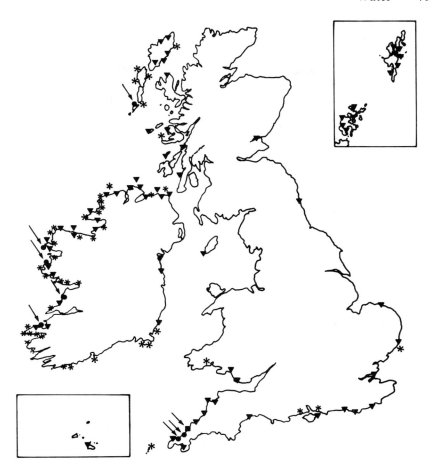

Fig. 17. Distribution of reported finds of drift seeds in the British Isles, including the Channel Isles and the Hebrides (insets). Frequency: ▼, single seeds; *, 2–10; → ●, 10–100; → ■, > 100. Note the preponderance of records from Cornwall, Ireland and Western Scotland, in accordance with prevailing currents (data of E. C. Nelson).

cause of seedling mortality, especially of coconuts (*Birgus latro*; Guppy, 1890; Hicks *et al.*, 1984) and mangroves (*Sesarma* spp.; Saenger, 1982).

The substrate may be unsuitable. Volcanic islands possess shorelines that are often inhospitable. The newly fractured edge of the surviving portion of Krakatau did not permit as extensive a regeneration of vegetation from drift seeds as the surviving beaches (Ernst, 1908). Tristan da Cunha is a volcanic island approximately 18 million years old (Wilson, 1963a), which is still almost totally unsuited for the growth of seedlings from seeds deposited on the rocky shoreline (Wheeler, 1962).

In the relatively young Kermadec Islands, very few species capable of wide dispersal by seed flotation have become established. Sykes (1977) describes in detail the distributions of *Ipomoea pes-caprae, Canavalia maritima* (Table II), *Lepturus repens* and *Calystegia soldanella*, emphasizing that the establishment of most of these species is both recent, and an ongoing, tenuous process. On Macauley Island, which emerged in the late Pleistocene, *Calystegia soldanella* became introduced at Sandy Bay beach between 1966 (seeds only present) and 1970 (an area about 2 m² covered). The main population of this species present in 1908, in Denham Bay of Raoul Island, had declined to just a few seedlings by 1966–1967 (p. 98 of Sykes, 1977). In this same locality, Sykes noted a solitary seedling of *Canavalia maritima*, "which would almost certainly have been washed away during the next storm" (p. 124 of Sykes, 1977). Thus another prerequisite for establishment is illustrated: the seed must be deposited on a substrate suitable for growth, and the seedling must escape or survive subsequent wave action.

The establishment of a single seedling and its survival to maturity may still not ensure the introduction of a particular species. This is pointedly true of dioecious species, such as the cycads (Section III, A) and palms (Section III, B) referred to previously. In such cases, seeds that will give rise to either male or female plants must arrive close together spatially and temporally, and both types of plant must become established. Ernst (1908) describes the discovery of a "fine specimen of a female plant of *Cycas circinalis*, bearing a handsome crown of fronds at the apex of a trunk 1 m 65 cm (5 feet 9 inches) high and 80 cm in circumference" in the littoral zone of the new island of Krakatau. However, there were no male plants at all on either Krakatau or Verlaten Island, and no other female plants (pp. 29 and 72 of Ernst, 1908).

Reproductive failure might also occur for some species if specific pollination vectors are absent from a new island habitat, but this is not a problem for legumes (Table II), which are often self-fertile (cleistogamous), nor is it usually a problem for strand species from other families (Carlquist, 1970, 1974; Porter, 1983).

VI. CONCLUSIONS

"An insular biota is itself a collection of evidence in support of long-distance dispersal of plants and animals, offering a living archive of dispersal events." (Carlquist, 1981).

As a means of seed dispersal, oceanic drift has succeeded in the past, providing an important contribution to the total vegetation of many islands (Sections III, IV). The exact rate of introduction of species to a particular locality can never be quantified, but even in islands with existing vegetation, the introduction of new species by this means is still happening. This is illustrated by the recent discovery of *Haloragis prostrata* in the New Hebrides (Green, 1979) and observations in the Kermadec Islands (Section V).

With sufficient time, species that have become established on an island can become modified to the extent that they are considered to be distinct endemic species (Sections III, IV). This is also possible for species brought by oceanic drift to mainland habitats (Section III, C, *2*). A readier appreciation of the diagnostic potential of biochemical analyses (Section III, C) will widen the scope of their application, and bring to light many more examples than those considered here.

In conclusion, it is appropriate to remember that disturbances to remaining natural areas of vegetation caused by human activity, with attendant plant and animal escapes, pose a serious threat to continuing studies of plant migration and speciation. This was recognized by Carson: "The tragedy of the oceanic isles lies in the uniqueness, the irreplaceability of the species they have developed by the slow processes of the ages. In a reasonable world, men would have treated these islands as precious possessions ..." (p. 119 of Carson, 1965). We do not yet live in a reasonable world.

ACKNOWLEDGEMENTS

I thank Dr T. R. Grant and Dr D. H. Ashton for their help in obtaining seed samples, Prof C. Donoso-Zegers for providing seed samples and Figure 14, and Mr J. B. Murray for drawing Figures 13 and 17. I also thank Mr C. Freeman for assistance with measurements of buoyant density, Dr F. G. Lennox for obtaining the data included as Figure 16, Dr E. C. Nelson for providing data incorporated into Figure 17, and all others who provided figures or information included in this chapter.

Note added in proof: Readers will be interested in the illustrated article 'Krakatau' by Professor Ian Thornton in *Australian Geographic* Volume 1, pp. 40–54 (1986).

REFERENCES

Ashcroft, W. J., and Murray, D. R. (1979). *Aust. J. Bot.* **27**, 343–352.

Attenborough, D. (1984). *In* 'The Living Planet'. W. Collins and British Broadcasting Corporation, London.

Bailey, K. (1976). *Science* **192**, 465–467.

Banus, M. D., and Kolehmainen, S. E. (1975). *In* 'Proceedings of the International Symposium on Biology and Management of Mangroves' (G. E. Walsh, S. C. Snedaker and H. J. Teas, eds.), pp. 370–384. Institute of Food and Agricultural Sciences, University of Florida, Gainsville, U. S. A.

Baretta-Kuipers, S. T. (1981). *In* 'Advances in Legume Systematics' (R. M. Polhill and P. H. Raven, eds), Part 2, pp. 677–705. Royal Botanic Gardens, Kew.

Bauman, A. J., and Yokoyama, H. (1976). *Biochem. Syst. Ecol.* **4**, 73–74.

Beadle, N. C. W. (1952). *Ecology* **33**, 49–62.

Beaglehole, J. C. (1962). 'The Endeavour Journal of Joseph Banks 1768–1771'. Volume II. Angus and Robertson, Sydney.

Bean, W. J. (1980). 'Trees and Shrubs Hardy in the British Isles. Volume IV, Ri-Z'. Eighth edition revised. J. Murray, London.

Bell, E. A. (1984). *In* 'Seed Physiology. Volume 1. Development' (D. R. Murray, ed.), pp. 245–264. Academic Press, Sydney.

Blaschke, P. M., and Grant-Mackie, J. A. (1976). *N. Z. J. Geol. Geophys.* **19**, 933–941.

Braithwaite, C., Taylor, J., and Kennedy, W. (1973). *Phil. Trans. Royl. Soc.* **266**, 307–340.

Bridgewater, P. B. (1982). *In* 'Mangrove Ecosystems in Australia — Structure, Function and Management' (B. F. Clough, ed.), pp. 111–120. Australian Institute of Marine Science and Australian National University Press, Canberra.

Burbidge, A. H., and Whelan, R. J. (1982). *Aust. J. Ecol.* **7**, 63–67.

Cambage, R. H. (1919). *J. Proc. Roy. Soc. N.S.W.* **LII**, 410–434.

Cambage, R. H. (1923). *J. Proc. Roy. Soc. N.S.W.* **LVII**, 283–299.

Carlquist, S. (1970). 'Hawaii — A Natural History'. The Natural History Press, New York.

Carlquist, S. (1974). 'Island Biology'. Columbia University Press, New York and London.

Carlquist, S. (1981). *Am. Sci.* **69**, 509–516.

Carlquist, S. (1983). *Sonderbd. naturwiss. Ver. Hamburg* **7**, 37–47.

Carson, R. (1965). 'The Sea Around Us'. Panther Books, London.

Clarke, L. D., and Hannon, N. J. (1969). *J. Ecol.* **57**, 213–234.

Clough, B. F., Andrews, T. J., and Cowan, I. R. (1982). *In* 'Mangrove Ecosystems in Australia — Structure, Function and Management' (B. F. Clough, ed.), pp. 193–210. Australian Institute of Marine Science and Australian National University Press, Canberra.

Cookson, I. C. (1954). *Aust. J. Bot.* **2**, 52–59.

Dalrymple, G. B., Silver, E. A., and Jackson, E. D. (1973). *Am. Sci.* **61**, 294–308.

Darwin, C. (1839). 'Narrative of the Surveying Voyages of His Majesty's Ships *Adventure* and *Beagle*, Between the Years 1826 and 1836, Describing Their Examination of the Southern Shores of South America, and the *Beagle*'s Circumnavigation of the Globe. Vol. 3. Journal and Remarks. 1832–1836'. Henry Colburn, London.

Darwin, C. (1845). 'Journal of Researches into the Natural History and Geology of the Countries Visited During the Voyages of H. M. S. *Beagle* 'Round the World, Under the Command of Capt. Fitz Roy, R. N.' John Murray, London.

Darwin, C. (1859). 'On the Origin of Species by Natural Selection, or the Preservation of Favoured Races in the Struggle for Life'. John Murray, London.

Dehgan, B., and Yuen, C. K. K. H. (1983). *Bot. Gaz.* **144**, 412–418.

Donoso-Zegers, C. (1978). *In* Manual No. 2, Fac. Ciencias Forestales Universidad de Chile, Talleres Gráficos Facultad de Agronomía Universidad de Chile, pp. 82–84. Santiago, Chile.

Donoso-Zegers, C., and Landrum, L. R. (1975). *In* Boletín Técnico No. 30, Fac. Ciencias Forestales Universidad de Chile, Talleres Gráficos Facultad de Agronomía, Universidad de Chile, pp. 12–19. Santiago, Chile.

Dowling, R. M., and McDonald, T. J. (1982). *In* 'Mangrove Ecosystems in Australia — Structure, Function and Management' (B. F. Clough, ed.), pp. 79–93. Australian Institute of Marine Science and Australian National University Press, Canberra.

Drinnan, A. N., and Chambers, T. C. (1985). *Aust. J. Bot.* **33**, 89–100.

Duke, N. C., Bunt, J. S., and Williams, W. T. (1984). *Aust. J. Bot.* **32**, 87–99.

Eames, A. J., and MacDaniels, L. H. (1947). 'An Introduction to Plant Anatomy'. Second edition. McGraw-Hill, New York and London.

Ernst, A. (1908). 'The New Flora of the Volcanic Island of Krakatau'. Cambridge University Press, Cambridge, U. K.

Evans, C. S., Qureshi, M. Y., and Bell, E. A. (1977). *Phytochemistry* **16**, 565–570.

Fleming, C. A. (1979). 'The Geological History of New Zealand and its Life.' Auckland University Press, Auckland.

Galloway, R. W. (1982). *In* 'Mangrove Ecosystems in Australia — Structure, Function and Management' (B. F. Clough, ed.), pp. 31–54. Australian Institute of Marine Science and Australian National University Press, Canberra.

Giddy, C. (1974). 'Cycads of South Africa'. Purnell, Cape Town.

Gill, A. M. (1975). *Proc. Ecol. Soc. Aust.* **8**, 129–146.

Gilmore, R. (1978). *Jetaway (Air New Zealand)*, **Part 37**, 36–37.

Given, D. R., and Williams, P. A. (1984). 'Conservation of Chatham Island Flora and Vegetation'. Botany Division, D.S.I.R., Christchurch, New Zealand.

Godley, E. J. (1975). *New Zealand's Nature Heritage* **5** (part 65), 1804–1806.

Good, R. d'O. (1951). *Nature* **167**, 518.

Gore, R. (1977). *Nat. Geog.* **151**, 668–689.

Green, P. S. (1970). *J. Arn. Arb.* **51**, 204–220.

Green, P. S. (1979). *In* 'Plants and Islands' (D. Bramwell, ed.), pp. 41–53, Academic Press, London.

Guinet, Ph. (1981). *In* 'Advances in Legume Systematics' (R. M. Polhill and P. H. Raven, eds), Part 2, pp. 835–857. Royal Botanic Gardens, Kew.

Gunn, C. R., and Dennis, J. V. (1976). 'World Guide to Tropical Drift Seeds and Fruits'. New York Times Books, New York.

Guppy, H. B. (1887). 'The Solomon Islands'. London.

Guppy, H. B. (1890). 'The Dispersal of Plants, as Illustrated by the Flora of the Keeling or Cocos Islands'. Reprint, London.

Guppy, H. B. (1906). 'Observations of a Naturalist in the Pacific Between 1896 and 1899. 2. Plant Dispersal'. Macmillan, London.

Guppy, H. B. (1917). 'Plants, Seeds and Currents in the West Indies and the Azores'. London.

Heyerdahl, T. (1950). 'Kon-Tiki'. Rand McNally & Co., Chicago.

Heyerdahl, T. (1958). 'Aku-Aku — The Secret of Easter Island'. George, Allen and Unwin, London.

Hicks, J., Rumpff, H., and Yorkson, H. (1984). 'Christmas Crabs'. Christmas Island Natural History Association.

Hooker, J. D. (1847a). *Trans. Linn. Soc. Bot.* **20**, 163–233.

Hooker, J. D. (1847b). *Trans. Linn. Soc. Bot.* **20**, 235–262.

Johri, B. M. (1984). 'Embryology of Angiosperms'. Springer-Verlag, Berlin.

Kenneally, K. F. (1982). *In* 'Mangrove Ecosystems in Australia — Structure, Function and Management' (B. F. Clough, ed.), pp. 95–110. Australian Institute of Marine Science and Australian National University Press, Canberra.

Knox, R. B., and Kenrick, J. (1983). *In* 'Pollen: Biology and Implications for Plant Breeding' (D. L. Mulcahy and E. Ottaviano, eds.), pp. 411–417. Elsevier Biomedical, New York.

Labouriau, L. G. (1978). *Rad. Environ. Biophys.* **15**, 345–366.

Laseron, C. F. (1955). 'Ancient Australia'. Reprinted edition. Angus and Robertson, Sydney.

Liddle, M. J., and Elgar, M. A. (1984). *Bot. J. Linn. Soc.* **88**, 303–315.

Löve, D., and Dansereau, P. (1959). *Can. J. Bot.* **37**, 173–208.

Maiden, J. H. (1920). 'The Weeds of New South Wales. Part I'. W. A. Gullick, Government Printer, Sydney.

Markham, K. R., and Godley, R. J. (1972). *N.Z. J. Bot.* **10**, 627–640.

Mildenhall, D. C. (1972). *N.Z. J. Bot.* **10**, 485–494.

Mildenhall, D. C. (1975). *N.Z. J. Geol. Geophys.* **18**, 209–228.

Murray, D. R. (1979). *Z. Pflanzenphysiol.* **93**, 423–428.

Murray, D. R., and Porter, I. J. (1980). *Plant System Evol.* **134**, 207–214.

Murray, D. R., and Weder, J. K. P. (1983). *Aust. J. Bot.* **31**, 119–124.

Murray, D. R., Ashcroft, W. J., Seppelt, R. D., and Lennox, F. G. (1978). *Aust. J. Bot.* **26**, 755–771.

Murray, D. R., Mackenzie, K. F., Vairinhos, F., Peoples, M. B., Atkins, C. A., and Pate, J. S. (1983). *Z. Pflanzenphysiol.* **109**, 363–370.

Nelson, E. C. (1978). *Watsonia* **12**, 103–112.

Pedley, L. (1978). *Austrobaileya* **1**, 77–234.

Pena, P., and de L'Obel, M. (1570). 'Stirpium aduersaria noua', London.

Pickard, J. (1983). *Cunninghamia* **1**, 133–265.

Polhill, R. M. (1981). *In* 'Advances in Legume Systematics' (R. M. Polhill and P. H. Raven, eds), Part I, pp. 213–230. Royal Botanic Gardens, Kew.

Porter, D. M. (1976). *Nature* **264**, 745–746.

Porter, D. M. (1979). *In* 'Plants and Islands' (D. Bramwell, ed.), pp. 225–256. Academic Press, London.

Porter, D. M. (1983). *In* 'Patterns of Evolution in Galapagos Organisms' (R. E. Bowman, M. Berson and A. E. Levitson, eds), pp. 33–96. American Association for the Advancement of Science, Pacific Division, San Francisco.

Rao, A. R. (1974). *In* 'Aspects and Appraisal of Indian Paleobotany' (K. R. Surange, R. N. Lakhanpal and D. C. Bharadwaj, eds). Birbal Sahni Institute of Paleobotany, Lucknow.

Rao, A. R. (1981). *Paleobotanist* **28–29**, 207–209.

Raven, P. H. (1973). *N.Z. J. Bot.* **11**, 177–200.

Raven, P. H., and Axelrod, D. I. (1974). *Ann. Missouri Bot. Gard.* **61**, 539–673.

Renvoize, S. A. (1979). *In* 'Plants and Islands' (D. Bramwell, ed.), pp. 107–129. Academic Press, London.

Ridley, H. N. (1930). 'The Dispersal of Plants Throughout the World'. Reeves, London.

Rix, M. (1981). 'The Art of the Botanist'. Lutterworth Press, Guildford and London.

Robinson, A. R. (1963). 'Wind-Driven Ocean Circulation'. Blaisdell Publishing Company, New York.

Rodd, A. N., and Pickard, J. (1983). *Cunninghamia* **1**, 267–280.

Saenger, P. (1982). *In* 'Mangrove Ecosystems in Australia — Structure, Function and Management' (B. F. Clough, ed.), pp. 153–191. Australian Institute of Marine Science and Australian National University, Canberra.

Sloane, H. (1696). *Phil. Trans.* **19**, 298–300.

Sloane, H. (1725). 'Natural History of Jamaica, 2'. London.

St. John, H. (1980). *Pac. Sci.* **33**, 357–367.

Sullivan, W. (1977). 'Continents in Motion — The New Earth Debate'. Macmillan, London.

Sykes, W. R. (1977). 'Kermadec Islands Flora'. New Zealand Department of Scientific and Industrial Research Bulletin 219. Government Printer, Wellington.

Sykes, W. R., and Godley, E. J. (1968). *Nature* **218**, 495–496.

Tran, V. N., and Cavanagh, A. K. (1980). *Aust. J. Bot.* **28**, 39–51.

Tran, V. N., and Cavanagh, A. K. (1984). *In* 'Seed Physiology. Volume 2. Germination and Reserve Mobilization' (D. R. Murray, ed.), pp. 1–44. Academic Press, Sydney.

Tunstall, B. R., and Connor, D. J. (1975). *Aust. J. Plant. Physiol.* **2**, 489–499.

Urzúa, A., and Cassels, B. K. (1970). *Phytochemistry* **9**, 2365–2367.

van der Pijl, L. (1957). *Acta. Bot. Neerl.* **6**, 291–315.

Vassal, J. (1972). *Bull. Soc. Hist. Nat. Toulouse* **108**, 125–247.

Walkom, A. B. (1918). *Proc. Linn. Soc. N.S.W.* **43**, 37–115.

Walkom, A. B. (1919). *Proc. Linn. Soc. N.S.W.* **44**, 180–190.

Walters, G. A., and Bartholomew, D. P. (1984). *Bot. Gaz.* **145**, 351–357.

Wapshire, A. J. (1974a). *Aust. J. Agr. Res.* **25**, 275–292.

Waspshire, A. J. (1974b). *Aust. J. Agr. Res.* **25**, 775–781.

Weder, J. K. P. (1978), *Z. Pflanzenphysiol.* **90**, 285–291.

Weder, J. K. P., and Murray, D. R. (1981). *Z. Pflanzenphysiol.* **103**, 317–322.

Wells, A. G. (1982). *In* 'Mangrove Ecosystems in Australia — Structure, Function and Management' (B. F. Clough, ed.), pp.57–78. Australian Institute of Marine Science and Australian National University, Canberra.

Wheeler, P. J. F. (1962). *National Geographic* **121**, 678–695.

Wilson, J. T. (1963a). *Nature* **197**, 536–538.

Wilson, J. T. (1963b). *Nature* **197**, 925–929.

Winter, K., Osmond, C. B., and Pate, J. S. (1981). *In* 'The Biology of Australian Plants' (J. S. Pate and A. J. McComb, eds), pp. 88–113. University of Western Australia press, Perth.

Yakovka, G. P. (1972). *Taxon* **21**, 716.

CHAPTER 3

Seed Dispersal Syndromes in Australian *Acacia*

DENNIS J. O'DOWD and A. MALCOLM GILL

	I.	Introduction	87
	II.	Definitions and Limitations	88
	III.	Inference of Dispersal Syndromes	89
	IV.	Principal Components Analysis	91
	V.	Seed Dispersal Syndromes	101
		A. Arillate *Acacia*	101
		B. Non-arillate *Acacia*	107
	VI.	Characteristics of Seed Dispersal Agents	108
	VII.	Comparison of *Acacia* with other Ant- and Bird-dispersed species	111
	VIII.	Ecological Consequences of Seed Dispersal	111
	IX.	Evolutionary Derivation of Dispersal Syndromes	113
	X.	Comparison with American and African *Acacia*	116
	XI.	Conclusions	117
		Acknowledgements	118
		References	118

I. INTRODUCTION

The diversity of dispersal structures associated with seeds may provide one of the classic examples of the evolution of adaptations, yet few examples of adaptive radiation involving entire genera or families are available (Stebbins, 1974). The genus *Acacia*, one of the largest genera of flowering plants, furnishes an excellent natural experiment for investigations on the evolution of seed dispersal characteristics. It is distributed pantropically and has diversified on each of the southern hemisphere continents (South America, Africa, and Australia) having been exposed for millions of years to distinct

SEED DISPERSAL
ISBN 0 12 511900 3

sets of potential dispersal agents on each continent. In Australia, *Acacia* is the largest genus of flowering plants (*ca.* 835 spp.) and is present in most plant communities (including *A. bakeri* in tropical rainforest) in all major climatic regions. Little is known about its seed dispersal. Our objectives here are two: (1) to determine the diaspore (i.e., dispersal unit) characteristics of a large sample of Australian *Acacia* spp.; and (2) to establish whether diaspore attributes within this heterogeneous natural grouping of plants correspond with general classes of dispersal agents, i.e., do dispersal syndromes occur? A 'yes' answer implies that certain dispersal agents play a principal role in selecting for co-occurring diaspore traits. These data then provide a baseline for further consideration, including: (1) the biochemical, morphological, and ecological correlates of the proposed dispersal syndromes; (2) the evolutionary derivation of seed dispersal traits in Australian *Acacia*; and, (3) the comparative ecology and evolution of seed dispersal in African, American and Australian *Acacia*.

II. DEFINITIONS AND LIMITATIONS

Dispersal is defined here as the transport of diaspores away from the parent plant. Dispersal agents are organisms with the ability to transport and handle diaspores, and whose characteristics are likely to leave a significant proportion of seeds viable following dispersal. In most Australian *Acacia*, the diaspore (the dispersal unit) is composed of the mature seed plus an expanded and elaborated portion of the funicle, which connects the ovule to the carpel wall during seed development (Fig. 1). We term this appendage hereafter an aril, although its homology with arils in other taxa is questionable (van der Pijl, 1955). Its important function, irrespective of structural homology with other taxa, is its role as a food which induces the dispersal of seeds by animals.

Our definition of dispersal systems among Australian *Acacia* stresses the non-random occurrence of combinations of diaspore traits related to the nature of dispersal agents. This 'syndrome' concept (Grant and Grant, 1965) has several advantages over initially relating single attributes to seed dispersal: it considers a number of correlated characters simultaneously, and therefore should have stronger predictive abilities. The syndrome concept further implies that some diaspore features are selectively moulded and integrated by the dispersal agents which most effectively disperse seeds. Under most circumstances, natural selection should be parsimonious: diverse selective pressures can often be accommodated by a single structure. While we emphasize dispersal as a strong selective force moulding diaspore attributes, this is not to deny that there are other selective pressures, such as

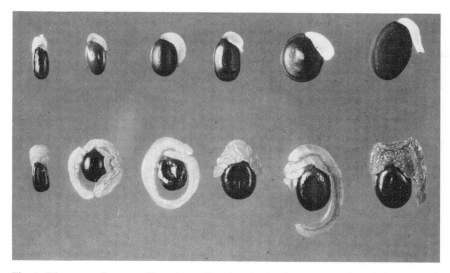

Fig. 1. Diaspores of some arillate Australian *Acacia* showing a range of seed sizes and aril structures. Top row shows species defined as 'ant-syndrome' in the text (from left to right, *A. mucronata* var. *mucronata, A. myrtifolia, A. terminalis, A. obliquinervia, A. elata,* and *A. falcata*). All arils are white to light brown. Bottom row shows 'bird-syndrome' species, *A. cowleana* (orange aril), *A. melanoxylon* (pink), *A. tetragonophylla* (yellow), *A. ligulata* (orange), *A. cyclops* (red), and *A. salicina* (red). See Table I for further diaspore characteristics.

germination and establishment demands, which must also be consistent with diaspore features.

Classifying a given *Acacia* species into a dispersal syndrome does not necessarily (or even probably) mean that its diaspores are exclusively transported by a single group of dispersal agents, but suggests that one class of the potential dispersal assemblage has been a major selective force in the evolution of its diaspore characteristics. Since we build these syndromes partially upon anecdotal accounts of seed dispersal rather than systematic study, the listed dispersal agents should be regarded as the most probable dispersal agents. This list will remain tentative until more thorough field studies are undertaken (c.f. Gill, 1985; N. Reid, pers. comm., 1983). The syndrome concept is used here to identify correlates of dispersal mode. A comprehensive understanding of the patterns described awaits concerted field study of the dispersal process.

III. INFERENCE OF DISPERSAL SYNDROMES

At least three general features of diaspores can be used to describe dispersal characteristics in Australian *Acacia*: (1) the nature of the reward provided to potential dispersal agents; (2) attributes of the diaspores which attract potential dispersal agents; and, (3) presentation patterns of the diaspores. We emphasize aril traits in our analysis, consistent with the notion that a major cost of dispersal is related to structures important in attraction and reward for potential seed vectors (van der Pijl, 1972; Howe, 1977).

To classify *Acacia* by dispersal attributes, we chose six easy-to-measure characteristics of diaspores for each of 92 species (93 'types') all associated with either the attraction or reward structure of the diaspore: dry seed mass, dry aril mass, aril length, aril contortion, seed lipid content, and aril lipid content. Dry mass of seeds and arils was determined after drying each sample of 50 diaspores (separated into seeds and arils) at 80^0C for 24 h. To measure aril length and contortion, arils from each arillate species were enlarged photographically and their length traced using a map measure. The number of complete 180 degree turns along the aril length was used as a measure of aril contortion. Four arils were selected at random for each *Acacia* species and the means were calculated.

Percent lipid content (seed and aril) was quantified for each species by wide-band nuclear magnetic resonance in a quantity analyser (Newport Instruments) with reference to standards of extracted seed and aril lipids from both *Acacia cyclops* and *A. terminalis*. While aril lipids act as nutrient rewards, certain fatty acid (FA) components may function as a trigger to release specific behaviours in dispersal agents and induce seed collection and removal (Bresinsky, 1963; Marshall *et al.*, 1979). We compared the FA composition of arils from 20 *Acacia* species (ten ant-syndrome and ten bird-syndrome species) and seeds from two *Acacia* species by trans-methylation of FA without prior extraction (Welch, 1977). Fatty acids were analysed using a Varian Aerograph 204B gas chromatograph equipped with a hydrogen flame ionization detector and columns of Silar 10C on Gas Chrom Q. Peak areas were calculated with reference to standard mixtures (Applied Sciences).

An additional variable, percent investment in dispersal per diaspore, i.e., 'reward-to-bulk-carried' ratio (Herrera, 1981a), was derived by dividing aril mass by total diaspore mass for each *Acacia* species. The level of investment in dispersal may reflect the disparate energy requirements of different classes of dispersal agents, and determine which agents transport seeds and thus the efficiency of seed removal.

Diaspore analyses of other species have shown that aril mass, lipid content and other attributes vary between seed crops and individual plants (Howe and Vande Kerckhove, 1980; Foster and McDiarmid, 1983). In some cases our diaspore samples come from single *Acacia* plants, but usually they represent population samples from a single locality taken from an unknown number of individuals. Since both aril and seed attributes are under genetic and environmental control (Foster and McDiarmid, 1983), considerable variability may occur within species that we do not describe here.

A Principal Components Analysis (PCA) was used to arrange species in two dimensions by summarizing the suite of original variables into the major dimensions (axes) of variation (Sneath and Sokal, 1973). Each succeeding dimension maximized the residual variance after extraction of the variance attributable to the previous dimension, so that the initial two dimensions often describe the major relationship between traits. Further, each dimension consists of 93 components conveying information about relationships between the diaspore characteristics of each *Acacia* species (Table I). For example, if component scores in the first dimension are similar for two species, then they have similar diaspore features as described by that dimension.

Once the PCA display was generated, two types of dispersal evidence were superimposed on it: (1) aril colour (for all arillate *Acacia* — 85 of 93 species), a known but imperfect correlate of seed dispersal by birds (Turcek, 1963; Willson and Thompson, 1982); and, (2) actual observations of seed dispersal by animals (for 22 of 93 species). Subsequent grouping into dispersal 'regions' was inferred from the relationship between the PCA display and the above criteria. Once groupings were defined, syndromes were characterized by examining the original six variables to determine any metric, morphological, and biochemical correlates of seed dispersal.

To determine whether diaspore presentation and display patterns were consistent with syndrome characteristics as defined from the PCA, we selected several *Acacia* spp. (*A. melanoxylon, A. cyclops, A. pycnantha* and *A. myrtifolia*) to measure diaspore retention times in the tree or shrub canopy. For bird-syndrome species, we predicted that fruits would be retained and diaspores displayed in the canopy. In ant-syndrome species, diaspores should be presented to dispersal agents on the soil or litter surface, through either shedding of diaspores or dehisced fruits. For each species, five shoots with mature pods were bagged in mosquito netting and retention times of fruits and diaspores in the canopy were determined following pod dehiscence.

IV. PRINCIPAL COMPONENTS ANALYSIS

In the PCA, 69% of the variance in diaspore attributes was explained by the first two axes. Coupling the Principal Component co-ordinates of *Acacia* spp. (Table I) with the weighting of diaspore traits on axes 1 and 2 (Table

Table I. Seed and aril attributes of the Australian *Acacia* spp. (92 species, 93 'types') available for the Principal Components Analysis: (1) PCA scores for each species on axes 1 and 2; (2) taxonomic groupings (Ac = subgenus *Acacia*, section *Acacia*; Bo = subgenus *Phyllodineae* (P), section *Botryocephalae*; Ph = *P*, section *Phyllodineae*; Pl = *P*, section *Plurinerves*; Ju = *P*, section *Juliflorae*; Pu = *P*, section *Pulchellae*; (3) seed attributes (dry mass and percent lipid content); (4) aril attributes (colour, dry mass, length, contortion, and percent lipid content). Aril colour defined as W = white to brown; 0 = orange, R = red, and Y = yellow; and, (5) percent of total diaspore invested in aril (dry aril mass/total dry diaspore mass).

Species	PCA scores		Taxonomic section	Seed			Aril				
	axis 1	axis 2		Dry mass (mg)	% Lipid	Colour	Dry mass (mg)	Length (mm)	Contortion	% Lipid	% Investment
Acacia acinacea[a]	0.039	1.528	Ph	12.5	16.2	W	1.3	3.7	0.00	43.1	9.2
A. acuminata	0.084	1.823	Ju	16.5	17.5	W	0.6	3.1	0.00	44.0	3.4
A. adunca	0.301	-0.519	Ph	56.2	14.0	W	4.0	5.2	0.00	37.3	6.7
A. alleniana	2.010	-1.437	Ph	22.7	5.3	W	0.4	4.1	0.25	8.6	1.8
A. aneura	-0.166	0.705	Ju	11.4	13.1	W	0.2	6.9	4.00	37.3	1.9
A. aspera	-0.967	1.382	Ph	17.9	17.4	W	2.1	8.7	2.50	50.5	10.5
A. aulacocarpa	-1.890	-0.867	Ju	29.9	12.6	W	1.6	20.5	11.25	34.5	5.2
A. auriculiformis	-3.769	-0.433	Ju	11.2	13.1	0	2.0	23.6	15.50	54.7	15.0
A. baileyana	0.311	0.601	Bo	19.2	13.2	W	1.7	4.8	0.00	39.5	8.3
A. beckleri	0.866	-0.273	Ph	17.8	9.2	W	1.2	6.0	0.00	31.1	6.1
A. bidwillii	3.926	-4.616	Ac	169.2	6.6	-	-	-	-	-	-
A. binervata	0.972	-0.253	Pl	19.0	10.3	W	1.7	4.6	0.00	24.1	8.2
A. brachystachya	-0.537	-0.037	Ju	19.2	11.6	W	0.5	13.2	4.75	38.8	2.4
A. buxifolia	0.740	0.247	Ph	14.0	11.8	W	1.3	5.3	0.00	27.0	8.7
A. calamifolia	0.523	0.283	Ph	14.5	9.8	W	1.2	3.7	0.00	45.6	7.5
A. cardiophylla	0.282	0.960	Bo	16.4	13.3	W	1.0	3.5	0.00	45.9	5.9
A. complanata	1.138	-1.239	Pl	51.5	10.3	W	2.5	5.1	0.00	22.1	10.3
A. conferta	1.204	-0.015	Ph	11.7	10.2	W	0.8	3.7	0.00	20.0	6.2
A. cowleana	-2.302	0.180	Ju	8.7	12.2	0	1.1	18.0	10.50	55.9	10.8
A. craspedocarpa[b]	2.836	-2.027	Ju	84.8	9.6	-	-	-	-	-	-
A. cultriformis	0.445	0.531	Ph	13.9	12.3	W	1.4	5.2	0.00	36.5	9.1

A. cyclops	-4.991	-3.521	Pl	24.8	12.1	R	14.4	55.0	4.75	40.0	36.8
A. cyperophylla	-0.541	0.212	Ju	11.9	12.6	W	1.4	13.1	3.50	36.2	10.6
A. dawsonii	-0.158	1.248	Pl	5.1	14.8	W	0.5	5.6	2.25	38.0	8.7
A. dealbata	0.445	0.975	Bo	9.1	11.9	W	0.4	3.1	0.00	46.1	4.0
A. deanei subsp. paucijuga	0.629	0.465	Bo	18.6	12.0	W	1.0	4.0	0.00	36.9	5.1
A. decora	0.623	0.541	Ph	8.9	11.1	W	0.4	5.7	0.00	36.3	4.2
A. decurrens	0.805	0.517	Ph	14.6	12.3	W	0.8	3.6	0.00	29.7	5.4
A. doratoxylon	-0.142	-0.154	Ju	15.4	10.5	W	1.5	11.3	2.50	37.9	8.7
A. drummondii	-0.319	1.941	Pu	3.2	17.1	W	0.5	4.5	1.75	43.4	13.4
A. elata	1.092	-1.170	Bo	31.5	7.7	W	2.2	5.4	0.00	25.5	6.6
A. excelsa	2.426	-1.071	Pl	23.8	7.1	—	—	—	—	—	—
A. extensa	0.697	0.623	Ph	11.6	14.2	W	1.5	4.0	0.50	19.4	11.2
A. falcata	0.724	0.212	Ph	11.7	11.0	W	1.6	3.8	0.00	30.4	12.3
A. falciformis	0.884	-1.012	Ph	44.4	10.1	W	2.7	6.1	0.00	28.5	5.8
A. farnesiana	3.326	-3.201	Ac	105.4	6.1	—	—	—	—	—	—
A. filicifolia	0.818	0.409	Bo	11.5	10.7	W	0.9	3.2	0.00	34.6	5.4
A. fimbriata	0.805	0.196	Ph	12.0	11.0	W	1.2	4.7	0.00	28.1	9.2
A. floribunda	-0.377	1.611	Ju	7.1	13.2	W	0.2	4.0	1.00	65.7	3.4
A. hakeoides	0.454	0.571	Ph	24.9	13.7	W	1.2	5.4	0.00	36.7	4.6
A. havlandii	-1.933	2.131	Pl	6.1	17.3	W	0.7	12.4	4.00	74.6	10.9
A. hemiteles	1.966	-1.942	Ph	50.1	5.1	W	0.9	3.7	0.00	20.5	1.8
A. implexa	-2.087	-0.732	Pl	17.8	10.2	W	2.4	19.8	9.50	51.4	11.8
A. iteaphylla	0.136	0.306	Ph	32.8	10.2	W	0.7	5.9	0.00	64.4	2.0
A. ixiophylla	-0.485	1.647	Pl	6.2	14.7	W	0.6	5.8	1.25	57.7	8.5
A. juncifolia	2.343	-0.874	Ph	13.8	6.9	—	—	—	—	—	—
A. kybeanensis	0.563	0.002	Ph	15.8	11.2	W	2.5	4.7	0.00	30.6	13.7
A. lanigera	-0.074	0.761	Pl	10.2	11.9	W	0.4	7.4	1.75	47.6	3.5
A. lasiocalyx	-2.226	-1.061	Ju	23.2	10.8	W	3.0	24.0	9.25	46.2	11.5
A. lateriticola	0.406	1.786	Pu	4.3	17.2	W	0.5	2.4	0.25	28.9	10.1
A. leprosa	-0.804	1.019	Ph	5.7	16.1	W	0.4	14.1	4.50	31.2	6.7

Table I (continued)

Species	PCA scores		Taxonomic section	Seed		Colour	Aril				
	axis 1	axis 2		Dry mass (mg)	% Lipid		Dry mass (mg)	Length (mm)	Contortion	% Lipid	% Investment
A. ligulata											
var. "red aril"	−2.212	−1.688	Ph	34.2	10.0	R	7.9	21.5	4.25	56.9	18.7
var. "orange aril"	−1.755	−0.862	Ph	28.4	10.9	O	4.4	20.4	4.50	55.5	13.5
A. linifolia	0.288	0.383	Ph	26.9	13.9	W	2.4	5.1	0.00	36.5	8.3
A. longifolia	−1.926	−0.310	Ph	16.0	12.5	W	2.8	17.3	8.75	44.7	14.8
A. maidenii	−2.462	0.546	Ju	14.9	13.9	Y	3.0	16.9	7.00	69.7	16.5
A. mangium	−2.993	−0.874	Ju	10.2	11.9	O	2.2	26.2	14.50	38.2	17.4
A. mearnsii	0.343	0.758	Bo	11.2	11.4	W	0.8	4.3	0.00	47.5	6.4
A. melanoxylon	−2.222	−3.178	Pl	9.6	5.9	R	3.9	47.6	7.25	13.1	28.7
A. microcarpa	0.624	1.332	Ph	6.1	15.5	W	0.5	3.3	0.00	26.0	7.7
A. montana	−0.575	1.157	Pl	12.2	16.6	W	1.4	10.6	2.25	37.5	10.1
A. mucronata											
var. mucronata	−0.634	1.689	Ju	4.7	17.8	W	0.4	8.8	3.50	35.8	7.5
A. myrtifolia	−0.073	2.003	Ph	11.3	18.9	W	1.3	3.5	0.00	38.8	10.2
A. neriifolia	0.566	−0.558	Ph	39.5	13.4	W	4.0	5.7	0.00	21.9	9.3
A. obliquinervia	0.361	0.578	Ph	26.7	14.6	W	2.3	3.4	0.00	35.6	8.0
A. obtusata	0.436	0.801	Ph	14.9	13.2	W	1.1	4.2	0.00	38.3	6.6
A. oxycedrus	−1.120	−0.840	Ju	31.8	13.1	W	4.8	15.9	4.25	34.2	13.2
A. pendula	2.086	−0.250	Pl	20.6	10.9	—	—	—	—	—	—
A. penninervis	0.598	−0.891	Ph	52.6	11.0	W	3.1	6.0	0.00	37.9	5.5
A. podalyriaefolia	0.296	−1.227	Ph	24.6	10.2	W	3.9	11.8	1.75	14.9	13.6
A. polybotrya	0.541	−0.241	Bo	32.7	12.1	W	2.8	5.2	0.00	32.5	7.8
A. pravissima	0.608	1.001	Pl	7.2	12.9	W	0.6	2.7	0.00	35.9	7.2
A. prominens	0.777	0.293	Ph	11.4	10.1	W	0.9	3.2	0.00	36.5	7.3
A. pulchella	0.189	1.901	Pu	3.7	15.9	W	0.3	2.0	0.00	44.3	7.3

A. pycnantha	0.028	0.996	Ph	14.9	14.5	W	1.6	6.0	0.00	42.7	9.8
A. quornensis	0.196	0.557	Ph	17.5	12.3	W	1.4	5.6	0.00	44.5	7.2
A. salicina	-4.552	-2.977	Ph	45.7	13.9	R	18.1	30.3	4.50	59.0	28.4
A. saligna	0.784	0.586	Ph	14.1	13.5	W	1.1	4.2	0.00	23.8	7.1
A. schinoides	1.166	-0.497	Bo	15.6	8.5	W	1.4	4.4	0.00	23.1	8.2
A. semirigida	1.196	-0.116	Ph	27.8	12.9	W	1.9	2.6	0.00	14.0	5.3
A. siculiformis[c]	1.934	0.111	Ph	7.9	11.2	—	—	—	—	—	—
A. sophorae	-2.516	-0.505	Ph	18.2	14.8	W	5.5	23.5	7.00	41.8	23.2
A. spectabilis	0.615	-0.287	Bo	30.0	10.6	W	2.0	5.6	0.00	36.8	6.3
A. stricta	-1.203	1.785	Ph	6.0	15.9	W	0.5	9.8	3.25	62.6	7.2
A. suaveolens	-0.920	0.642	Ph	28.9	13.5	W	1.4	9.1	3.50	62.5	4.6
A. terminalis	0.482	0.099	Bo	20.1	12.7	W	2.6	5.7	0.00	28.1	11.5
A. tetragonophylla	-4.118	-1.688	Ph	10.5	10.5	Y	6.1	44.0	8.75	57.5	36.5
A. ulicifolia[c]	2.266	-0.689	Ph	11.0	7.5	—	—	—	—	—	—
A. uncinata	1.200	-1.377	Ph	62.5	10.5	W	2.4	5.8	0.00	24.8	3.7
A. venulosa	-0.222	0.919	Pl	6.9	10.9	W	0.3	5.8	1.75	58.5	4.3
A. verniciflua	-0.393	0.955	Pl	9.2	15.9	W	1.1	9.8	3.25	29.4	10.8
A. vestita	0.786	-0.162	Ph	29.8	11.4	W	1.4	5.7	0.00	30.9	4.6
A. viscidula	-0.020	0.528	Pl	7.6	10.8	W	0.3	7.2	2.50	44.5	3.8

[a] taxonomy follows Pedley (1978).
[b] *A. craspedocarpa* defined as non-arillate according to Vassal (1971).
[c] *A. ulicifolia* and *A. siculiformis* have thread-like, funicular appendages which are non-arillate.

(A)

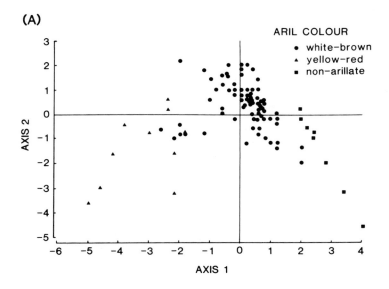

(B)

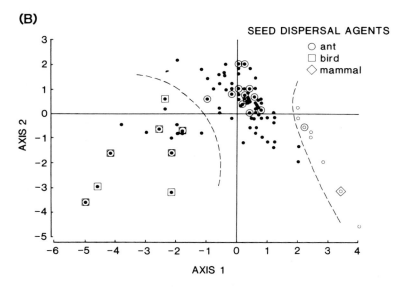

Fig. 2. Principal Components Analysis of diapsores from 92 Australian *Acacia* species (93 'types') on the first two principal axes. Symbols represent component scores on the first two axes for each *Acacia* species (component scores are given in Table I). Distance between points is a measure of oversall similarity in diaspore traits. (A) overlay of aril colour and (B) overlay of dispersal evidence from Table III. Dotted lines separate species' groups. Note that symbols used in (B) do not have the same meaning as in (A), although the points plotted are identical.

Table II. Principal Components Analysis of Australian *Acacia* diaspores. Rankings of component scores for each variable on the first and second axes are shown in parentheses.

	Axes					
	1	2	3	4	5	6
% variance explained	41.69	27.32	13.01	9.29	7.06	1.63
Component scores:						
seed dry mass	0.228(6)	-0.538(2)	0.408	0.683	0.097	0.129
seed lipid content	-0.235(5)	0.569(1)	0.437	-0.156	0.627	0.106
aril dry mass	-0.395(4)	-0.384(3)	0.566	0.423	-0.016	-0.444
aril lipid content	-0.443(3)	0.332(4)	0.205	-0.382	-0.710	0.031
aril length	-0.544(1)	-0.328(5)	-0.132	0.152	0.099	0.739
aril contortion	-0.495(2)	-0.149(6)	-0.511	-0.402	0.288	-0.477

II) allows a general description of variation in diaspore traits (Fig. 2). On axis 1, aril traits (% lipid content, mass, length, contortion) explain much of the total variance (Table II) and tend to decrease from left to right. For example, *A. cyclops* (extreme left in Fig. 2A) has an elaborate aril, whereas *A. bidwillii* (extreme right in Fig. 2A) is non-arillate. On axis 2, seed traits (mass and % lipid content) were most important: seed size generally decreased going up axis 2 while % lipid content of seeds generally increased. For example *A. bidwillii* (extreme bottom of Fig. 2), has a massive seed with very low % lipid content, whereas *A. havlandii* (uppermost coordinates in Fig. 2) has a small seed with high % lipid content. There was considerable range in component scores among *Acacia* spp. on the first two axes (Table I, Fig. 2). Overall, diaspore traits of most species clustered around the origin in Figure 2 but some deflected away in opposing directions, reflecting marked dissimilarities in overall characteristics. Non-arillate *Acacia* grouped to the extreme right of the display while arillate species spread from the origin to the lower left.

Overlaying aril colour on to the PCA display further aided discrimination among arillate *Acacia* (Fig. 2A). Those with brightly coloured arils (yellow, orange or red) extend to the lower left of the display. *Acacia* spp. with dull-coloured arils (white to brown) tend to cluster centrally, although a few 'outliers' mix with species with brightly coloured arils.

Evidence for seed dispersal in Australian *Acacia* is scant, and with a few exceptions, is primarily anecdotal. Observations, however, are sufficient to identify major classes of seed dispersal agents (Table III). Arillate *Acacia* appear to be primarily associated with dispersal by two distinct taxa: ants and birds. When these dispersal observations are superimposed onto the PCA display (Fig. 2B), they are largely consistent with the overlay of aril colour (Fig. 2A), enhance definition of these groups, and provide further support for diaspore groupings. Ants are the most generalized collectors

Table III. Putative dispersal agents for the Australian *Acacia* spp. used in the Principal Components Analysis. Three criteria for dispersal are presented: (1) transport of diaspores by the agent; (2) evidence of the agent feeding on arils, ingesting diaspores or defecating intact seeds; and, (3) observation of a concentration of intact diaspores or seedlings in or around nests of agents or below perches. Diaspore groupings (I-III) were defined from the PCA (Table I, Table II, Fig. 2). Trophic specializations (in parentheses) of various dispersal agents are as follows: F = frugivorous; I = insectivorous; N = nectivorous; O = omnivorous; S = granivorous; and, H = herbivorous.

Species	Dispersal Agents	Syndrome	Criteria	Source
Acacia aneura	ants	II	1,3	Davidson and Morton (1984)
	Melophorus sp. (O-S)[a]			
	Pheidole sp. (O)			
A. baileyana	ants	II	3	D. J. O'Dowd (pers. obs.)
	Iridomyrmex sp. (I,N)			
A. cyclops	ants	I	1,3	Gill (1985)
	Camponotus sp. (I,N)			
	Iridomyrmex sp. (I,N)			
	Melophorus spp. (O-S)[a]			
	Pheidole sp. (O)			
	Rhytidoponera spp. (O)[b]			
	birds		1,2	Gill (1985)
	Anthochaera carunculata (F,I,N)			
	Gymnorhina dorsalis (O)			
	Meliphaga virescens (F,I,N)			
	Strepera versicolor (O)			
	Zosterops lateralis (F,I,N)			
A. dealbata	ants	II	1	Ashton (1979)
A. extensa	ants	II	1,3	Majer (1980)
	Rhytidoponera violacea (O)			
A. farnesiana	mammals	III	2	Basáñez (1977); Pedley (1978)
	Bos taurus (H)			
	Capra hircus (H)			
	Equus cabalus (H)			

Species	Dispersers		Number	Reference
A. ligulata	ants	Iridomyrmex purpureus (I,N) Pheidole sp. (O)	I	1	Davidson and Morton (1984); N. Reid (pers. comm. 1983)
	birds	Acanthagenys rufogularis (F,I,N) Artamus cinereus (F,I,N) Corvus coronoides(O) C. bennetti (O) Cracticus torquatus (F,I,N) Gymnorhina tibicen (O) Manorina flavigula (F,I,N)		2	N. Reid (pers. comm. 1983)
A. linifolia	ants	Aphaenogaster longiceps (I,N) Pheidole sp. (O)	II	1,3	Berg (1975)
A. maidenii	birds	Trichoglossus chlorolepidotus (F,N,S?)	I	2	Pedley (1978)
A. mearnsii	ants		II	3	D. J. O'Dowd (pers. obs.)
A. melanoxylon	birds	Meliphaga chrysops (F,I,N)	I	2	Forde (1986)
A. myrtifolia	ants	Pheidole sp. (O) Rhytidoponera metallica (O)	II	1,3	D. J. O'Dowd (pers. obs.)
A. pulchella	ants	Rhytidoponera inornata (O)	II	1,3	Majer (1982)
A. pycnantha	ants	Camponotus sp. (I,N) Chelaner sp. (O-S)[a] Dolichoderus sp. (I,N) Iridomyrmex sp. (I,N) Pheidole sp. (O) Rhytidoponera metallica (O)	II	1,3	D. J. O'Dowd (unpubl. data)

Table III (continued)

Species	Dispersal Agents	Syndrome	Criteria	Source
A. salicina	birds Meliphaga virescens (F,I,N)	I	2	Forde (1986)
A. saligna	ants Iridomyrmex sp.(I,N)	II	1	Andersen (1982)
A. sophorae	ants Rhytidoponera metallica (O)	I	1	P. Bernhardt (pers. comm.)
	birds Meliphaga chrysops (F,I,N) M. virescens (F,I,N)		2	Forde (1986)
A. suaveolens	ants Aphaenogaster longiceps (I,N) Rhytidoponera metallica (O)	II	1,3	Drake (1981)
A. terminalis	ants Aphaenogaster longiceps (I,N)	II	3	Berg (1975)
A. tetragonophylla	ants Melophorus sp. (O-S)[a] Rhytidoponera sp. (O) birds Acanthagenys rufogularis (F,I,N) Meliphaga virescens (F,I,N)	I	1,3 2	Davidson and Morton (1984) Forde (1986)
A. ulicifolia	ants Aphaenogaster longiceps (I,N) Rhytidoponera metallica (O)	III	1	Drake (1981)

[a] *Melophorus* and *Chelaner* have diets which are variable with species ranging from extreme generalists to specialized seed-eaters (P.J.M. Greenslade, pers. comm. 1984).

[b] *Rhytidoponera* is not generally recognized as a seed-eater, but smaller species (e.g., *Rhytidoponera metallica*) both collect and eat some seeds (D.J. O'Dowd, unpubl. data).

and *Acacia* spp. with a broad range of diaspore characteristics are taken (large circles, Fig. 2B). Observations of bird dispersal are restricted to species with component co-ordinates skewed to the lower left of the origin (large open squares, Fig. 2B). Diaspores of *A. farnesiana*, with non-arillate seeds, are ingested by large, introduced mammals and this species has a component co-ordinate at the extreme lower right of the origin (large open diamond, Fig. 2B).

V. SEED DISPERSAL SYNDROMES

Coupling of the PCA to direct and indirect evidence of seed dispersal identifies three general groupings of Australian *Acacia* diaspores (Fig. 2B) from which syndromes of dispersal traits can be described: (1) a 'region' of arillate *Acacia* to which bird-dispersal is restricted, comprising 17% of the species in our sample; (2) the remaining 'ants only' region of arillate *Acacia* species, totalling 74% of the sample; and, (3) a group of non-arillate species whose dispersal criteria are unclear (9% of our sample).

A number of diaspore variables combine to delineate groupings in the PCA. These individual variables can be compared between diaspore groups to determine any correlates with each dispersal syndrome (Table IV).

A. Arillate Acacia

1. Aril Colour
Although no single diaspore character decisively segregates *Acacia* into dispersal groupings, aril colour is probably the best single discriminator between ant- and bird-dispersed species (Fig. 2A). Brightly-coloured arils occur more frequently in the bird-restricted region of the PCA display while arils of diaspores dispersed by ants alone are white to brown ($G = 55.1$, df $= 1$, $p < 0.001$, goodness-of-fit test). The colourful arils of these *Acacia* species are likely to attract avian dispersal agents. Insect dispersal agents typically forage on the ground and red, a common colour in these *Acacia* diaspore displays, is not included in the colour-vision spectrum of most insects (Wigglesworth, 1972). Birds generally have well developed colour vision (Polyak, 1957; Sillman, 1973) and Turcek (1963) has shown that red and black are common colours in European fruits whose seeds are dispersed by birds. Bicoloured diaspores, such as occur in most bird-syndrome *Acacia* (Fig. 1), are more attractive than monochrome displays in fruiting species of the deciduous forests of North America, especially when bird populations are low (Willson and Melampy, 1983).

As in pollination systems (Faegri and van der Pijl, 1971), the association between bright colours and bird visitation is not absolute and some bird-syndrome *Acacia* have dull-colour arils. *Acacia sophorae*, with light brown arils, is dispersed by honeyeaters (Forde, 1985) and in other diaspore and display characteristics it groups with bird-syndrome *Acacia* (Table I, Table IV). Other *Acacia* species with white-brown arils, including *A. implexa*, *A. longifolia* and *A. oxycedrus*, fall into this bird-restricted region of the PCA display. We would predict on this basis that they are primarily bird-dispersed but their inclusion remains tentative until field studies determine diaspore display patterns and seed removal agents.

2. Aril Mass, Length, Contortion, and Investment in Dispersal

Quantitative differences in aril mass, length and contortion are consistent with the general qualitative pattern in aril colour which distinguishes between dispersal syndromes (Table IV). On average, aril mass is over three times greater in bird-syndrome *Acacia* than in ant-syndrome *Acacia*. This greater reward reflects both the necessity of providing a signal sufficient to attract vision-oriented birds, and the greater metabolic requirements of large endothermic birds. This dual requirement can be met in several ways where aril size is organized into various combinations of length and contortion to yield a display. In bird-syndrome species, arils average almost five times the length of arils in ant-syndrome species and are about ten times more contorted (Table IV). Since the aril is elaborated along the length of the funicle, the display is arranged in two basic ways: the aril encircles the seed longitudinally (e.g., *A. cyclops* in Fig. 1) or subtends the seed forming a convoluted, turbinate appendage (e.g., *A. salicina* in Fig. 1).

Investment in dispersal per diaspore corresponds to the differences in aril mass averaging 7.2% for ant-syndrome and 18.9% for bird-syndrome *Acacia* (Table IV). Variation in investment per diaspore can strongly influence the removal rate of diaspores (Howe and Vande Kerckhove, 1980; Herrera, 1981b). In our sample of Australian *Acacia*, a significant relationship occurs between seed and aril masses in both ant- and bird-sydrome species (Fig. 3). A higher investment in aril tissue is maintained over the range of seed sizes in bird-syndrome *Acacia* and the slope of aril-to-seed mass is greater than in ant-syndrome species. Since diaspore display is an important component attracting visually-oriented birds, the maintenance of aril mass in proportion to seed size may be necessary to maintain an adequate signal. For foraging ants, visual display is less important, since most species rely on chemical rather than visual cues (Wilson, 1971).

3. Aril Lipids

Aril quality as well as quantity differs with dispersal syndrome. The average

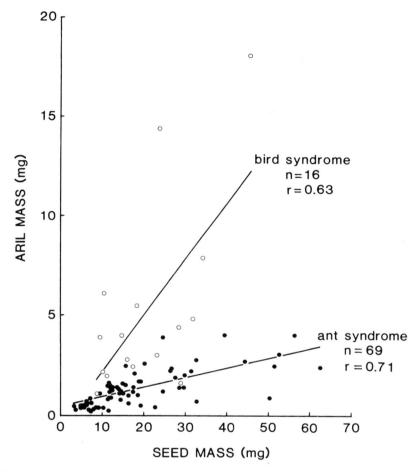

Fig. 3. Relationship between seed and aril masses among 85 arillate *Acacia* species. For analysis, species were divided into 'ant syndrome' (N = 69) and 'bird syndrome' (N = 16) as in Figure 2. Correlations between seed and aril masses were significant (for ant syndrome, r = 0.71, p < 0.001; for bird syndrome, r = 0.63, p < 0.01). Both slope and elevation of lines-of best-fit (for ant, y = 0.05x + 0.43; for bird, y = 0.28x - 0.01) differ significantly (slope, $F^{1,81}$ = 31.3, p < 0.001; for elevation, $F_{1,81}$ = 64.4, p << 0.001, analysis of covariance).

% lipid content of arils was significantly greater in bird-syndrome *Acacia* than in ant-syndrome *Acacia* (Table IV). Lipids are a primary constituent of most *Acacia* arils and in some species attain 70% of the aril dry mass (Table I). This value is comparable to the highest lipid contents observed in seeds (Earle and Jones, 1962; Trelease and Doman, 1984) and fruits (Mazliak, 1970). Lipids provide a greater energy reward per unit mass of

Table IV. Analysis of variance on diaspore characteristics of 92 species (93 'types') of Australian *Acacia* when grouped by dispersal syndrome. Numbers in parentheses are total *Acacia* spp. in each category. Percent investment is the dry mass of aril divided by the total dry mass of the diaspore.

Character	Syndrome			F	P
	Ant	Bird	'Other'		
Seed mass (mg)	18.5(69)	21.0 (16)	54.6 (8)	11.0	<0.001
% Seed lipid	12.7	11.8	8.2	10.4	<0.001
Aril mass (mg)	1.3	5.2	—	41.6	<0.001
% Aril lipid	36.5	47.1	—	8.9	<0.001
% Aril length (mm)	5.8	26.9	—	181.3	<0.001
Aril contortion	0.8	8.3	—	181.2	<0.001
% Investment	7.2	18.9	—	79.7	<0.001

aril, since they yield about twice as much energy upon catabolism as carbohydrates (Klieber, 1961; Slack and Browse, 1984).

The fatty acid compositions of lipids from arils of ant- and bird-syndrome *Acacia* did not differ strongly, although the relative amounts of both stearic and linolenic acids were significantly greater in ant-syndrome arils (Table V). Any ecological role for these differences remains unknown, although oleic acid (Marshall *et al.*, 1979) or its congener, ricinoleic acid (Bresinsky, 1963), may trigger specific behaviours in ants and stimulate seed collection. There were, however, no significant differences in oleic acid representation in lipids from arils of *Acacia* spp. assigned to bird- and ant-dispersal groupings. If FA's are important only in a nutritive sense, the requirements of ants and birds may be sufficiently alike that no strong selection for FA differences occurs.

Both lipid quantity (Table IV) and quality (Table VI) differ between arils and seeds, similar to differences found between pericarp and seed lipids of fruits (Mazliak, 1970). This may reflect the different roles of lipid stored to induce dispersal (aril lipids) and that stored to assist in seedling establishment (seed lipids). The average lipid content of arils was about three times greater than that of seeds and was paralleled by marked differences in FA composition. In *A. terminalis* and *A. cyclops*, ant- and bird-dispersed species respectively, seed FA was less saturated than aril FA (Table VI). Both the higher lipid content and higher levels of FA saturation in arils may reflect the role of aril tissue as animal food. Since most of the known dispersal agents of *Acacia* are generalist feeders and incorporate insects in their diets (Table III), the higher proportion of saturated FA may 'mimic' the FA composition typically found in animal prey. The proportion of saturated FA in the *Acacia* arils examined in Table V (29.6% saturated) is more similar to the saturation level in insects (29.7%)

Table V. Fatty acid (FA) composition of arils from *Acacia* species separated into ant- and bird-dispersed syndromes. N is the sample of *Acacia* species under each category.* FA components are expressed as percent of total FA and identified by the number of carbon atoms followed by the number of double bonds, e.g., 16:0 (palmitic acid) is saturated while 18:3 (linolenic acid) is unsaturated. Significance of F-values as follows: NS = not significant; † = $p < 0.01$. ANOVA was not done on palmitoleic acid (16:1) because arils of only three species contained this FA.

Syndrome	Fatty acids							
	14:0	16:0	16:1	18:0	18:1	18:2	18:3	N
Ant								
X̄	0.0	23.7	0.5	6.7	55.3	12.7	1.2	10
SD	—	10.8	1.7	5.2	14.8	8.1	1.2	
Bird								
X̄	0.0	27.3	1.9	1.5	59.2	10.1	0.0	10
SD	—	6.0	4.2	1.1	19.9	16.8	0.1	
$F_{(1,18)}$	—	0.9^{NS}	—	$11.9†$	0.3^{NS}	0.2^{NS}	$8.8†$	

* Species defined as ant-syndrome include *A. cyperophylla, A. decurrens, A. elata, A. myrtifolia, A. penninervis, A. prominens, A. pycnantha, A. stricta, A. suaveolens,* and *A. terminalis.* Bird-syndrome species include *A. auriculiformis, A. cyclops, A. ligulata* var. "red aril", *A. ligulata* var. "orange aril", *A. sophorae, A. maidenii, A. mangium, A. melanoxylon, A. salicina* and *A. tetragonophylla.*

reported by Gilmour (1961) than average leaf (16.3%) or seed (16.1%) values (Hitchcock and Nichols, 1971). Fatty acid saturation values of *Acacia* arils are also similar to the average saturation level (28.3%) reported for the pericarp of oil-containing fruits typically associated with bird-dispersal (Mazliak, 1970).

4. Seed Characteristics

To the extent that seed morphology and size have evolved independently of dispersal exigencies (Harper *et al.*, 1970; Baker, 1972), they represent initial constraints upon the evolution of attraction and reward structures of diaspores. While dispersal agents have undoubtedly shaped the characteristics of the aril, it is less clear how they might have affected the seed itself. Seed masses and % lipid content of ant- and bird-syndrome *Acacia* overlap broadly and no significant difference occurs between the two groups (Table IV). The size range of ant-syndrome species (3.2-62.5 mg) is actually broader than that of bird-dispersed *Acacia* (8.7-45.7 mg) in our sample. Nevertheless, dispersal agents may act in several ways to modify seed size, shape, and seedcoat texture. First, they may place both an upper and lower limit on seed size. The upper size range of both ant- and bird-dispersed *Acacia* may be limited if the diaspores become too massive to be handled efficiently. In bird-syndrome *Acacia*, large seed size may restrict the range

Table VI. Fatty acid composition of seeds and arils from *Acacia terminalis* (ant-dispersed) and *A. cyclops* (bird-dispersed). Fatty acids are expressed as proportion of total FA and identified by the number of carbon atoms followed by the number of double bonds, e.g., 16:0 (palmitic acid) is saturated, while 18:3 (linolenic acid) is unsaturated. Dry masses and lipid content of seeds and arils of the two species are shown in Table I.

	Fatty Acids							
	14:0	16:0	16:1	18:0	18:1	18:2	18:3	22:0
A. terminalis								
seed	0.0	9.4	0.0	0.9	19.7	70.0	0.0	tr
aril	0.0	23.6	0.0	5.6	50.7	18.9	1.2	0.0
A. cyclops								
seed	0.0	9.1	0.0	0.7	11.8	78.4	0.0	tr
aril	0.0	32.7	0.0	1.2	63.4	2.8	0.0	0.0

tr = trace.

of bird species available to disperse seeds. Gape size of birds is correlated with the ability to ingest diaspores (Herrera, 1981a). Large seed size may restrict the dispersal coterie to birds with large gape size and result in more 'seed wastage' (Howe, 1980) because birds with smaller gape size may feed on the arils alone and discard seeds below the parent plant. For example, N. Reid (pers. comm. 1983) has observed that chestnut-rumped thornbills (*Acanthiza uropygialis*) with a relatively small gape size feed selectively on the red arils of *A. ligulata* and discard seeds below the parent canopy. For ants, large seed size should also restrict the array of dispersal agents to large ants that can individually manipulate large seeds, or to those that recruit to and collectively handle single seeds. The overall efficiency of dispersal should also decrease, since smaller ants can still feed on the aril without transporting the seed (Berg, 1975; Gill, 1985). The lower range of seed size, especially in bird-syndrome *Acacia*, may be limited by the degree to which the diaspore display is perceptible to potential dispersers.

5. Pod and Diaspore Retention

Pod dehiscence and diaspore presentation patterns are consistent with the dispersal syndromes based on diaspore features alone (Fig. 4). In two ant-syndrome species examined, diaspores are presented on the soil or litter surface in distinct ways: (1) in *A. myrtifolia*, pods dehisce explosively and propel arillate seeds to the soil or litter surface. This dehiscence occurs rapidly as the pods dry in warm weather, so the availability of diaspores is strongly pulsed; (2) in *A. pycnantha*, pods initially dehisce along one suture, but most diaspores fall to the soil surface while they are still attached to one of the pod valves. Consequently seedfall occurs over a longer period, but pulses in podfall occur following high winds. This is contrasted by two bird-syndrome *Acacia* species (Fig. 4) where pods generally dehisce

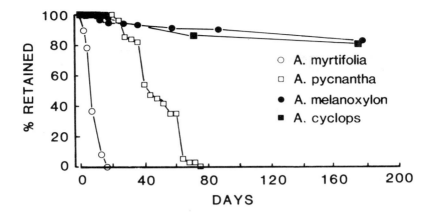

Fig. 4. Diaspore presentation patterns as indicated by per cent seed retention in the canopy of four *Acacia* species. Ant syndrome (open symbols) = *A. myrtifolia* (circles, number of seeds = 150) and *A. pycnantha* (squares, n=348); bird syndrome (filled symbols) = *A. melanoxylon* (circles, n=186) and *A. cyclops* (squares, n=101). Time zero is defined from first pod dehiscence in each bagged treatment.

along both sutures and most diaspores are retained and displayed in the plant canopy. Three basic display modes occur: (1) in *A. cyclops, A. tetragonophylla,* and *A. melanoxylon,* the brightly-coloured aril encircles the shiny, black seed longitudinally, forming a bicoloured 'bull's-eye'; (2) in the bicoloured diaspores of *A. salicina* and *A. ligulata,* seeds are subtended by a brightly-coloured, turbinate aril and retained in the canopy (Fig. 1); and (3) in other bird-dispersed *Acacia,* such as *A. auriculiformis,* shiny black seeds dangle from brightly-coloured arils following pod dehiscence (Corner, 1949). Diaspores can be displayed in the canopy of some bird-dispersed species for long periods — as long as 18 months in *A. melanoxylon* (D.J. O'Dowd, pers. obs.).

B. Non-arillate Acacia

The non-arillate group in our sample (n = 8) is a 'grab-bag' of species whose only unifying feature is the absence of an aril. In half of these species the funicle completely abscisses from the seed, although in *A. siculiformis* and *A. ulicifolia* a thread-like portion of the unelaborated funicle is retained, and in *A. craspedocarpa* and *A. bidwillii* small fragments of the unelaborated funicle sometimes remain attached to the seed. Both average seed mass and % lipid content depart significantly from those of arillate *Acacia* species. Seed mass (7.9-169.2 mg) averages almost three times that of arillate seeds,

108   *Dennis J. O'Dowd and A. Malcolm Gill*

but the average % lipid content of seeds is about 35% less than in arillate seeds (Table IV). Several factors may explain the larger seed size in non-arillate species of *Acacia*: (1) the difference in our sample may have a taxonomic basis in that the two species with the largest seeds (*A. bidwillii* and *A. farnesiana*) belong to a different section of the genus from the remaining species. In general, the non-arillate seeds of *Acacia* section *Acacia* are larger (Vassal, 1971; Janzen, 1977) and their inclusion may account in part for the larger average seed mass found in non-arillate *Acacia*; and (2) seed mass and size may be free to increase when no longer constrained by ant and bird dispersal agents. In abiotically-dispersed species, large seed size may increase the energy and nutrient reserves available to the embryo for establishment, potentially increasing the breadth of suitable safe sites available to the seed, but by foregoing animal dispersal and restricting seed number, decrease the probability of seeds reaching a suitable microsite. Furthermore, if dispersal agents restrict seed size, this may explain the greater % lipid content in arillate seeds than in non-arillate seeds. Since lipids are energy-rich relative to equivalent amounts of carbohydrate or protein, they are a more efficient storage compound if there are constraints placed on seed size.

Dispersal agents may also influence texture and thickness of the seed testa. Smooth seed coats characteristic of Australian *Acacia* may facilitate both internal passage through the guts of birds and external transport by ants through litter or on soil. The thick seed coats of most *Acacia* should augment the efficiency of dispersal by preventing many dispersal agents from becoming seed consumers (Janzen, 1983). If we take the intermediate values of the ranges in seed coat thickness presented by Cavanagh (1980) for 16 Australian *Acacia* spp. and divide those species into bird- and ant-dispersal syndromes as identified here, the overall average seed coat thicknesses are similar (ant syndrome = 192 μm $\pm$ 23 (SE), n = 13; bird syndrome = 200μm $\pm$ 28, n = 3). This suggests that there are no co-evolved traits unique to ants or birds involved in the protection of seeds during handling. Since natural selection favours parsimonious function, one diaspore trait is likely to function in several roles (e.g., protection from the physical environment and during handling by dispersal agents).

VI. CHARACTERISTICS OF SEED DISPERSAL AGENTS

While no quantitative data yet exist detailing the relative efficiency of various dispersal agents of any *Acacia* species, both birds and ants associated with *Acacia* diaspores range from seed-eaters to seed-dispersers. For most Australian ant species, whose trophic specializations are poorly understood, consumption of *Acacia* seeds following transport cannot be entirely ruled out. Most ants have a generalized diet and some, including species of *Pheidole, Melophorus* and *Rhytidoponera* can consume *Acacia* seeds (Table III). Because *Acacia* diaspores are characterized by thick, durable seed coats, (Tran and Cavanagh, 1984), most ant species probably act simply as dispersal agents for the majority of seeds transported. Some ants, such as small *Monomorium* and *Mayriella* species, act as 'parasites' on dispersal by consuming the aril piecemeal at the site where the seed is found; they only inadvertently move seeds and then only a few centimetres at most (Berg, 1975; Gill, 1985).

For birds, the definition of dispersal mutualists is probably clearer because information is often based on the recovery of intact seeds in faeces (Forde, 1986). A broad spectrum of bird species ingest *Acacia* diaspores, including strict seed consumers such as most parrots and many pigeons (Frith and Barker, 1974; Slater, 1974; Frith *et al.*, 1976), and a variety of 'soft-gutted' passerines (Table III). These passerines (primarily corvids, cracticids, melaphagids and zosteropids) either defecate seeds or regurgitate them intact (Gill, 1985; Forde, 1986; N. Reid, pers. comm. 1983). Like ant species that transport *Acacia* diaspores, these birds usually have broad diets, being omnivorous, or else having a fairly generalized diet that includes nectar, insects and fleshy fruits (Table III).

A general asymmetry occurs in the scope of diaspore 'types' collected by ants or birds. Ants collect and transport diaspores over almost the entire range of seed size and aril structure, but bird dispersal is limited to a subset of the total *Acacia* species considered in the PCA. In bird-syndrome *Acacia*, ants also transport diaspores (Table III; Fig. 2B): (1) some pods and diaspores fall to the ground or are dropped by birds and are available to ground-foraging ants (Gill, 1985; N. Reid, pers. comm. 1983) and (2) some arboreal ants detach and collect diaspores directly from the canopy (Davidson and Morton, 1984). Although the relative magnitude of dispersal by birds and ants in all species remain unquantified, diaspore retention and display patterns suggest that ants are less efficient in removing diaspores than are birds (Fig. 4). The presence of ants as transport agents in bird-syndrome *Acacia* may be explained in several ways: (1) they act as ancillary mutualists increasing the variety of microsites reached by diaspores (Davidson and Morton, 1984; Gill, 1985), particularly if bird numbers are variable in space and time so that a significant proportion of seeds may be transported by ants; or (2) conversely, since parasites are omnipresent in mutualistic systems, ants may be non-mutualists and reduce the benefits for plants and for birds. Bird-syndrome *Acacia* may not be able to provide access to their 'legitimate' dispersal agents and at the same time restrict access to all ants.

In ant-syndrome *Acacia*, the reverse pattern is not well-documented

(Fig. 2B). While parrots and pigeons have been reported to ingest seeds of ant-syndrome *Acacia*, e.g., *A. tricoptera* (Storr and Johnstone, 1979) and *A. terminalis* (D. O'Dowd, pers. obs.), given their gut and beak structures, it is likely that most seeds are crushed and killed. In only one instance have intact seeds of an ant-syndrome species (*A. victoriae*) been recovered from faeces of potential bird dispersal agents (Forde, 1986). Three factors may explain the asymmetry of diaspore generalism between birds and ants: (1) ant-syndrome *Acacia* do not present a potent visual signal to potential bird vectors; (2) diaspores are not presented at appropriate locations that overlap the foraging areas of the most likely bird dispersal agents (Fig. 4); and, (3) investment in reward (or the 'reward-to-bulk carried' ratio) is too small (Table IV, Fig. 3) or the handling time is too great for the diaspore to be included in the diets of large animals with high metabolic rates.

Little is known about biotic seed dispersal of non-arillate *Acacia* in Australia, although abiotic dispersal by water (chapter 2, III, C, *1*) or wind may be important. *Acacia ulicifolia* seeds can be collected by *Rhytidoponera metallica* (Drake, 1981) but this ant is a consumer of some seeds and often collects non-arillate seeds (C. Bonielle and D. O'Dowd, unpubl. data). Whether *Rhytidoponera* is attracted to the retained funicle or the seed itself is not known. Although Drake (1981) and Westoby *et al.* (1982) describe the seed of *A. ulicifolia* as arillate, its funicle lacks any feature consistent with aril structure (Berg, 1975). Both non-arillate *Acacia farnesiana* and *A. bidwillii* have tardily-dehiscent pods (Basáñez, 1977; L. Pedley, pers. comm. 1983). The pods of *A. farnesiana* are ingested by horses, goats and cattle both in America (Basáñez, 1977) and Australia (Pedley, 1978). Many seeds are voided intact.

Are Australian *Acacia* co-evolved with their dispersal agents? While these dispersal records are largely anecdotal (Table III), a lack of specificity of seed dispersal occurs within the proposed dispersal syndromes. Interactions between *Acacia* spp. and dispersal agents appear generalized: many species of ground-foraging ants transport *Acacia* seeds, and diaspores of a significant portion of *Acacia* spp. in Australia are ingested and voided intact by a variety of omnivorous birds (Table III). This generalism is consistent with the notion that diffuse co-evolution (Janzen, 1980) reflects adaptation to a variety of dispersal agents rather than to any particular species. Facultative mutualisms are likely to be readily established between *Acacia* spp. and dispersal agents, but like most mutualism involving seed dispersal (Wheelwright and Orians, 1982), direct co-evolution has not occurred.

VII. COMPARISON OF ACACIA WITH OTHER ANT- AND BIRD-DISPERSED SPECIES

The diaspores of most Australian *Acacia* (91% of species in our sample) have diverged to form two groups with signalling and reward structures necessary for inclusion in the menu of two separate types of dispersal agents, both with generalized diets. This conforms with two prevalent modes of dispersal in most Australian plant communities: (1) species with elaiosome-bearing seeds attractive to ants (Berg, 1975; Rice and Westoby, 1981) and (2) those with fleshy fruits attractive to birds (Clifford and Drake, 1981; Forde, 1986). For ant-dispersed *Acacia*, investment in dispersal per diaspore is similar to that of other elaiosome-bearing species, generally between 5–10% of total diaspore mass (Table VII). In bird-syndrome *Acacia*, the average investment is three times greater and approaches that of fleshy-fruited species in Australia (Table VII) and elsewhere (Snow, 1971; Stiles, 1980; Herrera, 1981a). Lipids, the primary class of storage compound found in the arils of most animal-dispersed *Acacia* seeds (Table IV), are usually elaborated both in elaiosomes of ant-dispersed species (Bresinsky, 1963) and the peri-carps or arils of a large number of temperate and tropical species dispersed by birds (Stiles, 1980; Herrera, 1981a).

Divergent patterns of display and presentation of diaspores in Australian *Acacia* also parallel the general characteristics of ant- and bird-dispersal modes. As in *A. myrtifolia* (Fig. 4), many ant-dispersed species have dia-spores whose initial dispersal is ballistic (Berg, 1975; Westoby and Rice, 1981) and all have diaspores that are presented on the soil or litter surface. Diaspore presentation and display patterns in bird-dispersed *Acacia* (Fig. 4) largely correspond to the in-canopy display patterns of diaspores in bird-dispersed plants, where, for example, many species also have bicoloured displays (Willson and Thompson, 1982).

VIII. ECOLOGICAL CONSEQUENCES OF SEED DISPERSAL

The basic structures of diaspores (funicle-seed) in Australian *Acacia* have provided scope for variation and selection of distinct character syndromes involving seed dispersal by agents from two disparate phyla — birds and ants. What are the selective bases for these syndromes? While no exper-imental studies demonstrate the benefits of seed dispersal for any *Acacia* species, at least two advantages have been suggested to accrue through local dispersal by birds and ants, consistent with several hypotheses put

Table VII. A comparison of average investment in dispersal (dry mass basis) for ant- and bird-syndrome *Acacia* with those of some other Australian species classified into dispersal categories. For *Acacia* species, dispersal syndromes are taken from Table III. Values for other species are data from K. French and D.J. O'Dowd.

Species	Family	Colour	Percent investment
Ant-dispersed Syndrome			
Acacia spp. (n = 12)	Mimosaceae	white/brown	7.6
			(1.9-11.5)
Bossiaea foliosa	Fabaceae	white	4.6
Daviesia mimosoides	Fabaceae	white	4.2
Dillwynia floribunda	Fabaceae	white	7.0
Hibbertia exutiaces	Dilleniaceae	white	11.9
Kennedia prostrata	Fabaceae	white	6.7
Pultenaea daphnoides	Fabaceae	white	10.7
Tetratheca pilosa	Tremandraceae	white	7.9
Bird-dispersed Syndrome			
Acacia spp. (n = 7)	Mimosaceae	yellow/orange/red	27.0
			(16.5-36.8)
Atriplex semibaccata	Chenopodiaceae	orange/red	54.2
Coprosma hirtella	Rubiaceae	orange/red	38.9
Elaeocarpus reticulatus	Elaeocarpaceae	blue	33.3
Enchylaena tomentosa	Chenopodiaceae	orange/red	43.1
Exocarpos cupressiformis	Santalaceae	red	59.9
Hymenanthera dentata	Violaceae	blue	55.1
Polyscias sambucifolia	Araliaceae	grey/blue	61.6

forward for the selective basis of seed dispersal in general (Howe and Smallwood, 1982). First, seeds may escape detrimental effects associated with the parent plant. Gill (1985) suggests that transport of *A. cyclops* diaspores away from parent canopies to open areas by birds (primarily grey currawongs, *Strepera versicolor*) decreases the probability of seed consumption by *Adrisa* sp., a cydnidid bug associated with litter below the canopies of *A. cyclops*. Second, both direct and indirect evidence shows that birds and ants can direct diaspores to specific microhabitats. Foraging ants relocate seeds of many *Acacia* spp. to ant nests (Table III), where they may be interred or discarded (Shea *et al.*, 1979; Majer, 1982). Glyphis *et al* (1981) and Davidson and Morton (1984) provide evidence that seedlings and saplings of three species of bird-syndrome *Acacia (A. cyclops, A. ligulata* and *A. tetragonophylla*) are non-randomly associated with the canopies of other tree or shrub species, suggesting that some bird species, by their perching and defecation behaviours, direct seeds to below-canopy micro-habitats. The consequences of directed dispersal in these *Acacia* have not been addressed, but in many cases the micro-environments to which the seeds are transported differ markedly from source areas (Beattie and Culver, 1981; Davidson and Morton, 1984). For example, increased levels of nutrients below tree or shrub canopies and at ant nests may increase the probability of seedling establishment at those sites. Alternatively, transport of seeds to ant nests may increase seed survivorship in fire-prone regions of Australia (Shea *et al.*, 1979; Majer, 1982). Although birds and ants may direct diaspores to specific microhabitats, these agents differ in many ways which may have major consequences for the seed dispersal shadows that they generate. These dispersal shadows may have a major impact on the subsequent structure and dynamics of plant populations, gene flow, and reproduction.

IX. EVOLUTIONARY DERIVATION OF DISPERSAL SYNDROMES

From the diversity of dispersal features in Australian *Acacia* it may be possible to construct a scenario of the evolution of *Acacia* dispersal systems. In the absence of any dispersal correlates with aril characteristics in *Acacia*, Corner (1949) viewed the aril structures we correlate with ant dispersal (Fig. 1) as non-functional vestiges from 'primitive' arillate seeds dispersed by birds. Van der Pijl (1972) also considered the inconspicuous arils found in most Australia *Acacia* as vestigial, but argued in general that the dispersal linkage of plants with ants was derived from bird-dispersed ancestors, primarily on the basis of the pre-eminence of birds and the paucity of ants in the fossil record. In his view and that of others (Berg, 1958; Bresinsky, 1963), the structure and phenology of the diaspore associated with ants

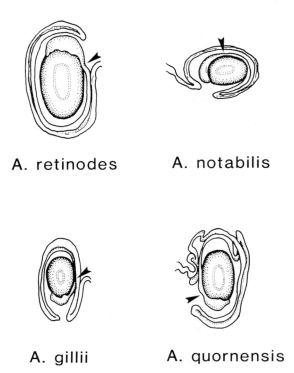

A. retinodes   A. notabilis

A. gillii   A. quornensis

Fig. 5. Funicle and aril structure of four *Acacia* species (redrawn from Whibley, 1980) indicating that the funicle encircles the seed longitudinally prior to terminating in a small, white-brown aril, typical of ant syndrome *Acacia*. The arrows indicate the point of funicle abscission from the diaspore.

in closely-related plant groups involved a transference of function from bird dispersal and reduction of the aril.

Our examination of funicle structure of some arillate *Acacia* supports this conclusion but for different reasons. First, the small arils of many Australian *Acacia* are not non-functional vestiges, but are responsible for the dispersal of seeds by ants. Second, the ultimate antiquity of either class of potential dispersal agents is a moot point since the radiation of *Acacia* in Australia has almost certainly occurred in the last ten million years (Barlow, 1981; Raven and Polhill, 1981; Lange, 1982), when suitable ant- and bird-dispersal agents were both abundant and diverse (R. W. Taylor, pers. comm. 1984; Sibley and Ahlquist, in press).

While the 'law of parsimony' would seem to suggest that a more elaborate and complex structure (bird aril) should be derived from a simpler one (the ant aril), selection should actually act along the lines of least resistance.

In eleven ant-syndrome *Acacia* (section *Phyllodineae*) from South Australia (Whibley, 1980), the unexpanded funicle encircles the seed in a double fold before ultimately enlarging into a small aril (Fig. 5). This parallels the encircling of the seed by the arillate funicle of some bird-syndrome Phyllodineae, including *A. cyclops, A. melanoxylon* and *A. tetragonophylla* (Fig. 1). The diaspores in these ant-syndrome species, however, absciss from the encircling funicle (Fig. 5) and retain a short, white-brown aril. In these species, at least, it appears that seed dispersal attributes associated with ants involve reduction of the aril from types correlated with bird dispersal. Ant- and bird-dispersal syndromes in *Acacia* involve the integration of a number of individual traits. Any complex of characters that has become developmentally canalized is likely to be retained in large part even after the conditions that originally exerted the selective pressure have disappeared. Alteration of integrated patterns, e.g., transference of dispersal function (Stebbins, 1970), is more likely to involve minimum modification from the existing structure (Fig. 5) rather than complete reorganization of the funicle.

If, as we suggest, bird dispersal preceded ant dispersal in Australian *Acacia*, what might have induced a shift to the present predominance of ant-syndrome *Acacia*? As the Australian plate drifted north into the sub-tropical arid belt during the Miocene, the coefficient of variation in rainfall increased dramatically over large parts of the continent (Barlow, 1981). Climatic uncertainty (e.g., patchiness and variability of rainfall) has been linked to the productivity and phenology of some *Acacia* species (Preece 1971, Wilcox 1974) and seed production in many *Acacia* spp. may have become increasingly variable in time and space during this period.

Birds, being large and endothermic, have higher energy requirements than ants. Schodde (1982) has suggested that with increasing uncertainty of food resources in time and space, a greater degree of nomadism developed in many bird species. For many *Acacia* species the consequence may have been that birds located and dispersed seeds less reliably. With increasing climatic uncertainty, ants may have been able to persist locally as dispersal agents for three reasons (Risch and Carroll, 1982): (1) they have relatively low energy requirements; (2) most are able to store food in their colonies; and, (3) they can cannibalize brood when other food is not available. Ants, continually sampling arillate *Acacia* diaspores (Fig. 2), may have become relatively more predictable and reliable dispersal agents of many *Acacia* species. If the ecological consequences of seed dispersal by ants and birds are similar (Davidson and Morton, 1984) then ants could have become the more consistent and important selective agent on diaspore attributes of many *Acacia* species, and this could account for the patterns observed in Figure 5.

The increasing importance of fire (chapter 6) during this epoch could further complement a transition to seed dispersal by ants. If ants, by directing seeds to their nests where they might survive high-intensity fires (Majer, 1982), increase the fitness of some *Acacia* relative to dispersal by birds, then there could have been a concomitant shift in diaspore attributes.

X. COMPARISON BETWEEN AMERICAN AND AFRICAN ACACIA

The dispersal assemblage of *Acacia* is likely to be among the most diverse and dichotomous known, involving active dispersal by ants, birds, and mammals. Seed dispersal characteristics of Australian *Acacia* are distinctive and little overlap occurs with those in African and American *Acacia*. Outside Australia, seed dispersal of *Acacia* by ants is unknown. Bird dispersal, however, has evolved independently in the polyphyletic swollen-thorn *Acacia* (12 species) of the American tropics (Janzen, 1969, 1974). Analogous rather than homologous diaspore features yield a display similar to that in bird-dispersed Australian *Acacia*. The seeds are embedded in a sweet white to yellow pulp, and in all species (except closely-related *A. cornigera, A. sphaerocephala* and *A. mayana*, which have indehiscent pods), the dark brown to black pods open to present a bicoloured display. In *A. collinsii*, for example, bright yellow seed clusters dangle from the dehisced pods. Orioles (*Icterus* sp.) ingest these seed clusters and pass the seeds intact. In species with indehiscent pods, the legume itself is attractively coloured either red, maroon, or yellow. Woodpeckers, orioles, saltators, and kiskadees split the pods and ingest the diaspores (Janzen, 1969).

The 'megafaunal' dispersal syndrome (Janzen and Martin, 1982), involving seed dispersal by large vertebrates, is well-developed in some of the remaining American *Acacia* and in many African *Acacia* (Lamprey, 1967; Gwynne, 1969; Janzen and Martin, 1982). Pods are the dispersal units in these species: the seeds are non-arillate, and the pods are indehiscent (or tardily dehiscent) and often leathery (Robbertse, 1975). They are ingested by large mammals, and a proportion of the seeds are passed intact. Dispersal agents include elephants, giraffe, rhinoceros, gazelle, cattle, and goats in Africa (Burtt, 1929; Tran and Cavanagh, 1984) and reintroduced horses and cattle in America (Basáñez, 1977). In Australia, only two species of *Acacia* are known to be routinely dispersed by large mammals, and these 'exceptions' may prove the rule. Pods of *A. farnesiana* (Fig. 2B, putatively of Central American origin) and *A. nilotica* (African origin) are ingested by large vertebrates (introduced cows and goats) in Australia (Pedley, 1978), parallel to the megafaunal dispersal described in their natural distributions (Basáñez, 1977; Ross, 1979).

Why is the megafaunal dispersal syndrome, which is well developed in African and American *Acacia*, so poorly represented among Australian *Acacia*? Three hypotheses may explain its absence: (1) a megafaunal dispersal component was never sufficiently developed to act as a strong selective agent on dispersal characteristics (Janzen and Martin, 1982); (2) the mega-faunal assemblage was declining during the relatively recent radiation of *Acacia* in Australia (Murray, 1984); or, (3) the importance of ants and birds as potential dispersal agents pre-empted major interactions between seeds and large vertebrates (c.f. Morton, 1985). Whatever the proximate causes of selection of dispersal attributes, the distinctive histories and faunal composition of the southern hemisphere continents were important in mould-ing the divergent diaspore characteristics seen in *Acacia*.

XI. CONCLUSIONS

The Principal Components Analysis presented here, coupled to indirect (diaspore colour) and direct (observation of dispersal agents) evidence of seed dispersal, segregates Australian *Acacia* into three dispersal groups: (1) arillate *Acacia* that are bird-dispersed (17% of species sampled) involving dispersal by a wide array of birds with generalized diets, including mela-phagids, cracticids, and corvids; (2) arillate species (74% of species sampled) that are dispersed by a variety of ant species, also with broad diets; and, (3) the remaining non-arillate *Acacia* species (9% of species sampled) which do not correlate well with active dispersal by ants or birds.

Three basic characteristics summarize ant- and bird-dispersal syndromes in Australian *Acacia*: display, attraction, and reward. In bird-syndrome species, bicoloured diaspores predominate and are displayed in dehisced legumes retained in the canopy. Investment in dispersal is high (aril mass, lipid content, length and contortion are significantly greater than in ant-dispersed *Acacia*), which may reflect the metabolic requirements of large, endothermic birds, and is similar to the investment shown by many Aus-tralian species with fleshy fruits. Diaspores of ant-dispersed species are presented inconspicuously on the ground and investment is low, similar to that of other species known to be dispersed by ants in Australia. The morphological and behavioural characteristics of many ants and birds sug-gest that they can generate 'directed' seed shadows which may have major consequences for seedling establishment and plant population structure.

No single selective pressure (e.g., nutrient-deficient soils, fire or distance-dependent seed consumption) is likely to have moulded the dispersal attrib-utes of Australian *Acacia*. Funicle structure, at least in some species, indicates that seed dispersal by ants has been evolutionarily derived from bird-dispersed progenitors. This within-continent analysis, as well as a comparison of dispersal attributes of *Acacia* in other continents in the southern hemisphere, implies that diaspore characteristics reflect the nature of the most effective group of dispersal agents.

ACKNOWLEDGEMENTS

Both CSIRO Divisions of Forest Research and Plant Industry, as well as the National Botanic Garden, Canberra, allowed us access to their seed collections of *Acacia*. We thank J. Green for help in measuring characteristics of *Acacia* diaspores and L. Tonnet for conducting the GC analyses for fatty acids and instructing us in the use of the quantity analyser. S. Shea and G. Folley collected seeds in the field for us. N. Forde, J. Majer, S. Morton and N. Reid provided access to their unpublished work and graciously allowed us to cite their field observations of seed dispersal in this study. A. O'Dowd drew Figure 5. We are particularly grateful to J. Wood for conducting the Principal Components Analysis. Portions of this study were conducted while DJO was a postdoctoral fellow at the CSIRO Division of Plant Industry.

REFERENCES

Anderson, A. (1982). *In* 'Ant-Plant Interactions in Australia' (R. C. Buckley, ed.), pp. 31–43. Junk, The Hague.
Ashton, D. H. (1979). *Aust. J. Ecol.* **4**, 265–277.
Baker, H. G. (1972). *Ecology* **53**, 947–1010.
Barlow, B. A. (1981). *In* 'Flora of Australia. Volume 1', pp. 25–76, Bureau of Flora and Fauna, Canberra. Griffin Press, Netley, South Australia.
Basáñez, M. (1977). *Biotica* **2**, 1–18.
Beattie, A. J., and Culver, D. C. (1981). *Oecologia* **56**, 99–103.
Berg, R. Y. (1958). *Norske Videnskaps-Academi Skirfter Matematisk-Naturvidenskapeligeklasse* **1**, 1–36.
Berg, R. Y. (1975). *Aust. J. Bot.* **23**, 475–508.
Berg, R. Y. (1981). *In* 'Ecosystems of the World, Volume 9B. Heathlands and Related Shrublands' (R. L. Specht, ed.), pp. 51–58. Elsevier Scientific Publishing Company, Amsterdam.
Bresinsky, S. (1963). *Bibliotheca Botanica* **126**, 1–54.
Burtt, B. D. (1929). *J. Ecol.* **17**, 351–355.
Cavanagh, A. K. (1980). *Proc. Roy. Soc. Vic.* **91**, 167–180.
Clifford, H. T., and Drake, W. E. (1981). *In* 'Ecosystems of the World, Volume 9B. Heathlands and Related Shrublands' (R. L. Specht, ed.), pp. 39–49. Elsevier Scientific Publishing Company, Amsterdam.
Corner, E. J. H. (1949). *Ann. Bot.* **13**, 367–414.
Davidson, D. W., and Morton, S. R. (1984). *Ecology* **65**, 1038–1051.

Drake, W. E. (1981). *Aust. J. Bot.* **29**, 293–309.

Earle, F. R., and Jones, Q. (1962). *Econ. Bot.* **16**, 221–250.

Faegri, K., and van der Pijl, L. (1971). 'Principles of Pollination Ecology'. Pergamon Press, Oxford.

Forde, N. (1986). *In* 'The Dynamic Partnership: Birds and Plants in Southern Australia.' (H. A. Ford and D. C. Paton, eds.), pp. 42–58. Government Printer, South Australia.

Foster, M. S., and McDiarmid, R. W. (1983). *Biotropica* **15**, 26–31.

Frith, H. J., and Barker, R. D. (1974). *Aust. Wildlife Res.* **1**, 129–144.

Frith, H. J., Wolfe, T. O., and Barker, R. D. (1976). *Aust. Wildlife Res.* **3**, 159–171.

Gill, A. M. (1985). *Proc. Roy. Soc. Western Australia.* **67**, 59–65.

Gilmour, D. (1961). 'The Biochemistry of Insects'. Academic Press, London.

Glyphis, J. P., Milton, S. J., and Siegfried, W. R. (1981). *Oecologia* **48**, 138–141.

Grant, V., and Grant, K. A. (1965). 'Flower Pollination in the Phlox Family'. Columbia University Press, New York.

Gwynne, M. D. (1969). *East Afr. Wildlife J.* **7**, 176–178.

Harper, J. L., Lovell, P. H., and Moore, K. G. (1970). *Annu. Rev. Ecol. Syst.* **1**, 327–356.

Herrera, C. M. (1981a). *Am. Nat.* **118**, 896–907.

Herrera, C. M. (1981b). *Oikos* **36**, 51–58.

Hitchcock, C., and Nichols, B. W. (1971). 'Plant Lipid Biochemistry'. Academic Press, London.

Howe, H. F. (1977). *Ecology* **58**, 539–550.

Howe, H. F. (1980). *Ecology* **61**, 944–959.

Howe, H. F., and Smallwood, J. (1982). *Annu. Rev. Ecol. Syst.* **13**, 201–228.

Howe, H. F., and Vande Kerckhove, G. A. (1980). *Science* **210**, 925–927.

Janzen, D. H. (1969). *Condor* **71**, 240–256.

Janzen, D. H. (1974). 'Swollen-thorn Acacias of Central America'. Smithsonian Contributions to Botany no. 13. Smithsonian Institution Press, Washington, D. C.

Janzen, D. H. (1977). *Ecology* **58**, 921–927.

Janzen, D. H. (1980). *Evolution* **34**, 611–612.

Janzen, D. H. (1983). *In* 'Physiological Plant Ecology III. Responses to the Chemical and Biological Environment'. Encyclopedia of Plant Physiology. New Series, Volume 12C (O. L. Lange, P. S. Nobel, C. B. Osmond and H. Ziegler, eds.), pp. 625–657. Springer-Verlag, Berlin.

Janzen, D. H., and Martin, P. S. (1982). *Science* **215**, 19–27.

Klieber, M. (1961). 'The Fire of Life'. Wiley and Sons, New York.

Lamprey, H. F. (1967). *East Afr. Wildlife J.* **5**, 179–180.

Lange, R. T. (1982). *In* 'A History of Australian Vegetation' (J. M. B. Smith, ed.), pp. 44–89. Griffin Press, Sydney.

Majer, J. D. (1980). *Reclamation Review* **3**, 3–9.

Majer, J. D. (1982). *In* 'Ant-Plant Interactions in Australia' (R. C. Buckley, ed.), pp. 44–60. Junk, The Hague.

Marshall, D. L., Beattie, A. J., and Bollenbacher, W. E. (1979). *J. Chem. Ecol.* **5**, 335–344.

Mazliak, P. (1970). *In* 'The Biochemistry of Fruits and Their Products, Volume 1' (A. C. Hulme, ed.), pp. 209–237. Academic Press, London.

Morton, S. R. (1985). *Ecology* **66**, 1859–1866.

Murray, P. (1984). *In* 'Quaternary Extinctions' (P. S. Martin and R. G. Klein, eds.), pp. 600–628. University of Arizona Press, Tucson.

Pedley, L. (1978). *Austrobaileya* **1**, 1–337.

Polhill, R. M., and Raven, P. H. (1981). 'Advances in Legume Systematics'. Parts I and II. Royal Botanic Gardens, Kew.

Polyak, S. (1957). 'The Vertebrate Visual System'. University of Chicago Press, Chicago.

Preece, P. B. (1971). *Aust. J. Bot.* **19**, 21–38.

Raven, P. H. and Polhill, R. M. (1981). *In* 'Advances in Legume Systematics, Part I' (R. M. Polhill and P. H. Raven, eds.), pp. 27–34. Royal Botanic Gardens, Kew.

Rice, B. L., and Westoby, M. (1981). *Aust. J. Ecol.* **6**, 291–298.

Risch, S., and Carroll, R. (1982). *Oecologia* **55**, 114–119.

Robbertse, P. J. (1975). *Bothalia* **11**, 481–489.

Ross, J. H. (1979). 'A Conspectus of the African *Acacia* Species'. Memoirs of the Botanical Survey of South Africa No. 44. Botanical Research Institute, Pretoria, South Africa.

Schodde, R. (1982). *In* 'Evolution of the Flora and Fauna of Arid Australia' (W. R. Barker and P. J. M. Greenslade, eds.), pp. 191–224. Peacock Publications, Adelaide.

Shea, S. R., McCormick, J., and Portlock, C. C. (1979). *Aust. J. Ecol.* **4**, 195–205.

Sibley, C. G., and Ahlquist, J. E. (*in press*). The phylogeny and classification of passerine birds, based on comparisons of the genetic material, DNA. Proc. XVIII Internat. Ornithol. Congr.

Sillman, A. J. (1973). *In* 'Avian Biology. Volume 3' (O. S. Farner and J. R. King, eds.), pp. 349–387. Academic Press, New York.

Slack, C. R., and Browse, J. A. (1984). *In* 'Seed Physiology. Volume 1. Development' (D. R. Murray, ed.), pp. 209–244. Academic Press, Sydney.

Slater, P. (1974). 'A Field Guide to Australian Birds. Volume II. Passerines'. Rigby Press, Adelaide, Australia.

Sneath, P. H. A., and Sokal, R. R. (1973). 'Numerical Taxonomy'. W. H. Freeman and Company, San Francisco.

Snow, D. W. (1971). *Ibis* **113**, 194–202.

Stebbins, G. L. (1970). *Israel J. Bot.* **19**, 59–70.

Stebbins, G. L. (1974). 'Flowering Plants. Evolution Above the Species Level'. Belknap Press, Cambridge, Massachusetts.

Stiles, E. W. (1980). *Am. Nat.* **116**, 670–688.

Storr, G. M., and Johnstone, R. E. (1979). *Western Australian Naturalist* **14**, 156–157.

Tran, V. N., and Cavanagh, A. K. (1984). *In* 'Seed Physiology. Volume 2. Germination and Reserve Mobilization' (D. R. Murray, ed.), pp. 1–44. Academic Press, Sydney.

Trelease, R. N., and Doman, D. C. (1984). *In* 'Seed Physiology. Volume 2. Germination and Reserve Mobilization' (D. R. Murray, ed.), pp. 201–245. Academic Press, Sydney.

Turcek, F. J. (1963). Proc. XIIIth Internat. Ornithol. Congr., pp. 285–292. The American Ornithologists' Union, Baton Rouge, Louisiana, U. S. A.

van der Pijl, L. (1955). *K. Nederlanse Akademie van Wetenschappen. Proceedings Series* C58, 154–161, 307–312.

van der Pijl, L. (1972). 'Principles of Dispersal in Higher Plants'. 2nd edition. Springer-Verlag. Berlin.

Vassal, D. J. (1971). *Bulletin de la Société d'Histoire Naturelle de Toulouse* **107**, 191–246.

Welch, R. W. (1977). *J. Sci. Food and Agric.* **28**, 635–638.

Westoby, M., and Rice, B. (1981). *Aust. J. Ecol.* **6**, 189–192.

Westoby, M., Rice, B., Shelley, J. M., Haig, D., and Kohen, J. L. (1982). *In* 'Ant-Plant Interactions in Australia' (R. C. Buckley, ed.), pp. 75–88. Junk, The Hague.

Wheelwright, N. T., and Orians, G. H. (1982). *Am. Nat.* **119**, 402–413.

Whibley, D. J. E. (1980). 'Acacias of South Australia'. D. J. Woolman, Government Printer, South Australia.

Wigglesworth, V. B. S. (1972). 'The Principles of Insect Physiology'. 7th edition. Chapman and Hall, London.

Wilcox, D. G. (1974). *In* 'Plant Morphogenesis as the Basis for Scientific Management of Range Resources'. Proceedings of the 1st Workshop of the U. S./Australia Rangelands Panel, Berkeley, California, 1971, pp. 60–71. USDA Misc. Publ. 1271.

Willson, M. F., and Melampy, M. N. (1983). *Oikos* **41**, 27–31.

Willson, M. F., and Thompson, J. N. (1982). *Can. J. Bot.* **60**, 701–713.

Wilson, E. O. (1971). 'The Insect Societies'. Belknap Press, Cambridge, Massachusetts.

CHAPTER 4

Seed Dispersal by Fruit-Eating Birds and Mammals

HENRY F. HOWE

I.	Introduction	123
II.	Definitions	126
III.	Foraging for Fruits	128
	A. Conceptual Framework	136
	B. Animal Correlates of Seed Dispersal	139
	C. Theory and Practice	146
IV.	Attracting Dispersal Agents	150
	A. Morphological Syndromes	150
	B. Ecological Adaptations	155
V.	Population Ecology of Seed Dispersal	162
	A. Consequences of Seed Dispersal	163
VI.	Community Effects	172
	A. Community Composition	172
	B. Co-evolution or Co-occurrence?	179
VII.	Summary and Conclusions	181
	Acknowledgements	182
	References	183

I. INTRODUCTION

Plants have an ancient and uneasy relationship with vertebrate animals which eat their fruits, and either digest or disperse their seeds. As early as 300 million years ago, Carboniferous progenitors of modern cycads bore fleshy fruits, which were apparently adapted for consumption by primitive reptiles (Sporne, 1965). By the Cretaceous, 65 to 125 million years ago, the lowland gymnosperm forests were rapidly infiltrated and ultimately replaced by flowering plants with fleshy fruits adapted for bird and mammal

SEED DISPERSAL
ISBN   0 12 511900 3

consumption (Stebbins, 1974; Regal, 1977). Relicts of the ancient fruit flora still exist. Possums now scatter the fruits of the Australian cycad *Macrozamia*, whose progenitors might once have been dinosaur food (Burbidge and Whelan, 1982). But most fascinating is the enormous diversity and abundance of modern fruits which are more or less specialized for consumption by bats, large and small birds, arboreal carnivores, monkeys, and ungulates (van der Pijl, 1972; Janson, 1983). Some examples are shown in Figure 1. In contemporary forests as many as 45 to 90% of the tree species bear fleshy fruits apparently evolved for consumption by birds and mammals (see Table 1 of Howe and Smallwood, 1982).

Given the prevalence of seed dispersal by animals, it is puzzling that neither the advantages of seed dissemination to the plant nor the means of bringing it about are well understood. One reason is that relevant questions have been approached from at least four independent traditions. First, an early interest in seed dispersal over long distances, often hundreds or thousands of kilometers (Ridley, 1930), eclipsed the appreciation of local consequences of seed dispersal which might favour some parent plants over others (Janzen, 1970, 1971). Second, autecologies of frugivores document diets, but not the effects of food choice on plant reproduction (Snow, 1962a; Milton, 1980). Third, more recent emphasis on 'dispersal systems' of particular plant species and their visitors document fruit removal without considering visitor diets (Howe, 1977, 1980). Fourth, a community approach paints a general picture of potential interactions between plants and animals without considering either plant or animal ecology in depth (Fleming and Heithaus, 1981; Herrera, 1984a). This chapter discusses the causes and consequences of frugivory and local seed dispersal in light of the last three traditions. Chance long-distance seed dispersal is critical for interpreting colonization of remote islands or continents (MacArthur and Wilson, 1967; Carlquist, 1981; Chapter 2), but local reproductive competition among fruiting plants may suffice to explain why berries evolved.

Once their importance is recognized, the descriptive natural history of frugivory and its consquences for local seed dispersal are surprisingly inadequate for answering basic questions (Howe and Smallwood, 1982). Why do some birds kill seeds with bills or gizzards, while others do not? Can a fruiting tree repel inefficient dispersal agents and encourage those that effectively disperse seeds? How beneficial to a tree is an elephant that eats thousands or even millions of seeds at one time, but leaves them in huge fecal masses in which all, or all but one, succumb to sibling competition, or to pathogen, rodent, or insect attack? Do dispersal agents ensure the escape of some plant offspring from insect or rodent attack under the crown, or do birds and mammals simply scatter seeds in especially favourable sites for seedling establishment? How can selection balance whatever advantage might exist in seed dissemination against the obvious risk that animal visitors will drop seeds under the parental crown, or kill them outright?

Fig. 1. Neotropical fruits dispersed by birds or mammals. (A) *Virola surinamensis* fruits dehisce to expose a seed (2 cm long) covered by a brilliant red aril rich in lipids. Primary dispersal agents are toucans, although monkeys also eat the fruits (from Howe and Vande Kerckhove, 1981). (B) *Tetragastris panamensis* fruits dehisce to expose one to six white, arillate seeds, dangling from a purple core. Monkeys, other arboreal mammals, and some birds eat the sugary fruits and disperse the seeds (from Howe, 1980). (C) *Stemmadenia donnell-smithii* capsules dehisce to expose an orange pulp in which seeds are embedded. Many birds quickly eat the oily pulp and disperse viable seeds (see McDiarmid *et al.*, 1977; photo by H. F. Howe).

Even such elementary questions, raised by casual natural history, indicate the complexity of issues that must be teased apart to understand contemporary forests in which most trees and shrubs bear fleshy fruits adapted to be animal food. At the very least, a perception of either an ancient or modern forest must be a warped shadow of reality unless the means and consequences of seed dispersal by fruit-eating birds and mammals are understood.

This chapter explores the ecology of seed dispersal by birds and mammals that consume fruits and defecate or regurgitate seeds in viable condition. The aim is to explore the uneasy mutualism between plants whose fitness is enhanced by seed dissemination and animals whose interests are furthered by digesting fruit pulp — and either digesting or discarding seeds. This exploration is prefaced by a brief discussion of terminology (Section II), followed by a consideration of the specific topics: (1) how animals choose fruits and what the potential consequences of that choice are (Section III); (2) how plants attract fruit-eating animals (Section IV); (3) consequences of seed dispersal for plant fitness (Section V); and, (4) consequences of seed dissemination for community composition (Section VI).

II. DEFINITIONS

Several terms are defined in Table I, either because they suggest multiple interpretations, or because they have become so confused in the literature that their use must be clarified.

Some terms, such as 'adaptation,' 'aril,' and 'fruit' have technical meanings which must be clarified in an ecological context. For instance, an adaptation is a behavior, physiological property, or structure that helps an organism survive and ultimately reproduce. This assumes that the adaptation has been shaped by natural selection; the term cannot be applied to a property with an accidental advantage, no matter how great that advantage might be (see Williams, 1966). Accidental dispersal of seeds consumed by finches, which are themselves eaten by owls, is not subject to selection (Grant *et al.*, 1975). An aril is an edible outgrowth of the seed coat modified for the attraction of dispersal agents. I use the term loosely, without distinguishing between 'true' and 'false' arils derived from different parts of the seed coat (see Corner, 1949; van der Pijl, 1972). Similarly, the term 'fruit' loosely refers to any fleshy structure of parental origin in which the seed is embedded. It is usually maternal tissues which give the fruit its taste, colour, nutritive reward, and hence opportunity for dispersal. Detailed homologies of fruit parts are of little interest to a monkey or bird, and are best left to a longer discussion (van der Pijl, 1972).

Table I. Definitions of terms[a].

---

adaptation: functional property of an organism evolved by natural selection
aril: an edible outgrowth of the seed coat
dispersal: departure of a seed from the parent plant; *secondary dispersal* is further movement
       of seeds following their initial departure
dispersal agent: an animal that moves viable seeds from one location to another
dispersal system: a plant species and the animals that eat its fruits
establishment: the process by which a seed germinates and assumes independent growth as
       a seedling
frugivore: an animal that eats fruit
fruit: the mature gynoecium; the ripened ovary including the seeds
sapling: a young tree at least 1 cm in diameter at chest height
seed: a *gymnosperm seed* contains the embryo embedded in the female gametophyte; an
       angiosperm seed is the embryo with or without endosperm or other storage tissue
       surrounded by one or more integuments
seedling: a young plant smaller than a sapling; an *early seedling* is a young plant still dependent
       on seed reserves
seed predator: an animal that eats a seed or part(s) of a seed, usually killing the embryo

---

[a] Modified from Howe and Smallwood (1982). For further explanation, see text.

Other important terms have been badly confused in the literature. 'Dispersal' of seeds is one. Plant geographers emphasize the movement of seeds over very long distances, as would occur in colonizing distant islands or continents (e.g. Ridley, 1930; Carlquist, 1974). Here the use is ecological, referring to movement of seeds in the immediate vicinity of parent plants, ranging in distance from centimeters to kilometers (c.f. Janzen, 1970, 1971). I avoid the connotation that dispersal is to a 'safe site' for germination and growth (Harper, 1977), because the notion of a 'safe site' is categorical, while the chance that a seed survives is predicted by a continuous probability distribution. Of course, some sites are 'safer' than others. The probabilistic nature of seed and seedling survival justifies a simple definition of seed dispersal coincident with abscission. This is preferable to the intuitive definition as 'movement away from the maternal parent plant' (c.f. Janzen, 1983a), because 'away' has different meanings for different plants. For *Virola surinamensis* (Myristicaceae; Fig. 1A) in the Panamanian rain forest, 40 m 'away' confers a dramatic advantage to a seed (Howe *et al.*, 1985); for the *Mirabilis hirsuta* (Nyctaginaceae) on a badger mound, 40 cm produces the same result (Platt, 1976). Of interest here are the adaptive means of ensuring that seeds are taken by animals, coupled with the demographic processes which make those adaptations important in plant reproduction.

    Among the most confused terms in dispersal ecology is 'frugivore,' with or without modifying adjectives. This chapter is concerned with animals that both eat fruits and consistently disperse seeds rather than destroy them.

I give only passing mention to frugivores such as peccaries (*Tayassu* spp., see Kiltie 1981, 1982) or parrots (*Brotogeris*, see Janzen, 1981a; *Amazona autumnalis*, see Howe, 1980), which usually kill the seeds that they consume. Minute seeds sometimes escape direct mandibulation by such 'seed predators', but few reports suggest that this is ecologically important (e.g. (*Brotogeris* and *Muntingia calabura*, Fleming *et al.*, 1985).

Of living fruit-eating vertebrates, few are 'total frugivores' in the sense that they restrict their diets to fruit (e.g. *Steatornis caripensis*, the oilbird; see Snow, 1962a). Most birds eat animal matter as nestlings, and most birds and mammals eat some animal matter, nectar, or foliage as adults. But some 'obligate frugivores' eat little but fruit. Among these some (e.g. large *Ramphastos* toucans [Ramphastidae]) seem to restrict their diet to a few favored fruits at any one time, whereas others (e.g. smaller *Pteroglossus* aracaris [same family]) have a more catholic diet (Howe, 1977; Wheelwright, 1985a). Representative families and genera illustrate these 'obligate' frugivores (Table II). Possibly different, from the perspective of a fruit-bearing plant, are 'opportunistic' frugivores which alternate fruits with insects, vertebrates, foliage, nectar or even digestible seeds, often choosing fruits when they are superabundant (McKey, 1975; Howe and Estabrook, 1977). Among the so-called opportunists are many migratory birds, non-breeding insectivorous birds, folivorous monkeys, or generally herbivorous or omnivorous terrestrial or climbing mammals (Tables II and III). In no sense is 'obligate' synonymous with highly effective seed dispersal, nor does 'opportunism' necessarily carry the connotation of poor seed dispersal. It remains to be seen whether frugivores accustomed to fruit are better dispersal agents, from the plant perspective, than animals that are not behaviorally or physiologically accustomed to a given fruit as food.

III. FORAGING FOR FRUITS

Ecological theory is largely concerned with competition, predation, or succession. Formal theory has existed for each of these topics for several decades with result that data collection in each area has both an end in view and a ready-made context once collected. The phenomena of interest here — foraging by frugivores, plant tactics for attracting animals, and the demographic consequences of seed dispersal — are almost untouched by the creators and testers of theory. Foraging tactics by frugivores are a convenient starting point because the conceptual framework is more developed than that of either fruit adaptation for dispersal or seed and seedling demography. The theory provides an explicit set of incentives for probing constraints of energy, time, and risk which influence frugivore choice and use of fruits.

Table II. Size and taxonomic diversity of selected bird genera known to eat fruits and disperse seeds. This listing does not include birds that nearly always digest seeds (e.g. parrots [Psittacidae] and finches [Fringellidae]).

Order Family (Common Name[s]; Family Distribution) Selected Genera	Weight (g)	Habit	Food
Primarily Frugivorous Birds			
Casuariiformes			
Casuariidae (Cassowaries; Asia, Australia)			
Casuarius	29 000-58 000[a]	Terrestrial	Fruits; digests some seeds
Cracidae (Currosows and Guans; Neotropical)			
Crax, Penelope	±1000-2000[b]	Aerial and Terrestrial	Fruits, insects, seeds
Caprimulgiformes			
Steatornithidae (Oilbird; Neotropics)			
Steatornis	415[c]	Aerial; Caves	Fruits
Columbiformes			
Columbidae (Doves, Pigeons; Cosmopolitan)			
Ducula, Ptilionopus	112-518[d]	Aerial	Fruits; many genera granivorous
Trogoniformes			
Trogonidae (Trogons; Pantropical)			
Pharomachrus, Trogon	50-206[c,e]	Aerial	Fruits and insects
Coraciiformes			
Bucerotidae (Hornbills; Africa, Asia)			
Aceros, Bycanistes, Ceratogymna	206-738[f]	Aerial	Fruits and some insects
Piciformes			
Ramphastidae (Toucans; Neotropical)			
Aulacorhynchus, Pteroglossus, Ramphastos	155-660[c,e]	Aerial	Fruits, insects, vertebrates

Table II (continued)

Order Family (Common Name[s]; Family Distribution) Selected Genera	Weight (g)	Habit	Food
Passeriformes			
Cotingidae (Cotingas; Neotropical)			
Carpodectes, Cotinga, Gymnoderus, Lipangus, Pterissocephalus, Procnias, Querula, Rupicola, Rhytipterna, Tijuca, Xipholena	41-395^{e,g}	Aerial	Fruits and some insects
Pipridae (Manakins; Neotropical)			
Manacus, Pipra	12-20^{c,h}	Aerial	Fruits and some insects
Tyrannidae (Tyrant-flycatchers; New World)			
Erator, Tityra (formerly in Cotingidae)	49-86^{e}	Aerial	Fruits and some insects
Bombycillidae (Waxwings; Northern Hemisphere)			
Bombycilla	33^{c}	Aerial	Fruits and some insects
Muscicapidae (Tanagers, Thrushes; Old World Warblers and others; Cosmopolitan)			
Catharus, Erithacus, Sylvia, Tangara, Turdus	12-33^{c,i,j}	Aerial	Fruits and invertebrates
Dicaeidae (Flowerpeckers; Asia)			
Dicaeum	4-10^{d}	Aerial	Fruits
Vireonidae (Vireos; New World)			
Vireo	18^{k}	Aerial	Fruits
Oriolidae (Orioles; Old World)			
Oriolus	60-81^{d}	Aerial	Fruits and insects
Paradisaeidae (Birds of Paradise; Asia)			
Diphyllodes, Epimachus, Lophorina, Manucodia, Parotia, Paradisaea	89-246^{l}	Aerial	Fruits
Casuariiformes			
Dromaiidae (Emu; Australia)			
Dromaius	±54 000^{m}	Terrestrial	Omnivorous; digests some seeds

Coraciiformes			
Bucerotidae (Hornbills; Africa, Asia)			
Tockus	±200[d]	Aerial	Insects, vertebrates, fruits
Piciformes			
Picidae (Woodpeckers; Cosmopolitan except Australia)			
Melanerpes, Phloeoceastes	54-240[c,e]	Aerial	Insects and fruits
Passeriformes			
Tyrannidae (Tyrant Flycatchers; New World)			
Elaenia, Legatus, Megarhynchus, Mionectes, Myiodynastes, Myiozetetes, Tyranniscus, etc.	9-71[c,e]	Aerial	Insects and fruits; perhaps dozens of species of partial frugivores
Vireonidae (Vireos; New World)			
Vireo	17-18[c,e]	Aerial	Mostly insects (many species)
Mimidae (Thrashers; Northern Hemisphere)			
Dumatella, Mimus	39-51[c]	Aerial	Insects and fruits
Muscicapidae (Tanagers, Thrushes, Old and New World Warblers, Blackbirds; Cosmopolitan)			
Above and dozens of genera, ranging in size from *Vermivora* to *Zarhynchus*	9-163[c,e]	Aerial	Omnivorous as a group; perhaps hundreds of species of partial frugivores
Zosteropidae (White-eyes; Africa, Asia)			
Zosterops	9-10[f]	Aerial	Buds, berries, insects, nectar
Corvidae (Crows, Jays, Magpies; Cosmopolitan			
Cissa, Corvus, Dendrocitta and others	100-960[f]	Aerial	Omnivorous; many digest seeds

[a] Crome (1976); [b] Howe (1980); [c] Moermond and Denslow (1985); [d] Crome (1975); [e] Howe (1977), Howe and De Steven (1979); [f] Ali and Ripley (1983); [g] Snow (1982); [h] Worthington (1982); [i] Herrera (1984a); [j] Snow and Snow (1984); [k] Morton (1977); [l] Beehler (1983); [m] Gould (1967).

Table III. Size and taxonomic diversity of selected mammalian genera known to eat fruits and thought to disperse seeds. Mammals known to digest virtually all seeds eaten, such as pigs (Suidae) and mice (Cricetidae) are not included.

Order Family (Common Name(s); Family Distribution) Selected Genera	Weight (g)[a]	Habit	Food
Primarily Frugivorous Mammals			
Chiroptera			
Pteropodidae (Old World Fruit Bats; Africa, Asia, Australia) *Cynopterus, Dobsonia, Epomophorus, Epomops, Hypsignathus, Pteropus*	16–900	Aerial	Fruits, nectar. Many genera are likely agents of seed dispersal.
Phyllostomatidae (New World Fruit Bats; Neotropics) *Ariteus, Artibeus, Carollia, Leptonycteris*	10–87	Aerial	Fruits, nectar, pollen. Many genera are likely dispersal agents.
Primates			
Cebidae (New World Monkeys; Neotropics) *Ateles, Brachyteles, Lagothrix*	680–12 000	Arboreal	Fruits, some flowers and seeds
Carnivora			
Procyonidae (Racoons, etc.; New World, Asia) *Bassaricyon, Potos*	1000–2700	Arboreal	Fruits and small animals
Viverridae (Civets; Old World) *Arctictis, Nandinia*	±1500–14 000	Arboreal	Fruits and small animals

Partially Frugivorous Mammals

Marsupialia			
Didephidae (Opossums; New World) *Caluromys*	1000–3000	Arboreal	Omnivorous
Phalangeridae (Phalangers; Australasian) *Cercartetus, Distoechurus, Phalanger, Pseudocheirus, Trichosurus*	15–5000	Arboreal	Omnivorous
Macropodidae (Kangaroos, Wallabies; Australasian) *Dendrolagus, Dorcopsulus, Macropus*	1500–27 000	Terrestrial	Grasses, leaves, fruits
Chiroptera			
Phyllostomatidae (New World Fruit Bats; Neotropics) *Glossophaga, Monophyllus, Phyllostomus*	9–100	Aerial	Fruits, insects, nectar, pollen
Primates			
Lemuridae (Lemurs; Vicinity of Madagascar) *Lemur, Microcebus*	85–2000	Arboreal	Fruits and insects
Lorisidae (Lorises, Galagos; Africa, Asia) *Nycticebus, Perodicticus*	500–1500	Arboreal	Omnivorous
Cebidae (New World Monkeys; Neotropics) *Alouatta, Aotus, Cebus*	600–9000	Arboreal	Fruits, leaves, seeds, some eat insects
Callithricidae (Marmosets; New World) *Callithrix, Cebuella, Saguinus*	70–900	Arboreal	Omnivorous
Cercopithecidae (old World Monkeys; Africa, Asia) *Cercopithecus, Macaca, Papio*	7000–41 000	Arboreal, Terrestrial	Omnivorous
Pongidae (Apes, gibbons; Africa, Asia) *Chimpansee, Gorilla, Hylobates, Pan, Pongo*	8000–275 000	Arboreal, Terrestrial	Omnivorous

Table III (continued)

Order Family (Common Name(s); Family Distribution) Selected Genera	Weight (g)	Habit	Food
Carnivora			
Canidae (Dogs, foxes; Cosmopolitan islands) *Canis, Chrysocyon, Urocyon, Vulpes*	2500–79 000	Terrestrial	Animals, sometimes many fruits
Ursidae (Bears; Northern Hemisphere) *Ursus*	150 000–780 000	Terrestrial	Omnivorous
Procyonidae (Racoons; New World and Asia) *Ailurus, Nasua, Procyon*	1500–4500	Arboreal	Omnivorous
Mustelidae (Weasels; Cosmopolitan except Australia) *Eira*	3000–4400	Arboreal	Omnivorous
Viverridae (Civets; Old World) *Arctogalidia, Crossarchus, Helogale, Paguma, Paradoxurus, Viverra, Viverricula*	450–11 000	Arboreal	Omnivorous
Proboscidea			
Elephantidae (Elephants; Africa, Asia) *Elephas, Loxodonta*	5000,000–7 500,000	Terrestrial	Foliage, fruits, grasses
Perissodactyla			
Equidae (Horses; Old World) *Equus*	260 000–350 000	Terrestrial	Foliage, fruits, grasses
Tapiridae (Tapirs; Asia, Neotropics) *Tapirus*	225 000–300 000	Terrestrial	Foliage, fruits, seeds
Rhinocerotidae (Rhinoceroses; Africa, Asia) *Diceros, Rhinoceros*	1 000 000–4 000 000	Terrestrial	Foliage, fruits, grasses

Artiodactyla			
Camelidae (Camels; now Africa, Asia, S. America)			
Camelus	450 000–690 000	Terrestrial	Foliage, fruits, grasses
Tragulidae (Chevrotains; Africa; Asia)			
Hyemoschus, Tragulus	2500–4500	Terrestrial	Foliage, fruits
Bovidae (Bison, Buffalos, Cattle, Impala, Cosmopolitan except Australia)	5000–9 000 000	Terrestrial	Vegetation; many genera probably dispersal agents.
Aepyceros, Bison, Bos, Cephalophus			
Cervidae (Deer, Neotropics, Asia)			
Mazama, Muntiacus			

[a] Weights for most groups from Walker (1975)

A. Conceptual Framework

The premise of foraging theory is that animals optimize the expenditure of energy and the taking of risk to maximize individual genetic fitness (Schoener, 1971; Pyke *et al.*, 1977). Energy expenditure, risk and genetic fitness are not easily measured in the field, but the premise that animals optimize trade-offs does lead to testable hypotheses. Those concerning profitability and choice in complex communities provide the most intriguing avenues of research at present.

1. Profitability

Profitability refers to the energetic or nutritional benefit gained from a food after the energetic or nutritional cost of its handling has been subtracted. 'Handling' includes search, manipulating the fruit once found, and processing components of the fruit in the gut. Because fruits are often, to a human observer, easy to find or even superabundant, many naturalists failed to recognize handling costs other than those imposed by overall fruit size. Recent recognition of the relative importance of different handling costs has catalysed an enormous amount of productive research.

In a seminal paper over 15 years ago, Snow (1971) recognized that fruit-eating birds must both process nutritious fruit flesh and contend with heavy seed ballast that displaces digestible food in the digestive tract. McKey (1975) and Howe and Estabrook (1977) further argued that search costs were likely to be substantially different for animals visiting fecund plants with superabundant fruits, as compared with visitors to plants producing smaller fruit crops. Qualitative reasoning could predict little more than the general expectation that birds (and other fruit-eating animals) should somehow balance their intake of digestible and indigestible fruit parts, and further that they should ignore small fruit crops unless they were of especially high quality.

Quantification of the relevant variables was the second major step. Herrera (1981a) used a reasonable adaptation of foraging theory to test the widely cited assumption that tropical fruits were intrinsically more nutritious than temperate fruits. In his argument, relative yield (RY) of dry nutritious matter could be described as:

$$RY = \frac{P(1-W)}{P + ns} = \frac{1-W}{1 + \dfrac{ns}{P}} \tag{1}$$

where n = number of seeds in a fruit, s = fresh weight of the seed, P

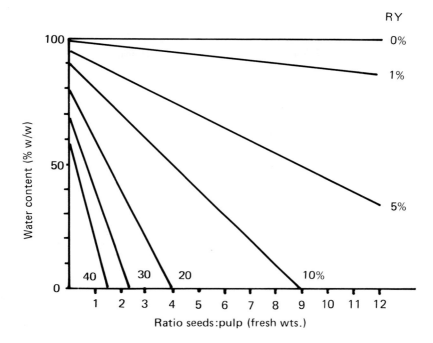

Fig. 2. Variation in the design' profitability (RY) of fruits changes with minor changes in water content. Note that fruits with a low seed-to-pulp ratio may become highly profitable even with watery pulps. The potential for modification of RY decreases substantially as the relative seed load per fruit increases. From Herrera (1981a).

= fresh weight of nutritive pulp, and W = percent of water in pulp. Yield is easily translated into overall profitability (OP) by calculating:

$$OP = RY \times d \tag{2}$$

where d = relative richness (per unit dry weight) of some valuable substance (e.g. lipid, mineral, protein). This model offers an estimate of 'design profitability', or that benefit derived from eating fruits of a given seed bulk (Fig. 2). This is a measure of fruit value *per se* which does not take the effort or risk of foraging into account.

A calculation of the intrinsic worth of a fruit offers some interesting comparisons. For a *Stemmadenia* fruit like that illustrated in Figure 1C, for instance, data from McDiarmid *et al.* (1977) show that 82% of a pulp averaging 10.5 g is water, and an average of 132.5 seeds weighing 0.094

g are embedded in each fruit. The total yield from equation (1) is 8%. The dry substance of the pulp is 63.9% lipid and 11% protein, offering one estimate of d of 74.9. A calculation of overall profitability from equation (2) is 6% (slightly lower than Herrera's [1981a] calculation). Similar calculations for other illustrated fruits give overall profitabilities of 1.3% for *Tetragastris panamensis* (data from White, 1974) and 11.0% for *Virola surinamensis* (data from Howe and Vande Kerckhove, 1981). Herrera (1981a) found that the nutritive value of pulp *(d)* was twice as high in fruits from the Neotropics as in fruits from southern Spain, but that large seeds of Neotropical plants reduced fruit 'yield' (RY). The consequence was that overall profitabilities (OP) between the two samples were indistinguishable: 3.3 ± 2.7 versus 2.9 ± 3.6 (Herrera, 1981a). The implication is that Neotropical and Spanish fruits have generally similar profitabilities, but that some tropical plants must invest heavily in pulp because of the large seeds selected by their competitive environment (Richards, 1952; Ng, 1978). Enormous variation in both samples suggests the potential for a wide range of fruit 'strategies' for encouraging dispersal by animals.

Actual profitability of a fruit depends on many factors. Herrera (1981a) notes that birds differ in their ability to assimilate different nutrients and in the ease with which they rid themselves of bulky seeds. Other foraging costs, such as those involved in searching for fruiting trees, are not included. A manakin may avoid unripe green fruits most of the year, for instance, but eat them in quantity during times of dearth (Foster, 1977). At an even more basic level, the profitability of a fruit from one community is not comparable to that of a fruit from another. For example, fruit-eating birds in southern Spain that weigh 12–20 g (Herrera, 1984a) could not possibly eat the 'highly profitable' arillate seed of *Virola surinamensis* (Fig. 1A) which usually weighs 4–5 g (Howe and Vande Kerckhove, 1981)! Constraints on frugivores obviously need to be considered.

Often the appropriate response to a very simple model which minimizes the number of interacting variables is a more complex version which summarizes all of the likely variables of any consequence. Martin (1985) has produced a fine example of the latter by including behavioral, ecological, and morphological constraints on frugivorous birds. He argues that the potential yield of a fruit to a bird is provided by the product of the pulp mass (F_m) and the nutritional quality of the pulp (F_n), discounted by body size (M) and the time spent finding and eating this fruit (T). The simplest relationship for this more complex profitability (P) is:

$$P = (F_m \, F_n / M^{.61}) / T \tag{3}$$

where $M^{.61}$ is the energy requirement of free-living birds (Kendeigh *et al.*, 1977). The improvement here is the inclusion of a term for search and eating time. The second major step is to explore the components of T, which can be summarized as handling time itself (T_h), search time (T_s), interference time due to competitors in the tree (T_i), and time spent watching for predators (T_p). Martin further had the foresight to note that the available time spent at a fruiting plant (T_t) is rarely the actual time (T_a) spent at it, for any number of reasons. All of these components of exploitation are self-explantory except handling time (T_h), which must incorporate fruit size (S) and some correlate of the handling efficiency of a given bird species. Martin defines handling time as:

$$T_h = b + kS^{c(S - a)} \tag{4}$$

where *a* is the critical fruit size above which handling is difficult for a bird species, *b* is the minimum handling time, and *k* and *c* are species-specific constants that modify the slope of the curve. The complete summary statement then is:

$$P = T_a\,[F_m\,F_n/M^{.61}) - E]\,/\,T_t\,(T_h + T_s + T_i + T_p) \tag{5}$$

where E is the energy spent finding the food. In this formulation the problem of seed ballast is not considered an energy expenditure because of weight (which it might be for an airborne creature), but as a component of handling time (T_h). The longer a bird holds indigestible seeds in the gut before either regurgitating them (often 10–20 min) or defecating them (as long as this or much longer), the less digestible pulp it can process.

The beauty of Martin's model is that the various components of exploitation time (T) can be identified with the natural history of very different foragers. For instance, a large bird will probably have little to fear from predators (Howe, 1979) or from competitive displacement, so T_p and T_i are likely to be inconsequential. But a large bird may have to search widely to find fruits worthy of eating, in which case E and T_s will loom in importance (Fig. 3). The model can easily be altered to assess foraging strategies of such mammals as fruit-eating bats and highly frugivorous monkeys (Table III), neither of which digest seeds. It is less likely to work for large terrestrial frugivores, such as large flightless birds or horses (*Equus cabalus)* which digest most seeds consumed (e.g. Noble, 1975; Janzen, 1982a).

B. Animal Correlates of Seed Dispersal

It is in the direct interest of plants to have their seeds dispersed, but effective seed dispersal is irrelevant to the immediate interests of a fruit-eating bird or mammal. A consequence of this difference in individual advantage is

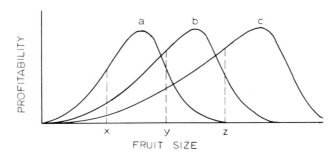

Fig. 3. Profitability as a function of fruit size for birds weighing (A) 17, (B) 250 and (C) 500 g. The curves are calculated from equation (3), assuming that the gapes of the birds are 14, 30 and 45 mm respectively, and that handling time increases with fruit size as predicted in equation (4). Profitability is defined here as the energy gain (estimated by pericarp mass) relative to metabolic requirements of the birds and their handling requirements (from Martin, 1985).

that dramatic adaptations for attracting fruit-eating animals occur among plants, but only adaptations for processing fruits exist among frugivores. Processing often kills seeds. Hence, fruit-eating birds and mammals have digestive physiologies, modifications of the digestive tract, and behavioral attributes which influence their abilities to handle fruits, and only coincidently the likelihood of dispersing seeds effectively. In the context of foraging theory, seeds that are not digested constitute a waste of time, energy, or feeding capacity for a fruit-eating bird or mammal. Similarly, the presence or absence of particular nutrients in a fruit pulp determines its quality of 'reward' which makes various handling costs worthwhile. Even the natural history now available makes it obvious that different plants exert different adaptive leverages on the many possible dispersal agents that may eat their fruits.

1. Birds

Aerodynamic considerations constrain the diets of fruit-eating birds. The first problem is that birds must minimize the burden of swallowing heavy, bulky, and often indigestible seeds when they eat fruits (Snow, 1971; McKey, 1975). Furthermore, fruits are generally low in protein and high in calories (Snow, 1962a; White, 1974). Birds require something in the order of 7–8% dietary protein, so a random selection of fruit species would probably not meet their needs (Moermond and Denslow, 1985). A bird must either eat an enormous quantity of fruits to achieve protein balance and dispense with excess calories, or it must eat a few fruits and find protein in other foods, such as insects.

A large and important fraction of fruit-eating birds bypass all of these constraints. Many opportunistic species, which use fruits primarily as an energy source, simply strip the pulp or aril off and drop the seeds. Seed-eating grosbeaks (Fringillidae) in a lowland forest in Costa Rica tear *Protium* (Burseraceae) arils off and drop the seeds under the crown (Howe and Estabrook, 1977), while in the scrub of southern Spain a variety of gran-ivorous finches (Fringillidae) and insectivorous tits (Paridae) and warblers (Muscicapidae) do the same things at many plants (see Table 2 in Herrera, 1984a). Even *Amazona* parrots, which normally eat seeds, may strip the arils from particularly noxious seeds and drop the latter under a fruiting tree (Howe, 1977). In these and many other cases, birds use fruits as a secondary food source and balance their diets elsewhere. Any positive influence on seed dissemination is strictly accidental.

Birds that depend heavily upon fruits must circumvent some challenges of frugivory. The problem is most acute for unusual 'total' frugivores, such as the oilbird (*Steatornis caripensis*) which eats nothing but fruits (Table II; Snow 1962a). Young oilbirds must be fed enormous quantities of fruits which are relatively poor in protein (5–14% of dry pericarp weight; mean of $10.5 \pm 2.9\%$ S.D.), but rich in lipid (19–44%; mean of $31.0 \pm 8.8\%$), and in carbohydrate (47–70%; mean of $58.5 \pm 9.7\%$) (Snow, 1962a). The young are so obese that they are incapable of flight when parents abandon them in nesting caves. To be such an extreme frugivore, the oilbird must raise young in the safety of ledges high in tropical caves — an option not open to most birds (Morton, 1973; Foster, 1978).

Other birds heavily dependent upon fruits have a variety of physiological and morphological adaptations which facilitate the frugivorous habit. In the south-western United States *Phainopepla nitens*, a silky flycatcher (Pti-logonatidae), processes mistletoe berries through an unusually short gut in 12–45 min (mean 29 min) (Walsberg, 1975). Herrera (1984b) finds that highly frugivorous European flycatchers (Muscicapidae) pass seeds much faster (29.9 ± 8.5 min) than birds of the same family that do not disperse seeds (64.0 ± 22.9 min), although the two groups do not differ in the sizes of the digestive organs. Many opportunistic and obligate frugivores regurgitate any but the smallest seeds after stripping off the edible aril; the entire process takes less than 15 minutes in birds as diverse as tyrannid flycatchers (*Megarhynchus pitangua, Myiozetetes granadensis, M. similis, Myiarchus nuttingi, Pitangus sulphuratus, Tityra semifasciata*), vireos (*Vireo flavoviridis, V. olivaceous*), and toucans (*Ramphastos sulfuratus, R. swain-sonii*) in the Neotropics (Howe, 1977, 1981; Howe and Vande Kerckhove, 1979, 1981). Sorensen (1981) reports that in temperate England, British thrushes (*Turdus*) also regurgitate large seeds. Morphological adaptations may involve modifications of internal organs (Pullainen *et al.*, 1981). The stomach is often reduced in size and capacity in birds that subsist largely on fruits, as compared with those which are primarily insectivorous (see Fig. 12 of Jenkins, 1969). The quetzal (*Pharomachrus mocinno*) has no crop (Wheelwright, 1983), and a number of species either lack a gizzard for grinding up food or have a non-muscular (reduced) gizzard (e.g. *Saltator,* Jenkins, 1969; *Phainopepla nitens,* Walsberg, 1975). Reductions of organ size increase the 'cost' of indigestible seeds in terms of the lower proportion of the gut volume devoted to processing digestible fruit pulp. A coincidental result is that birds with reduced gut size often void or regurgitate seeds quickly. Physiological adaptations are more subtle, but clearly exist. Moermond and Denslow (1985) find that granivorous and insectivorous birds often die or lose weight on fruit diets which are adequate for frugivorous species. Such a diet, for a 30 g tropical frugivore, consists of twice its weight daily or the equivalent of 171 berries 6 mm wide. These observations suggest promising lines of investigation for determining the physiological adaptations which distinguish animals that eat or do not eat fruits.

A popular misconception is that birds often scarify seeds, thereby assuring germination. Some bird-dispersed seeds cannot germinate without removal of an aril (Howe and Vande Kerckhove, 1981) others can (Howe, 1977). In the first case, untreated arillate seeds are attacked by mold within days. In the second, the seed germinates so rapidly (< 7 days) that mold does not set in. Occasionally, a seed with a hard coat benefits from scarification by small birds (e.g. *Acacia cyclops* by unspecified birds; Glyphis *et al.*, 1981). Few highly frugivorous birds scarify seeds; when they do so, the scarification is a matter of unsuccessful digestion rather than some mutualistic compromise with the plant, and usually involves large birds which eat and defecate large seeds or large numbers of seeds. Examples include cassowary (*Casuarius casuarius*) eating fruits of *Beilschmiedia* sp. (Lauraceae) and many other species (Stocker and Irvine, 1983; also see Crome, 1976; Pratt, 1983a), probably dodo (*Raphus cucullatus*) and fruits of *Calvaria major* (Sapotaceae) (Temple 1977, 1979; see Owadally, 1979), and emu (*Dromaius novae-hollandiae*) eating fruits of *Nitraria billardieri* (Zygophyllaceae) (Noble, 1975). Both the treatment and the effect of frugivory by these large flightless species seem far more typical of frugivory by large terrestrial mammals than by other birds.

Among most birds that void or defecate viable seeds, the attributes most influencing seed dispersal are behavioral rather than morphological or physiological (Howe and Estabrook, 1977; Herrera, 1984a,b,c, 1985b). Much more will be said about this below (see 'consequences of failure'). Suffice it to say here that the most important characteristic of an avian visitor to a fruiting shrub or tree is likely to be its predilection for sitting in place, or choosing a place in the open or in the shade to sit while processing seeds. These characteristics are not easily manipulated by a plant, nor are they very different between birds of remarkably diverse taxonomic affinities. Hence, the ecological relationship between a fruiting plant and its dispersal assemblage is likely to be rather general (Howe and Smallwood, 1982; Wheelwright and Orians, 1982; Janzen, 1983a).

2. Mammals

Fruit-eating mammals show interesting parallels and contrasts with avian analogues (Table III). Seed ballast is less of a problem for fruit-eating mammals, both because they are usually larger than fruit-eating birds and because most do not fly. Like birds, however, mammals may drop seeds without ingesting them (these are spat out rather than regurgitated), pass them through the gut, or digest them. Differences are that ingested seeds often remain in a mammalian gut for hours, days, or even weeks. If they survive, seeds are frequently left in large masses rather than scattered singly or in small piles, as is the case with flying birds.

Flying mammals should face the same aerodynamic difficulties as birds, but they seem better able to handle the problem of seed ballast. Fig (*Ficus;* Moraceae) seeds are minute, but the entire compound fruit of two species in Panama average 3.1 g (*F. yoponensis*) and 9.5 g (*F. insipida*), which amounts to 6% or 20% of the body weight of their primary dispersal agent, the medium-sized bat *Artibeus jamaicensis* (Morrison, 1978a). In Panama, as elsewhere (August, 1981), the bats carry an entire fig to a feeding roost located dozens to hundreds of meters from a feeding tree, apparently to avoid predators at the trees (Morrison, 1978b). There the pulp is chewed off, and the remainder is dropped. Many seeds fall under the roost, while many others are defecated in small splats as the animals forage. These small New World fruit bats offer some parallels to birds when they visit trees with large fruits, such as *Dipteryx panamensis* (Papilionaceae). Fruiting crops of this species are heavily visited by a large variety of mammals, most of which eat both the pulp and seed. Three bats visit *Dipteryx* trees in Panama, but two of them (including *Artibeus jamaicensis*) are too small to carry the 18–23 g fruit away (Bonoccorso *et al.*, 1980). One, *A. lituratus* (69 g), does take the entire fruit up to hundreds of meters away before it gnaws off the pulp and drops the seed. No bird is likely to fly carrying 29% of its body weight.

It is interesting that a bat can carry a far larger fruit than a bird of the same, or even much larger, size. But it would be a mistake to imply that bats forage primarily on fruits with large seeds. Among the best studies of frugivory are those of bats that feed on small-seeded fruits in the seasonal dry forest of Costa Rica. Fleming *et al.* (1977), for instance, found that radio-tagged bats (*Carollia perspicillata, Glossophaga soricina*), feeding on widely-dispersed *Piper* (Piperaceae) fruits (catkin-like stalks on small bushes), were on the alert for experimentally placed *Piper* stalks, whereas bats (*Glossophaga soricina, Artibeus jamaicensis*) commuting directly to much larger crops of *Ficus, Chlorophora* (Moraceae), or *Muntingia calabura* (Eleocarpaceae) ignored *Piper* fruits along the way. Heithaus and Fleming (1978) found that *Carollia perspicillata* commuted 4.7 km to feeding patches. The bats repeatedly made short visits to clusters of food plants, apparently to minimize predation by owls or carnivorous bats. Notably, *Carollia perspicillata leaves* > 90% of the *Piper* seeds that it consumes in piles under feeding roosts (Fleming, 1981). Presumably other seeds are scattered en route. Fleming and Heithaus and their collaborators have distinguished different feeding 'strategies' of bats in one community, and have documented the effects of that feeding on plant distribution (see below; Fleming and Heithaus, 1981; Fleming, 1982; Heithaus, 1982).

Old World fruit bats (Pteropodidae) lack echolocation and rely on fruit, nectar and leaves for food. They undoubtedly drop or defecate seeds of many of the fruits that they eat, at substantial distances from feeding trees. Ayensu (1974), for instance, notes that the bat *Epomophorus gambianus* is probably an effective dispersal agent of exotic *Azadirachta indica* (Meliaceae) fruits in West Africa. Despite the far longer history of frugivory in pteropids than phyllostomatids (Heithaus, 1982), little is known of their role in seed dispersal and seedling recruitment.

Arboreal mammalian frugivores include arboreal carnivores, marsupials, primates (lemurs, tarsiers, and monkeys) and rodents in several families. Many arboreal frugivores chew the pulp off at least some fruits and drop the seeds, but by and large they swallow both the pulp and seeds and defecate the seeds later. One difference between these animals and either birds or bats, from the plant perspective, is that seeds actually ingested take several to many hours to pass through the gut. Consequently, arboreal mammals are far less likely to defecate seeds directly under the crown or into piles under feeding roosts than are many birds or bats. However, these species often spit out or knock down more fruits than they actually ingest, reducing their reliability as agents of fruit removal (Howe, 1980; 1983).

Little is known about the influence of frugivory by such arboreal carnivores as the kinkajou, olingo, or tayra (Table III). The first two are nocturnal mammals, frequently noted in night censuses of fruiting trees otherwise visited by birds, monkeys, or bats (Howe, 1980, 1983; Charles-Dominique et al., 1981; Glanz, 1982). I discovered that *Virola surinamensis* seeds defecated by the kinkajou are viable, but that the animals defecate many (an unknown proportion) in heaps under nest hollows in trees (pers. obs. and C. K. Augspurger, pers. comm. 1980). One such heap accumulated over 500 seeds during one season, all of which were killed by weevils (*Conotrachelus*) (Howe, 1983). The cavity was in use and under observation for four seasons, yet no *Virola* seedlings survived under it. However, many *Tetragastris panamensis* (Burseraceae) seeds are scattered by kinkajous in fecal clumps averaging five seeds per clump (p. 952 of Howe, 1980), indicating potentially effective dispersal. Quantitative studies of seed dissemination by these animals would be difficult but most timely.

In the Neotropics, the primate family Cebidae includes approximately 30 species (Walker, 1975), of which several have been well studied. These animals include highly frugivorous forms such as the spider monkeys (*Ateles*), some folivores which eat substantial amounts of fruit such as the howler monkeys (*Alouatta*), and a few true omnivores which eat insects, seeds, and fruits (e.g. *Cebus*). Milton (1981) finds that *Ateles geoffroyi* and *Alouatta palliata* in Panama spend, respectively, 72% and 42% of their time eating fruits (also Klein and Klein, 1977). These genera have similar digestive tracts, except that the frugivorous *Ateles* has a colon with only half the surface area of that of folivorous *Alouatta* monkeys (see Hladik, 1967, for anatomical data on primates). Spider monkeys pass seeds through the digestive tract in 4.4 ± 1.5 h, while howler monkeys pass both foliage digesta and seeds through in 20.4 ± 3.5 h (Milton, 1981). Effects on seed and seedling survival are not easy to judge, since both the short and the long gut passage times would ensure seed dispersal well away (50–1000 m) from parent trees. In Mexico, *Alouatta palliata* spends 80% of its time eating fruits, and drops viable seeds of many species within a highly variable 10–893 m daily foraging range (Estrada and Coates-Estrada, 1984). Less is known of *Cebus* frugivory and its consequences for seedling survival. *Cebus capucinus* picks apart and kills seeds of *Gustavia superba* (Lecythidaceae) in its search for insects (Oppenheimer, 1982) and eats, chews up, and presumably digests seeds of *Virola surinumenesis* (Howe, 1983). In the same Panamanian forest, this monkey passes viable *Tetragastris panamensis* seeds (Howe 1980). The influence of arboreal mammals on seed dispersal has been suggested by many authors (Hladik and Hladik, 1969; Howe, 1980; Milton, 1980; Charles-Dominique *et al.*, 1981), but the details remain poorly known in the Neotropics (see Estrada and Coates-Estrada, 1984).

Studies of frugivory by arboreal mammals in the Old World usually consider fruit as food rather than propagules. A good reason for this is that several Old World genera eat and kill seeds rather than disperse them (e.g. *Colobus* in McKey *et al.* 1981; McKey and Waterman, 1982; *Macaca, Hylobates, Pongo,* and *Presbytis* in Leighton and Leighton, 1983). The roles of these animals in seed dissemination are not at all clear. Lieberman *et al.*, (1979) did find that Ghanan olive baboons (*Papio anubis*) passed viable seeds of 59 shrub and tree species, with the diversity in feces correlated with the number of plants in fruit ($r = 0.80$, $p < 0.01$). Germination trials showed that seeds from two berries (*Nauclea latifolia* [Rubiaceae], *Securingea virosa* [Euphorbiaceae]) and one drupe (*Azadirachta indica* [Meliaceae]) had enhanced germination after passing through the baboons, while those of another berry (*Diospyros mespiliformis* [Ebenaceae]) did not (also see Jackson and Gartlan, 1965). The role of Old World primates in seed dissemination would appear to be an open field for investigation.

Finally, terrestrial mammals ingest fruits, digest some or all the seeds and pass varying proportions of seeds in viable condition. Alexandre (1978) notes that many African seeds germinate in elephant dung, and shows that several species are likely to be dependent on elephants for recruitment. Lamprey *et al.* (1974) have similarly found that large African herbivores enhance the germination of some *Acacia* species. In the Neotropics, experiments with captive or domesticated animals have demonstrated considerable seed-dispersal potential. A captive tapir (*Tapirus bairdii*) readily ate legume seeds fed to it (Janzen, 1981b). Some such as carao seeds (*Cassia grandis* [Caesalpiniaceae]) apparently germinated in the gut, and were digested. The same fate befell 78% of the guanacaste (*Enterolobium cyclocarpum* [Mimosaceae]) seeds, but the remainder appeared in viable condition 9–28 days (mean 15 days) later. Similar trials showed that cows killed 14–21% *Enterolobium* seeds, while horses killed 44–83% (Janzen, 1982a). Cattle defecate most seeds within one week, while horses hold up to 25% of them for 15 to 60 days in intestinal caecae (also Janzen, 1981c,d). Retained seeds are likely to germinate and be digested. Once evacuated, viable seeds may be eaten or hoarded by rodents (Janzen, 1982b) or be left in place. Similar patterns are to be expected in Africa and Asia, where buffalo (Bovidae), antelopes (Antilocapridae), and elephants (Elephantidae) are known to use fruit heavily during at least part of the year (Alexandre, 1978; Jarman and Sinclair, 1979). Other large mammals may kill virtually all seeds that they consume. Peccaries in the Amazon (*Tayassu*) regularly crack and eat *Astrocaryum* and *Jessenia* palm nuts with break forces of 140 ± 30 and 100 ± 40 kgf, respectively, and the larger peccary species is capable of cracking *Iriartea* (Palmae) nuts which have break forces of 350 ± 70 kgf (Kiltie, 1982)!

C. Theory and Practice

The loose link between foraging theory and natural history of frugivory is not surprising, because the key theoretical developments are recent. Fruits are sedentary, attractive, and often available in quantity, and birds and mammals make use of them in a way quite different from their use of other foods, such as animal prey or foliage. The differences are in the relative importance of terms in the profitability equation (5) (Martin, 1985). For instance, most insects of a given size are of similar nutritional quality (White, 1974), but the time and effort birds and mammals require to find and subdue them differs greatly. In contrast, the nutritional quality of fruits varies enormously, and fruits occur in quantities far larger than most animals can eat. Search time within a clump is often minimal, while that between clumps may be considerable. Moreover, the locations of clumps of food may be as predictable to aggressive competitors or to predators as they are to a frugivore looking for a meal. Such factors affect components of search and handling differently when animals are foraging for fruits rather than searching for prey. Add to these complications the fact that frugivores range from 6–7 500 000 g in weight (Tables II and III), and the difficulties of a universal foraging theory are apparent. Some examples underscore the differences between frugivory and predation.

Diamond (1973) suggested that Asian fruit pigeons partition fruit resources on the basis of size. The suggestion would be reasonable for birds feeding on mobile, furtive, and cryptic insects intent on escaping predation. A detailed study of fruit and bird size in highland Costa Rica by Wheelwright (1985a) shows, however, that frugivore size is strongly correlated with maximum size of fruit consumed, but large birds also eat many small fruits (Fig. 3). Partitioning is rather loose, if present at all, because large birds can often profitably reduce handling time (T_s in Martin's models, above) by eating many small fruits rather than by searching far and wide for larger ones.

The issue of fruit accessibility has been explored in more depth in studies of food choice by Costa Rican tanagers, saltators, and manakins. Moermond and Denslow (1983) used cage experiments to show that manakins (*Manacus candei*) are choosier about the kinds of fruits they eat than tanagers (*Euphonia gouldi*), for which accessibility from perches is more important (Fig. 4). In both birds, however, small changes in accessibility can override choice of species or fruit size. A further exploration by Levey *et al.* (1984) shows that several tanagers ate *Miconia* (Melastomataceae) berries before *Urera* (Urticaceae) berries when both were presented 0.3 or 0.9 m apart, but took whatever was accessible when berries were placed 3.4 or 6.4 m apart. In these trials, two manakin species (*Manacus candei* and *Pipra mentalis*) switched to the most accessible *Urera* berries only when *Miconia* was at least 6.4 m away. Such experiments suggest that wild tanagers might feed in whatever bushes or trees they initially land, while manakins could be expected to be more selective in their feeding sites. Changes in the profitability of a fruit in these cases are certainly attributable to changes in

Fig. 4. Choice trials by small frugivorous birds of the Costa Rican understorey. All birds were selective when fruits of different species were equally accessible from the perch. Tanagers and saltators chose 'non-preferred' fruits that were closer to the perch than 'preferred' fruits 90% of the time, whereas manakins (shown) made this switch only 56% of the time (from Moermond and Denslow, 1983).

the handling component of the total time spent finding and eating fruits (T; see above). Differences among foraging tactics of bat species point to analogous issues for small mammalian frugivores (Fleming *et al.*, 1977; Heithaus and Fleming, 1978).

Other aspects of handling and search affect the natural history of frugivory and underscore the differences between the foraging of a predator searching for cryptic and furtive prey and a frugivore choosing between obvious, accessible, and abundant fruits. For instance, territorial birds may defend a fruiting tree to the detriment of seed dispersal (Howe and Estabrook, 1977). In New Guinea, Pratt (1983b, 1984) reported that one bird of paradise (*Paradisaea rudolphi*) defended a shrub (*Schefflera pachystyla* [Araliaceae]) against other birds of paradise, a large pigeon (*Reinwardtoena reinwardsti*) defended a similar shrub (*S. chaetorrhachis*) against a number of smaller birds, and a large cuckoo (*Eudynamis scolopacea*) chased large and small competitors out of one fruiting tree (*Chisocheton* sp. [Meliaceae]) which relies heavily upon paradise birds for dispersal (Beehler, 1983). In Central America, large toucans (*Ramphastos swainsonii*) chase virtually all other birds out of productive *Casearia corymbosa* (Flacourtiaceae) trees (Howe,

1977), and similarly defend *Virola sebifera* (Myristicaceae) trees against other birds during periods of fruit scarcity in the rain forest (Howe, 1981). Salmonson and Balda (1977) showed that in temperate climates solitaires (*Myadestes townsendi*) defend *Juniperus* fruits in western North America. Snow and Snow (1984) showed that mistle thrushes (*Turdus viscivorus*) defend at least four trees, a vine, and parasitic mistletoes for several months during the English winter. In such cases, the attentions of an aggressive frugivore both reduce the number of seeds taken out of a tree by other visitors, and increase the chance that seeds eaten by the aggressor are dropped under the crown of the fruiting tree.

A predictable source of fruits for frugivores may also be a predictable source of frugivores for predators. Predator avoidance may explain the peripatetic behavior of small birds in fruiting trees, as well as their tendency to take fruits to the cover of nearby underbrush between feeding bouts (Howe, 1979). Morrison (1978b) suggests that avoidance of predators is the prime reason why fruit bats (*Artibeus jamaicensis*) avoid fig trees on moonlit nights, and Heithaus (1982) suggests that most small frugivorous bats snatch fruits and carry them to feeding roosts in order to avoid climbing or flying predators at fruiting trees. From the perspective of a plant, territoriality and predation at trees influence terms in the profitability equation (5) (T_i and T_p) in ways that are likely to reduce seed dispersal (territoriality or extreme predation) or increase it (moderate predation). To a small forager, the difference between facing an aggressive competitor or a predator lies between inconvenience and death.

Most attempts to compare foraging strategies have been made between members of taxa and small frugivores that rarely, if ever, digest seeds. Comparisons between unrelated taxa, or those with substantially different means of handling fruits, highlighted the incompleteness of our existing theoretical structure. I have already mentioned that bats have a far greater ability to handle large fruits than do birds of the same size. One might expect that a 2 g 'bat fruit' eaten by a 50 g bat could 'get away with' much larger seeds or many more seeds in each fruit than a 2 g 'bird fruit' taken by a 50 g bird. Predictions from a model such as Martin's would have to be restricted to a particular taxon. Starker differences exist. Much of the 'handling time' for a frugivorous horse, for instance, is in carrying seeds in the gut for days or even weeks (Janzen, 1981b; 1982b). During that time, many seeds are digested. Fruit that is highly profitable to a tanager may achieve reasonably effective seed dispersal, but high fruit profitability to a horse is likely to be lethal to seeds. It will be a challenge to link the theory of foraging of animals that differ in size by six orders of magnitude with predictive theories of frugivore attraction and seed dispersal.

IV. ATTRACTING DISPERSAL AGENTS

A plant may attract dispersal agents with fruits that are colored, presented, or provisioned to appeal to fruit-eating animals. Although many, or even most, animals attracted to a fruiting tree are not effective dispersal agents, the general means by which plants encourage frugivory are probably indicative of long-term adjustments to mutualists. For convenience, I distinguish between morphological 'syndromes' and ecological adaptations.

A. Morphological Syndromes

Homologies are of tremendous use in illustrating the variety of ways in which plants respond to selection for vagility, but they are of no concern to a bird or monkey faced with the choice of equally obvious, accessible, and nutritious red berries or red berry-like figs. Selection for functional convergence is so strong that organs other than the fruit itself may take over both the attractive and nutritional roles of the pulp (e.g. the funicle of *Acacia cyclops* [Mimosaceae], Glyphis *et al.*, 1981; Table V of Chapter 3). Superficial fruit form, color, and function can be similar among distantly related plants.

Constellations of fruit attributes associated with particular categories of dispersal agents are called 'dispersal syndromes' (Ridley 1930; van der Pijl 1972; Chapter 3.II). A fruit dispersed by wind is easily identified by a feathery pappus or winglike structures, one disseminated by water is buoyant, and a seed carried by ants bears an oily elaiosome. Scentless black, blue, or red fruits eaten by birds are distinguished from aromatic green, yellow or white fruits taken by mammals (see Fig. 1). Recent field research has both suggested further refinement of these vertebrate dispersal syndromes, and pinpointed the shortcomings of the syndrome concept.

1. Bird Fruits

Early naturalists noticed that birds were partial to berries of some forms and colors rather than others (Ridley, 1930), but a seminal study of berry choice among European birds provided much of the impetus for specifying colors attractive to birds (Turcek, 1963). Tests of berry choice by 156 bird species for 251 plant species showed that birds preferred red and black diaspores over others, with a probable preference for blue fruits as well. A suggestive preference for white was due to heavy use of mulberries (*Morus nigra* [Moraceae]). The strength of this analysis was its experimental approach combined with a comparison of use by birds against the distribution of fruit colors (black, blue, brown, green, orange, yellow, red, white) in the environment. Because the choices of and for different bird and plant species are not distinguished, it is not possible to discern inter-actions among particular species or categories of species (e.g. dispersal agents *versus* granivores or berries *versus* achenes). What can be surmised without demanding statistical independence is that red and black fruits are more common than most, and that birds prefer them to others. Of some interest is the general observation that some colors near the center of the spectrum, (orange, yellow, and green) are usually rejected by European birds.

A number of observations and surveys have broadened our perception of the bird-fruit syndrome. In the Amazon, Janson (1983) noted that birds tend to disperse small fruits which are red, black, white, blue, purple, or mixed-color. A comprehensive survey of feeding records of Central Amer-ican frugivores showed that most fruits eaten by birds are black (or a bicolored combination of black and white), red, or blue (Wheelwright *et al.*, 1984). This study also revealed consumption of a number of green, white and orange fruits. In a study of several species of obligately frugivorous birds of paradise in Papua New Guinea, Beehler (1983) found that the birds most commonly ate fruit pulp that was red, but it could also be brown, orange, purple, white, or yellow. At the same site, Pratt (1984) observed a variety of birds eating orange, purple, and white fruits. In a broad overview of tropical foods of fruit-eating birds, Snow (1981) lists a large number of plant genera with fruits found in all parts of the color spectrum. These field studies and tabulations confirm the earlier impressions of Snow (1971) and McKey (1975) that obligate frugivores (the 'specialists' of Snow [1981]) often seek out large lipid-rich green or purple fruits to a degree not suspected in earlier work with European birds (Turcek, 1963). The same natural history suggests that other birds, many of them small opportunistic omnivores, eat black, blue, and red fruits, as observed by Turcek (1963), but also seek out white and orange fruits. A subdivision of the bird-fruit syndrome reflects these findings (Table IV). Obviously, experimental refinement would be most useful.

Other characteristics of fruits eaten by birds seem reasonable, in light of visual cues perceptible to the avian eye. For instance, many North American fruits have bicolored displays, with a red or black berry set against a bract, capsule, or pedicel of a different color. Experimental study of ripe and non-ripe cherries (*Prunus serotina* [Rosaceae]) with red or cryptic green pedicels confirms that bicolored displays are especially attractive to birds (Willson and Melampy 1983). Willson and Thompson (1982) found that such displays are more common when frugivore abundance is seasonally low, suggesting that competition for dispersal agents is more intense then than at other times. Bicolored displays also occurred in the tropics, both in bird-dispersed species (Wheelwright *et al.*, 1984) and in tree species taken by both primates and birds (e.g. a white arillate seed against a purple capsule: *Tetragastris panamensis*, Fig.1B.)

Table IV. Seed dispersal syndromes of vertebrate frugivores.

Subcategory	General syndrome Salient features
	Bird fruit syndrome[a]
Obligate frugivore fruits[b]	Large arillate seeds or drupes; seeds >10 mm; scentless; rich in lipid or protein; black, blue, green, purple, red
Opportunist fruits[b]	Small or medium-sized arillate seeds, berries, or drupes; seeds <10mm; no scent; rich in lipid, protein, sugar or starch; black, blue, orange, red, white
	Mammal fruit syndrome
Arboreal frugivore fruits[c]	Large arillate or compound; aromatic, rich in protein, sugar, or starch; brown, green, orange, yellow, white
Aerial frugivore (bat) fruits[a]	Large or small; odourless or musky, often rich in starch or lipid; green, yellow, white or whitish; often pendant
Elephant fruits[d]	Large, tough and indehiscent; fibrous pulp; dull colors; resistant seeds

[a] van der Pijl (1972); [b] Snow (1971), McKey (1975); [c] Janson (1983); [d] Gautier-Hion *et al.* (1985), Howe (1985)

2. Mammal Fruits

The same sorts of observations that identified the 'bird-fruit syndrome' have suggested that fruits either gnawed and dropped or ingested and passed by mammals are characteristically heavily scented, and may be green, orange, yellow or white (van der Pijl, 1972). Although extensive experimental studies of the sort done by Turcek (1963) have not been made on mammalian frugivores, a wealth of field information suggests that the 'mammal-fruit syndrome' can be subdivided.

3. Arboreal Mammal Fruits

Monkeys and arboreal carnivores such as *Potos* are important frugivores in tropical forests. Janson (1983) discovered that large (> 14 mm) brown, green, orange, or yellow fruits, often protected by husks, were favored primate foods in the Amazon, in contrast to smaller and more brightly colored fruits eaten by birds (Section IV. A,*1*). Essentially the same trends have been reported by Hladik and Hladik (1969) for Central Panama, by Charles-Dominique *et al.* (1981) in Guyana, and by Leighton and Leighton (1983) in Borneo. A separate subcategory appears to be appropriate for such fruits (Table IV) as long as the qualifications discussed below are kept in mind.

4. Aerial Frugivore (Bat) Fruits

Bat fruits are identified by a generally musky odor, by their green, yellowish, or whitish color, and often by presentation as a pendant or dangling fruit (van de Pijl 1972). Intensive field work in Central and South America confirms the general features of the syndrome for such diverse genera as *Andira* (Papilionaceae) (Janzen *et al.*, 1976), *Dipteryx* (Papilionaceae) (Bonaccorso *et al.*, 1980), green and yellow *Ficus* (Moraceae) (Morrison, 1978a,b; August, 1981), *Lecythis* (Lecythidaceae) (Prance and Mori, 1978), *Piper* (Piperaceae) (Fleming, 1981), and others (Fleming *et al.*, 1977; Heithaus and Fleming, 1978; Bonaccorso, 1979). Although rather varied, bat fruits appear to deserve a separate category now, as they did in van der Pijl's (1972) monograph.

5. Megafaunal Fruits

Ruminants, elephants, and other large terrestrial mammals are primarily grazers and browsers, but many do have a predilection for large, oily, indehiscent fruits (Gautier-Hion *et al.*, 1980, 1985) (Fig. 5). The reproductive ecology of these trees is poorly known, but there is good anecdotal evidence that many are effectively dispersed by elephants (*Loxodonta*) and other large mammals. In the Ivory Coast, West Africa, for instance, numerous seeds germinate in elephant dung, and some tree species seem to disappear when elephants are exterminated (Alexandre, 1978). In Tanzania, East Africa, large herbivorous mammals (*Loxodonta, Aepyceros, Madoqua, Gazella)* eat and disperse *Acacia* seeds (Lamprey *et al.*, 1974; Douglas-Hamilton and Douglas-Hamilton, 1970). It seems reasonable to tentatively identify a megafaunal fruit syndrome which includes large, oily, indehiscent fruits (Table IV). Because large mammals leave seeds in huge masses, species adapted for dispersal by large mammals might be expected to have seeds and seedlings resistant to insects, pathogens, and competition.

Janzen and Martin (1982) include many neotropical fruits in the megafaunal syndrome, proposing that such species have persisted with drastically reduced seedling recruitment since the extinction of American ground sloths, elephant-like gomphotheres, camels, and horses at the end of the Pleistocene 10 000 years ago. This hypothesis may have a core of truth for plants with highly resistant seeds and seedlings. But adult trees die at a rate of about 1% per year, and several trees included in the megafaunal syndrome suffer devastating seed and seedling mortality if fruits are undispersed (e.g. *Ficus, Brosimum, Scheelea*) (Howe, 1985). Such plants probably could not survive for 500 years without dispersal, much less 10 millenia. Moreover, at least half of the species listed by Janzen and Martin seem to conform to other syndromes, and in fact have effective living dispersal agents in protected forests. Just the same, this intriguing hypotheses deserves continuing attention as more becomes known about undisputed 'megafaunal' plants in Africa and Asia.

Fig. 5. The fruit of *Baionella taxisperma* (Sapotaceae) is an 'elephant-fruit' from West Africa. This fruit superficially resembles some neotropical fruits, such as that of *Gustavia superba* (Lecythidaceae), which might once have been dispersed by large mammals that became extinct during the Pleistocene (photo by L. Emmons).

6. Limitations

Useful as they are in broad floristic comparisons (see below), dispersal syndromes are, at best, marginally predictive. Some criteria (e.g. green fruit color) span all subcategories of syndromes, at least occasionally (Table IV). Exclusive criteria are elusive. One reason is that a given fruit, such as a green fig (*Ficus*), may be eaten by monkeys, bats, tapirs or horses, and even now and then by both large birds which swallow the whole thing and smaller birds which peck out pieces (Morrison, 1978a). Gautier-Hion *et al.* (1980) find equally broad overlap in fruit use by West African mammals. Even a classic bird fruit with a very limited assemblage, such as *Virola surinamensis* (Fig. 1A) is also visited by some monkeys, such as *Ateles geoffroyii* (Howe, 1983; also Charles-Dominique *et al.*, 1981). Most sobering, a classic red 'bird berry' (*Muntingia calabura*) is now known to be dispersed primarily by bats of several genera (*Artibeus, Carollia, Glossophaga*) rather than by birds in the seasonal dry forest of Costa Rica (Fleming *et al.*, 1985). Finally, the sorts of observations used to describe fruit syndromes rarely determine whether frugivores disperse, or kill, the seeds. Given the imprecision of the syndrome concept when applied to plants and living dispersal agents, vague syndromes relying entirely on conjecture and anecdote should be avoided (Howe, 1985). Syndromes remain a useful ecological taxonomy, but not a substitute for actual study of the dispersal process and its consequences. Perhaps the concept serves best as a reminder of the remarkable evolutionary flexibility of plants faced with the challenge of reproduction in forests populated by an enormous diversity of animals that eat their fruits, for better or worse.

B. Ecological Adaptations

Plants ensure seed dispersal by fruit-eating animals with more than the traditional adaptations of fruit structure. Presentation, nutritional content, seediness, schedules of fruit production, toxins and taste may influence which animals visit a plant, and what they do once they arrive. A comparative approach points out the ways in which different species secure fruit removal, while studies of intraspecific variation in seed dispersal and destruction probe the subtle choices of animals that favour some parent plants over others.

1. Comparisons among species
Several attempts have been made to define reproductive characteristics that appeal to different sets of dispersal agents. Most such work has concerned bird-dispersed fruits in either neotropical forests or in temperate forests in Europe and North America.

Neotropical forests commonly harbour dozens of species of fruit-eating animals, and characteristically have floral compositions in which 51–93% of the canopy trees and 77–98% of the subcanopy trees and shrubs bear fleshy fruits adapted for animal consumption (Howe and Smallwood, 1982). McKey (1975) recognized in the immense diversity of fruit adaptations a possibility of very different tactics for attracting very different frugivores. Among trees with large fruits and consequently relatively few seeds, he suggested that it would be in the interest of plants to ensure efficient removal of fruits by making their crops most desirable and most accessible to a limited 'gallery of connoisseurs' (Howe, 1982). Predicted qualities include highly nutritious fruits, presented over a long season, that would be unlikely to satiate a limited number of regular visitors. On the other hand, McKey (1975) recognized the rather different potentials for highly fecund trees, many with small seeds enveloped in sugary or starchy pulp which might serve as an ancillary energy source for virtually any bird able to swallow the fruit. One expects highly visible displays of superabundant fruits, presented to a practically limitless number and variety of frugivores, omnivores, and herbivores of quite variable efficiency as dispersal agents (see Howe and Estabrook, 1977).

First, phenological patterns are clearly influenced by many ecological factors, of which visitation by frugivores is only one. Perhaps a majority of plants bearing fleshy fruits in the Neotropics are understorey shrubs or subcanopy trees which may lack the light necessary to produce bursts of superabundant fruits, or of extremely lipid or protein-rich fruits. Most such species produce a few 'low quality' fruits at a time over extended periods out of physiological necessity (Howe and Smallwood, 1982). Salient exceptions to extended fruiting seasons prove the rule. *Hybanthus pru-nifolius* produces a sharp burst of explosively dispersed seeds shortly after rains during extended dry seasons in which many deciduous forest trees are barren or sparsely foliated (Augspurger, 1981; Leigh and Windsor, 1982) and the forest floor is relatively well-lighted. Both the explosive dispersal and the abrupt phenology of this shrub are odd. *Casearia corymbosa* is a common understorey tree in the seasonal dry forest of western Costa Rica, and it produces an extremely oily fruit (Howe and Vande Kerckhove, 1979). But it lives in open sunlit woods. Probably the physical gloom of their environment forces most understorey trees and shrubs of the Neotropics to 'dribble' relatively low quality berries or drupes out over fruiting seasons lasting many weeks or months. Shrubs or small trees that occupy gaps may not be so constrained (e.g. *Muntingia;* Fleming *et al.*, 1985).

Among neotropical canopy trees, the plant forms of special interest to McKey (1975) and Howe and Estabrook (1977), fruiting seasons are more variable in shape. Sugary fruits of *Tetragastris panamensis* (Fig. 1B) occur in more peaked fruiting seasons than oily arillate seeds of *Virola sebifera* and *V. surinamensis* (Fig. 1A) but all three fruiting seasons are peaked (kurtosis $\gg$ 1) rather than extended (Howe, 1981, 1982). Such data are rare, but seem to conform to a general pattern of abrupt fruiting seasons at the start of rains (Foster, 1982a). Most trees in this forest lack dormancy in seeds, and produce fruits at the beginning of wet weather when conditions for germination and establishment are best (Garwood, 1982; *V. sebifera* is exceptional because it bears fruit in the late wet season and has seed dormancy). It also seems likely that the shape of the flowering curve influences the shape of the fruiting curve. Suffice it to say that many possible factors influence schedules of fruit production, of which disperser attraction may be far from the most important.

Secondly, visitation and fruit depletion sometimes support and some-times fail to support McKey's (1975) dichotomy. Both species of *Virola* mentioned above produce oily fruits in a forest inhabited by 80 species of fruit-eating birds and mammals, yet both are dispersed by essentially the same seven or eight species of birds, primarily the toucans *Ramphastos swainsonii* and *R. sulfuratus* (Howe, 1981, 1983). Most birds, bats, and monkeys in the forest that are large enough to swallow the arillate seeds simply ignore them. On the other hand, 'cheaper' *Tetragastris* rewards draw at least 22 species of birds and mammals (Howe, 1980). In these three instances, the high average fruit removal from the two *Virola* species (76 ± 4% and 62 ± 5% SE, respectively) and the low removal from *Tetragastris* (28 ± 3% SE) offer some support for the hypothesis. In this case *Tetragastris* may have a peaked production of superabundant fruits to attract a diversity of visitors, but the predominant dispersal agent is actually one monkey, *Alouatta palliata* (Howe, 1980). Perhaps trees can only attract large (8 kg) monkeys to rather large crops, capable of sustaining a troop. Another 'generalized' fruit dispersed primarily by these monkeys in Central Panama, *Trichilia tuberculata* (Meliaceae), is most heavily used when it is available in large quantity (Leighton and Leighton, 1982). Other exceptions exist. Almost complete removal of extremely rich *Stemmadenia* (Apocynaceae; Fig 1C) fruit pulp and the small seeds embedded in it by a large variety of birds in Costa Rican pastures suggests an alternative life history for one tree with an extended fruiting season (McDiarmid *et al.*, 1977). Tiny lipid and protein-rich fruits in superabundance in South African *Trema* (Ulmaceae), *Apodytes* (Icacinaceae), and *Vepris* (Rutaceae) trees show another pattern in a subtropical forest in which only four of 35 frugivorous birds actually rely on fruits (Frost, 1980). Finally, specialization on Panamanian *Lindackeria* by normally insectivorous North American migrant wood warblers (Parulidae) suggests another plant 'tactic' which falls outside McKey's (1975) framework (Greenberg, 1981). Estimates of fruit production and depletion, and careful observations of frugivore activity, are still so rare in the literature on tropical frugivory that any conclusions must be tentative. But it is clear that many means of adaptation for and to frugivory are possible. Most relevant patterns have probably not been clearly documented, much less understood.

Another effect of phenology, thought to be important in tropical ecology, is competitive partitioning of the 'frugivore resource' through displacement of fruiting times. Snow (1965) was correct in pointing out that there is no necessary connection between timing of flower and fruit production, but his suggestion that different species of *Miconia* (Melastomataceae) partition the frugivore community on Trinidad is difficult to interpret because quantitative estimates were not made of fruit production (Howe, 1982; Worthington, 1982). The apparent spacings of qualitative phenologies were not distinguished from random (Poole and Rathcke, 1979). In Panama, Worthington (1982) found dramatic seasonal variation in both the number of species and the quantities of fruits of shrubs and small trees available in the forest. Notably, many 'manakin fruits' reached their peak productivity at the same time. Foster (1982a) showed that there was at least as much seasonal synchrony among canopy trees. A review of tropical fruiting phenologies shows dramatic seasonality from otherwise 'aseasonal' rain forests in Colombia (Hilty, 1980a) to seasonal forests of Sarawak (Fogden, 1972) and northern Australia (Crome, 1975; see Howe and Smallwood, 1982). Birds in these forests may eat quite different things (Crome, 1975; Frith *et al.*, 1976), but they may do so even when many species are in fruit. Wheelwright (1985b) looked for and failed to find displaced fruiting seasons among bird-dispersed trees in highland Costa Rica. Janzen and Martin (1982) suggest that Costa Rican trees thought to be adapted for dispersal by large extinct mammals stagger fruiting seasons as anachronistic adaptations for partitioning a long-extinct frugivore assemblage, but evidence for this seems to be weak or non-existent (see Poole and Rathcke, 1979; Gleason, 1981).

In short, tropical forest plants show an enormous variety of patterns of fruit provisioning and quantity of production that might be tied to disperser attraction. Rich fruits ensure better removal than less nutritious ones, but the extent to which plants bearing such fruits rely on a few or many visitors may depend largely on seed size. Phenological adaptations to frugivory are possible, but have not been clearly demonstrated in tropical communities.

Temperate forests lack both the diversity of plants bearing fleshy fruits and the variety of birds and mammals largely dependent on fruits for food. North American thrushes, vireos, and flycatchers eat fruits and disperse seeds, but only waxwings (Bombycillidae) and a silky flycatcher (Ptilogonatidae) even approach the reliance on fruits that is characteristic of many neotropical birds. In Europe, little morphological specialization characterizes the many birds that eat fruits (Herrera, 1984b), although at least one flycatcher (*Sylvia atricapilla*) subsists largely on fruits and has an enormous impact on seed dispersal (Herrera and Jordano, 1981; Jordano and Herrera, 1981; Herrera, 1984a). Both Europe and North America lack obligately frugivorous mammals, such as fruit bats, which are so common in tropical forests. Interestingly, it is temperate zone fruiting phenologies that are adaptively adjusted to a limited frugivore resource.

Patterns of investment, fruit production, and use by animals (almost all birds) do vary seasonally in both North America and Europe. On both continents few fleshy fruits are produced in midsummer, while the greatest number and variety ripen during fall bird migration (Thompson and Willson, 1979; Stiles, 1980; Herrera, 1981a, 1982a, 1984a). A small minority of fruits ripen after most migrants leave, and persist throughout the winter. Nutritional analyses of North American fruits seem to suggest higher profitabilities during the fall than winter (Stiles, 1980), although this pattern may be more apparent than real if fruit pulp and seeds are analysed together (Herrera, 1982b). Herrera (1982a) found that persistent winter fruits in southern Spain have higher lipid contents (19.7 ± 18.7%), than those which ripen and are dispersed in either the summer (2.5 ± 1.2% lipid; 0.4 ± 0.2 profitability for lipid) or the autumn (7.4 ± 13.7% lipid; 1.6 ± 3.0 profitability for lipid). High winter profitabilities probably ensure continual seed use by birds during times of food scarcity, when nomadic species might be tempted to wander. Ripening of berries and drupes in both Europe and North America in fall, and high investment in persistent winter fruits, are easily explicable as adaptations to either fall migrants or wintering birds.

European studies of fruit production and removal additionally show wide species variation in fruit removal. An exemplary study in England showed that tits (*Parus*) took fruits of several shrubs or vines (e.g. *Sambucus* and *Lonicera* [both Caprifoliaceae], *Rubus* [Rubiaceae], *Bryonia* [Cucurbitaceae], *Tamus* [Dioscoreaceae]) early in the autumn, whereas thrushes (*Turdus*) ate some of these and others (*Rosa, Prunus* [both Rosaceae], *Hedera* [Araliaceae], *Euonymus* [Celastraceae]) later (Sorensen, 1981). Most interesting, some shrubs were entirely depleted by birds, but the extent to which they were depleted did not depend on pulp quality alone. Further experimental work showed that thrushes have taste preferences when offered 'synthetic' fruits of flour dough with flavours added (Sorensen, 1983). Such taste preferences should be explored further. They might indicate species-specific adaptations for attracting some birds rather than others, or they might suggest that secondary compounds in fruits discourage some potential dispersal agents more than others. A restriction of the disperser assemblage could result from either an adaptive means of excluding destructive or wasteful visitors (see Howe, 1981, 1983; Howe and Vande Kerckhove, 1981), or as an epiphenomenon of plant defence against insects that eat fruits (Herrera, 1982c). Finally, Sorensen (1984) showed that thrushes prefer fruits with the most favorable balance of energy gain from pulp and passage time of indigestible seeds. This series of studies on English birds certainly gives a strong impetus to profitability analyses in frugivore foraging ecology. Such a system is 'ripe' for preference-accessibility tests of the sort that have been so useful in documenting berry preference in tropical tanagers (Moermond and Denslow, 1983).

An extraordinary series of studies in scrublands of southern Spain has documented, among other things, dramatic differences in the efficiency with which temperate berries are dispersed by fruit crop depletion of a summer-fruiting cherry (*Prunus mahaleb*) by thrushes (*Turdus merula*) and the blackcap (*Sylvia atricapilla*). Among fall fruits, removal was very high (89–100%) for fruits smaller than the gape width of common small dispersal agents, but was very low (< 20%) for fruits with widths larger than the gape widths of the rather small birds (12–18 g body mass) common in the scrub (Herrera, 1982a, 1984a). In both the English and Spanish plants, the influences of fruit size and seediness on handling costs for birds appeared to play crucial roles in the extent to which different species achieve fruit removal. Both studies also showed, with admirably thorough direct field observation, that 'efficient' fruit removal is not the prerogative of tropical trees visited by 'specialized' frugivores (cf McKey, 1975; Howe and Estabrook, 1977).

2. Comparisons within Plant Species

Data may conform to the expectations of a prediction for the 'right' or 'wrong' reasons. Comparisons of removal efficiency among different plant species may help suggest fruit attributes which are important to different dispersal agents, but inferences are likely to be equivocal unless the number of species sampled is substantial. One might find, as expected, that sugary fruits of *Tetragastris panamensis* attract a large variety of wasteful animals while the rich oily fruits of *Virola surinamensis* draw efficient specialists (McKey, 1975; see Howe, 1982). But *Tetragastris* and *Virola* are so distantly related that the reason for this concordance with 'theory' might be due more to different smells, tastes, secondary compounds, or fruiting seasons rather than to the variables — seed size and nutritional content of the pulp — which were identified in the model being tested. Discordant results could likewise be spurious. Intraspecific studies of dispersal success avoid the confounding variables endemic in comparisons of widely divergent plant taxa (Howe and Estabrook, 1977).

Variation exists in both absolute and relative dispersal of fruits eaten by birds and mammals. The absolute number of fruits eaten is directly related to the number available (see Howe and Smallwood, 1982, and references therein). Equally important, different individual trees disperse more or fewer seeds than conspecifics; the lifetime success of a tree, *vis a' vis* fruit removal, is the sum of the proportion of seeds taken each year times each annual seed crop size. If such differences persist throughout the lives of the plants, they may result in dramatic variations in fitness. Relevant questions here are whether intraspecific variations in (a) fruit quality and (b) fruiting phenology influence the extent to which individual crops are depleted by dispersal agents, as suggested by comparisons between species.

Fruit quality may be defined in terms of the nutritional composition of the pulp, or better, in terms of both the reward and cost of eating a fruit. Aril quality among Panamanian *Virola surinamensis* trees showed nearly a twofold range of lipid and protein contents, but neither was associated with variance in seed removal for any of three seasons studied ($p > 0.1$, Howe, 1983). The assay of nutritional components is admittedly coarse, but it appears that birds do not detect differences in aril content from one tree to another (also see Sorensen, 1981).

The ratio of aril to seed is a better approximation of 'profitability' because it represents both the benefit in digestible pulp and 'cost' in bulk or weight of indigestible seed that a visitor must consume. Spanish *Smilax aspera* (Liliaceae) fruits are 'seedier' in populations with few competing fruits of other species than in populations with many fruiting plants competing for dispersal agents (Herrera, 1981b). Geographic adaptations to very different edaphic conditions have not been excluded in this study, but the pattern is consistent with the hypotheses that plants enhance the profitability of fruits when they must vie for the attentions of fruit-eating birds. Finally, Jordano (1984) has found that several bird species take fruits from different *Rubus ulmifolius* clones according to the degree of 'seediness'. This might suggest 'partitioning' of a fruit resource.

Seediness also counts in tropical forests. In *V. surinamensis*, as much as 52% of the variation in seed removal (range 13-91%, $\bar{x} = 62 \pm 5\%$ taken) is attributable to the aril-seed ratio of individual trees (Howe, 1983). Interestingly, selectivity increases as the fruit abundance decreases, suggesting that specialized fruit-eating birds range widely for *Virola* trees when fruit is scarce, but eat whatever is close at hand when fruit is ubiquitous (see Tables 9 and 10 in Howe, 1983). As in aviary trials with much smaller birds in a confined space, accessibility overrides preference (Moermond and Denslow, 1983). The *Virola* example is especially interesting because dispersability tends to be highest from trees with small seeds, which tend to have high aril-seed ratios, and lowest from trees with large seeds, which tend to have a low aril-seed ratio (78% of the variation in seed size is between crops, 22% within crops; Howe and Richter, 1982). In this case small seeds may have the advantage in dispersability, but large seeds the advantage in seedling survival. Perhaps the usual range of intermediate seed and aril sizes represents a 'trade-off,' due to normalizing selection, which makes the relevant alleles effectively neutral until strong directional selection (e.g. frugivore extinction, a dramatic shift in disperser size) imposes a clear advantage to one end of the aril to seed ratio continuum (Milkman, 1982).

Intraspecific tests of phenological hypotheses have been less successful. Fleming (1981) found that the bat *Carollia perspicillata* not only found and ate all *Piper amalago* fruits normally made available during an extended Costa Rican season, but it consumed as many additional fruits as were experimentally provided. The pace of *Piper* fruit production may be due more to light constraints in its undergrowth habitat than to selection by frugivores. Higher proportions of fruits seem to be taken from *Casearia corymbosa* and *Tetragastris panamensis* trees of intermediate size than from either smaller or larger plants, but the decrease in fruit removal from large individuals may be due to predator avoidance (Howe and Vande Kerckhove, 1979) or to seed consumption by parrots (Howe, 1980), not to satiation of the visitors. In both species, individual fruiting phenologies were generally synchronized with conspecifics, but were rather irregular. A stronger test of the notion that plants extend fruiting seasons for a limited disperser assemblage is a correlation of the percentage of fruits removed against the platykurtosis (flatness) of individual *Virola surinamensis* fruiting curves; one expects the flattest curves to promote the most efficient fruit removal. None of the tests approach significance ($p > 0.2$) in any of three seasons (Howe, 1983). The length of an individual fruiting season may depend largely on climatic constraints on seed germination (Garwood, 1982) or on flower and fruit production (see Herrera, 1984a) for a given species, while its 'shape' could reflect either the physical constraints of light-poor environments or the vagaries of pollination success or predispersal seed consumption. There is as yet no evidence that tropical plants adjust fruiting phenologies to match the needs of dispersal agents.

3. Summary

Comparisons of the efficiency of fruit removal between and within taxa confirm the expectation that birds balance the reward gained from digestible fruit pulp against the cost of ingesting bulky and heavy seeds. Further evaluation of the high dispersability of small seeds compared with enhanced seedling vigour from large seeds of the same species shows that such choices pose adaptive dilemmas for the plant. But most questions are still unanswered. More factors than pulp composition and seed size influence fruit choice. Fruit accessibility, both on the scale of placement within a bush and distribution of bushes within a forest, appears to be critical. Other issues may be important. For instance, casual observation might suggest that Costa Rican shrubs present berries on the ends of branches because their avian dispersal select for distal presentation. Yet Denslow and Moermond (1982) found that birds actually prefer berries in the axils of branches. Distal berries are ubiquitous because rodents, which eat and kill seeds of many shrubs, cannot easily harvest berries on the tips of branches! Other aspects of fruit choice have only begun to be explored. Thrushes may reject some tastes because a fruit is not evolved to attract thrushes, because the fruit must discourage seed-eating insects and coincidentaly repel some potential dispersal agents, or because a plant evolves means of repelling inefficient dispersal agents in favour of others. Most questions concerning a plant's means of attracting dispersal agents have not yet been answered, nor even properly addressed.

V. POPULATION ECOLOGY OF SEED DISPERSAL

The different influences of birds, bats, monkeys and ungulates in either promoting or precluding effective seed dissemination are at an exciting phase of discovery. At one level, recruitment of some plants seems to match activities of some dispersal agents. Shrubs with berries favored by West Indian manakins were abundant around traditional courting leks (Snow, 1962b,c), *Prunus mahaleb* saplings were most common in brushy tangles visited by *Sylvia atricapilla,* its primary dispersal agent (Herrera and Jordano, 1981), and a variety of tropical trees had distributions consonant with the movements of fruit-eating bats (Fleming and Heithaus, 1981). Such associations lead to the more basic question of whether dispersal agents are efficient or merely moderately destructive, from the plant perspective. The key points are that dispersal efficiency is always relative, and that efficiency may be more apparent than real.

A. Consequences of Seed Dispersal

Several ecological factors might select for seed dispersal. These include: (1) escape of progeny from disproportionate mortality near parent plants (escape hypothesis); (2) occupation of open habitat (colonization or disturbance hypothesis); and, (3) occupation of rare special microhabitats which are necessary for seedling establishment (directed dispersal hypothesis). These hypotheses are not exclusive, and may all apply to dispersal advantage in some plants (Howe and Smallwood, 1982). The challenge is to discriminate between the relative importance of escape, colonization, and directed dispersal.

1. Seed Escape

Whatever the mode of seed dispersal, seeds of terrestrial plants usually fall in continuous leptokurtic distributions with the mode under or near the parent and a steadily declining number further away (Levin and Kerster, 1974; recent example: Howe *et al.,* 1985; Fig. 3). Janzen (1970) and Connell (1971) suggested that disproportionate mortality of seeds and seedlings near parent trees might allow propagules of competing species to establish, thereby promoting species diversity among adult trees. Density-dependent or distance-related seed and seedling mortality is probably not sufficient to explain species diversity. Animal species show parallel patterns of diversity without the requisite mechanism, and Hubbell (1979, 1980) and others have criticized the adequacy of the model for tropical trees, which are usually clumped, not spaced as implied by Janzen and Connell. But disproportionate mortality of seeds and young plants near parents could enhance the fitness of parent plants with dispersible seeds, compared with plants which fail to disseminate seeds. Put another way, a seedling may be the sole survivor of 1000 seeds that fall under a parental crown, or a sole survivor of 100 that fall 30 m away.

The escape hypothesis assumes, in its strictest form, that microhabitats

under and away from fruiting plants are identical except for biotic interactions brought about by heavy seed fall near the parent (Howe and Smallwood, 1982). Disproportionate mortality under the parent is due to seedling
competition, pathogens or seed consumers that affect seeds or seedlings
in high rather than low densities, and not to more favorable edaphic or
canopy conditions away from the parent. The hypothesis can be rejected
if distributions of juvenile plants reflect seed distributions, indicating that
mortality is random with respect to density of siblings or distance from
the parent. The hypothesis may be confirmed, but inconsequential, if the
density of seeds, seedlings, or juveniles never falls below that at which
competition limits the use of space in the canopy (Harper, 1977), or if
canopy openings which are unpredictable in space and time determine
seedling survival and consequently adult recruitment (see 'colonization
hypothesis', Section V.A,*2*).

High mortality of seeds or seedlings is common under tropical trees
(reviewed by Janzen, 1969, 1970, 1971; also Howe, 1977; Estrada and Coates-
Estrada, 1984), but this is not evidence. Variance in mortality is very high
(Hubbell, 1980) and high seed and seedling mortality is to be expected
in organisms that produce hundreds to millions of offspring each year (e.g.
Williams, 1975). Neither can clumped distributions of adult trees or adults
surrounded by saplings and young adults necessarily refute an advantage
to seed dispersal (cf. Hubbell, 1979, 1980). Many more seeds fall close
to most fruiting trees than are carried away; it may take hundreds or
thousands of seeds to produce a sapling under an adult, but only dozens
a few meters away (see Howe and Smallwood, 1982; Clark and Clark,
1984).

A number of studies report differences in survival under and away from
parent trees, but there is a good deal of confusion in the literature about
what constitutes a test. The most definitive tests are with short-lived plants,
mostly dispersed by wind or ants, for which per-capita survival can be
estimated from seed dispersal through age of first reproduction. Platt (1976)
has produced one solid confirmation with a prairie herb, *Mirabilis hirsuta*
(other data reviewed by Howe and Smallwood, 1982). Most partial tests
with long-lived plants either compare distributions of seeds with those of
older juveniles, or follow seed or seedling cohorts. In general, direct tests
show that there is an 'escape' from seed or seedling mortality near the
parent (Table V). The sample shown de-emphasizes anecdotes of doubtful
generality by including only reports of seed or seedling survival around
four or more adults. Mapping data also show disproportionate mortality
near parent plants; Connell *et al.* (1984) demonstrate that density sharply
reduces seedling and sapling survival among Australian trees. However,
some caveats are important.

Table V. Evidence for the hypothesis that disproportionate seed and seedling mortality occurs near parent trees. The sample consists of studies reporting patterns of seed or seedling mortality around four or more trees.

Species (Family)	Adult trees (N)	Dispersal agent	Evidence
Bursera graveolens (Burseraceae)	28	Probably birds	Most seeds fell under adults, but most juveniles were more than 3 m from adults. Sources of mortality were unspecified. Site on Galapagos Islands.[a]
Dipteryx panamensis (Leguminosae)	6	Probably mammals	No seedlings survived within 8 m of an adult over a two year period. Sources of mortality were unspecified. Site in Costa Rica.[b]
Platypodium elegans (Leguminosae)	4	Wind	Seedling cohorts move progressively farther from adults from 1–12 months; saplings are at a greater distance yet. Early mortality is from fungal pathogens. Site in Panama.[c]
Scheelea zonensis (Palmae)	24	Rodents	Bruchid weevils (unspecified) are more likely to attack seeds under than away from trees; highest juvenile density was 8–10 m from adults. Site in Panama.[d]
Virola surinamensis (Myristicaceae)	6–17	Toucans	Higher curculionid weevil (*Conotrachelus*) attack on seeds and young seedlings under and near than far from adults. Site in Panama.[e]

[a] Clark and Clark (1981); [b] Clark and Clark (1985); [c] Augspurger (1983a, b); [d] Wright (1983); [e] Howe *et al.* (1985)

First, a strong advantage to seed dispersal early life is likely to favor trees with dispersable fruits, but this cannot be assumed. Most seeds may be taken to places unsuitable for seedling growth. Oilbirds, for instance, drop tens of thousands of seeds in caves (Snow, 1962a). More likely, the parent may occupy a relatively favorable habitat, so that seedlings out from under the crown but only a short distance away stand a better chance of surviving than those taken much longer distances. Maps of tropical trees suggest that some tree species have strong habitat limitations (e.g. Hubbell, 1979; Hubbell and Foster, 1983). These issues have not been explored thoroughly.

Secondly, many things may kill a seed or seedling between dispersal and maturity. The reasons for dispersal advantage can be complex. For instance, rodents and weevils together kill over 98% of *Virola surinamensis* seeds during the first 12 weeks of seed life. The rodent depredations are random with respect to distance from *Virola* crowns, while weevils kill a disproportionate number of seeds under the crowns. Howe *et al* (1985) showed that there was a fortyfold advantage to seeds dropped only 45 m from the trees, as compared with those 5 m from the trees, during the first 12 weeks after fruit fall. Browsing mammals continue to kill seedlings for at least several months; herbivore damage may accentuate the advantage imposed by weevils (Howe, unpub. data). It is even possible that browsing by rodents, deer, and tapirs on *Virola* seedlings under the crown is so intense that the mammals would produce a dramatic advantage to seed dispersal even if weevils were absent. In the *Virola* example, seed and seedling mortality is due to a wide variety of animals, including a weevil (*Conotrachelus* spp.; Curculionidae), a bark beetle (undetermined, Nitidulidae), at least two rodents (*Agouti paca* and *Dasyprocta punctata*), deer *(Odocoileus virginiana)*, peccary (*Tayassu tajacu*), and tapir (*Tapirus bairdii*). Even without seedling competition or a high incidence of pathogen attack, it would take an extraordinary experiment to separate the influences of all real sources of mortality. If all the animals were locally extinct, as may be the case in much of Central America, competition and pathogens *might* kill a disproportionate number of progeny under the trees!

Thirdly, ecologists are only now paying much attention to the empirical sufficiency of tests of the escape hypothesis. Most relevant papers concern seed or seedling survival around one or two trees in neotropical forests (reviewed by Janzen 1969, 1970, 1971; Howe and Smallwood, 1982; Clark and Clark, 1984). Such work is not useless. For instance, a pilot study in a Malaysian forest shows higher seed and seedling mortality under one *Aglaia* individual than at distances several to 50 m away, and further shows that most seed dispersal is by hornbills ((Bucerotidae) which are at least superficially similar to neotropical toucans in habit and behavior (Becker and Wong, 1985). Such data, even anecdotal, raise the possibility of parallel dispersal phenomena among Old and New World canopy trees whose seeds are dispersed by large birds (Howe, 1983; Howe *et al.* 1985). In general, however, extremely high variance in mortality under and near trees makes it clear that multiple observations at one tree are, at best, an unreplicated set, at worst an 'inflated N' of one sample (Hubbell, 1980; Hurlburt, 1984; Sarukhan *et al.*, 1984). For this reason, much of the published literature is uninterpretable.

2. Colonization or Disturbance Hypothesis

A second major advantage to seed dissemination might be occupation of habitats quite different form that in which the parent grows. The colonization hypothesis presumes that habitats change with time; the tactic of a parent plant is to broadcast so many seeds so widely that some are likely to encounter a favorable situation such as a gap in the forest canopy or disturbed ground (Howe and Smallwood, 1982). Colonists may occupy the open habitat as it occurs or, as with many herb seeds, lie dormant and wait for a treefall, fire, or landslide to come to them (for reviews of seed banks, see Cook, 1980; for case histories, see Werner, 1976; Gross and Werner, 1982), while larger seeds germinate and wait in the understorey as repressed juveniles ready to take advantage of any opening as it occurs (Hartshorn, 1978; Brokaw, 1982). Seeds without dormancy must be dispersed because establishment is impossible in the shade of the parent or in the shade of competitive species which encroach on the parental site. For parent plants with dormant seeds or repressed seedlings, dispersal is highly advantageous because it maximizes the number of open sites that might eventually be encountered. The number of usable disturbances or open sites increases as π distance^2 from the parent tree in the appropriate habitat. The more widely broadcast the seeds, the more likely one or many of this infinite potential of spaces will be occupied by progeny. Effective distance for seed dispersal under this hypothesis depends on crop size. A tree with a small crop will not have to disperse its seeds widely to use all of the 'potential' sites for establishment in its vicinity, while a tree with a large crop must distribute seeds far more extensively to take maximal advantage of all of the possible sites available.

The disturbance hypothesis has a long history in the study of herbaceous plants (see references above, and citations therein), but it is far less explored in the biology of large plants likely to be dispersed by fruit-eating animals. Most examples must be inferred from indirect evidence because seed and seedling demographies are either unknown or cannot exclude alternative hypotheses. But evidence suggests that the 'colonization hypothesis' will be the most important in explaining the evolution of adaptations for dispersal.

The natural history of successional habitats and the prevalence of seed dormancy are instructive. Many shrubs and trees of forest gaps and early succession produce bird-dispersed berries or drupes with seeds capable of extended dormancy (e.g. Richards, 1952; Snow, 1971; Denslow, 1980a), as do bat-dispersed plants with small seeds (Fleming and Heithaus, 1981). Marks (1974) reports a classic example in pin cherry (*Prunus pensylvanica*; Rosaceae), a small tree of north-eastern North America which produces a drupe eaten by dozens of bird species. Cherry pits may be viable for up to 50 years in the soil. A major windfall or forestry operation can produce a virtually solid stand of this early successional tree. In this example, 'escape' from parents by seeds or seedlings is inconsequential. One might expect that most seeds capable of extended dormancy derive their benefit in dissemination from an ability to colonize future 'gaps'.

The applicability of the colonization hypothesis may even be more important than the widespread phenomenon of seed dormancy implies. Hartshorn (1978), for instance, reports that the mean turnover rate, or time required for a place in the forest to experience a gap, in a Costa Rican rain forest is 118 ± 27 years. While some species seem to be dependent on very large ($> 500 m^2$) gaps for regeneration (e.g. animal-dispersed *Cecropia* [Moraceae] and *Trema* [Ulmaceae] and wind-dispersed *Ochroma* [Bombacaceae]), most gaps are smaller, ranging from 8–376 m^2 (mean 88 $\pm$ 89 m^2). Hartshorn argues that the majority of tree species in this forest, perhaps 75%, require exposure to periodic canopy gaps of small or moderate size to regenerate.

Plants with rather different seed and seedling demographies may take advantage of chance breaks in a forest canopy. Tropical trees often have large seeds, which produce large seedlings capable of astonishing persistence in the understorey (Richards, 1952; Ng, 1978). One experimental study showed that two-year old *Shorea leprosula* and *S. maxweliana* (Dipterocarpaceae) seedlings could survive 100% defoliation, and that 25% defoliation produced no discernible difference in mortality over a two year period (Becker, 1983). The first of these species appeared far more dependent on gaps than the second. In Central America, survival of *Gustavia superba* (Lecythidaceae) seeds and seedlings may be as high as 80% for the first six months after fruit fall (Sork, 1985), and survival of *Tetragastris panamensis* seedlings directly under the parental crown ranges from 30–40% during the first year (Howe, 1980). 'Escape' from the parent, for such species, is probably far less important than occupation of sites where growth is feasible.

On the other hand, both 'escape' and occupation of contemporary or future gaps in the canopy may be critical for other species. Howe *et al.* (1985) found a decisive more than fortyfold advantage to seed dispersal away from *Virola surinamensis* trees, but the vigour of the few surviving seedlings at 18 weeks was definitely enhanced by slight variations in canopy conditions which were independent of distance from *Virola* trees. Canopy photographs showed that seedlings with three leaves were in situations with 3.5% of the canopy overhead open, whereas those with two leaves had 2.3% of the canopy open ($p < 0.05$). Light conditions did not affect the chance to survive to independence from seed reserves (at about 12 weeks), but such conditions may be critical after independence. The influence of minor breaks in the canopy is all the more interesting because *Virola* is relatively shade-tolerant; saplings are less likely than most species to be in large gaps (Howe *et al.*, 1985). 'Escape' from pathogens near the parent is well-documented in wind-dispersed *Platypodium elegans* (Papilionaceae), but is far less important for seedling recruitment than is occupation of a large light gap (Table V; Augspurger, 1983a,b).

3. Directed-dispersal hypothesis

Dispersal agents may take seeds to special microhabitats which are required for establishment and growth (Howe and Smallwood, 1982). The implication is that effective dispersal is not simply 'away' from the parent. The special site is predictable, and perhaps permanent. A convincing demonstration would show that dispersal agents take seeds to non-random places which are well suited, by virtue of edaphic or other conditions, for establishment and growth (see also Chapter 6,V,B).

'Directed dispersal' is unusual among plants dispersed by vertebrates, perhaps because a plant can exert so little control over the behavior of a fruit-eating animal (Wheelwright and Orians, 1982). The phenomenon is well documented in ant-dispersed plants. Culver and Beattie (1980) showed that two English violet species (*Viola odorata* and *V. hirta* [Violaceae]) germinate and establish far better in ant mounds than elsewhere, and Davidson and Morton (1981) showed that several Australian shrubs grow on ant mounds, but that ant-dispersed congeners have demonstrably poor growth off the mounds. The clearest examples among vertebrate-dispersed plants seem to be Asian mistletoes. Docters van Leeuwen (pp. 124 ff, 1954) found that several flowerpeckers (Dicaidae) consumed the berries of several genera of mistletoe (e.g. *Dendrophthoe, Viscum* [Loranthaceae]) and defecated the seeds in sticky strings, which were wiped off on bark. The birds consequently left the seeds of these epiphytic plants in the best possible place for establishment (also see Keast, 1958; Davidar, 1983a). Davidar (1983b) further showed that some bird species produced less mistletoe seedling mortality than others. Another intriguing example is dispersal of juniper and cedar (*Juniperus communis* and *J. virginiana*) seeds in New England pastures. Livingstone (1972) found that American robins (*Turdus*) tend to defecate seeds on rocks and the seeds wash into cracks in soil and rubble; this minimizes desiccation and allows seedling establishment. Seeds of *J. communis* dropped on flat ground are killed and eaten by mammals. Seeds left under juniper trees die (Salmonsen, 1978).

4. Special Cases
The distinctions between escape, colonization, and directed dispersal hypotheses are not always clear. For instance, as many as 85% of the flora in true Old World deserts either possess no adaptations for dispersal, or possess adaptations to interfere with dispersal. Ellner and Shmida (1981) argue that in such severe ecological circumstances there is no advantage to dispersal, or to investment in dispersal structures. Conditions are so severe that density-independent mortality kills virtually all seeds or seedlings, precluding competition among siblings near the parent. In such circumstances adaptations are favored that reduce density-independent mortality rather than promote dispersal. This has some relevance to less severe deserts, where many plants bear fruits eaten and dispersed by fruit-eating animals (Chapter 5). In the south-western United States, the seeds of cacti (e.g. the saguaro, *Cereus giganteus* [Cactaceae]) are dispersed by animals (Steenbergh and Lowe, 1969, 1977; Turner *et al.,* 1969). Seeds falling in the open or among rocks are essentially doomed, as seeds or seedlings, to rapid density-independent desiccation or to death from mammalian browsers. Only those lodging in the shade or protection of other plants have a chance, albeit small, of survival. The movement of dispersal agents may be more likely to promote dispersal to clumps of plants which provide cover for the animals than to open areas. If so, this might be interpreted as 'directed dispersal'.

5. Distinguishing Alternatives
The foregoing discussion shows that the selective advantages of seed dispersal have not been well measured. There has been a reluctance by field workers to distinguish alternatives rather than to advocate an argument. For instance, if the purpose of dispersal is seen to be occupation of a 'safe site' (Harper 1977), then seedling recruitment due to the activities of an animal is simply defined as the fulfilled purpose. The reasoning is circular. Other patterns of seed dissemination might have been far superior from the plant perspective.

    Virola surinamensis (Myristicaceae) bears a classic bird-dispersed fruit (Fig. 1A) which is eaten by nine of 80 possible fruit-eating birds and mammals in the Panamanian rainforest (Howe, 1983). As reported above, escape from weevils confers a dramatic advantage on seeds taken even a few meters from the *Virola* crown, and a more than fortyfold advantage to seeds dropped 45 m away (Howe *et al.*, 1985). Two interesting points must be made about the phenomenon of 'escape' during the first twelve weeks after fruit drop. First, rodents eat as many seeds as *Conotrachelus* weevils, but have no influence whatever on the 'escape' effect; if anything, they kill more seeds and young seedlings away from trees than near them. The lesson here is that seed consumption by itself, may be irrelevant to any advantage to dispersal. Secondly, an earlier phase of the *Virola* study had shown that different birds take seeds different distances (Howe and Vande Kerckhove, 1981). Trogons (*Trogon massena*) and motmots (*Baryphthengus martii*) virtually always drop seeds within 20 m of the *Virola* crown, and the motmots leave most *under* the crown where seed and early seedling mortality approaches 100%. Toucans (*Ramphastos swainsonii* and *R. sulfuratus*) and guans (*Penelope purpurascens*) regurgitate or defecate the vast majority of seeds that they eat well beyond the crown, where survival is demonstrably higher (Howe *et al.*, 1985). Just as it is the weevils and not the rodents which impose the necessity of 'escape' on *Virola* populations, it is the large frugivorous birds and not the other common visitors which make seed escape possible. Continuing studies show that 'colonization' of favorable light environments provides the advantage to dispersal of those seedlings that survive the first few weeks, even though this plant is more shade-tolerant than most, and is not notably reliant on large gaps (Howe *et al.*, 1985). An important message from the *Virola* study seems to be that some birds, but not others, assist seeds in 'escaping' almost certain death immediately after fruit drop. But canopy conditions rather than proximity to the parent tree determine survival among plants that reach independence from seed reserves. Most animals which eat *Virola* fruits are not effective dispersal agents (Howe, 1977, 1980; Howe and Vande Kerckhove, 1979).

6. Summary

This review of the effects of seed dispersal on seed and seedling survival illustrates the promise of comprehensive approaches to dispersal ecology. It also underscores deficiencies of piecemeal studies. An investigator is not safe in simply counting the visits of different birds or mammals to fruiting trees. Many visitors do not remove seeds. Of those that do, some take seeds substantial distances, others do not. The importance of distance in seed dispersal depends entirely on the biology of seed and seedling survival of the plant in question. Birds or bats which take seeds long distances will have a different effect on a plant occupying a rare microsite, than on a plant with persistent seedlings occupying a common habitat, than on a plant with vulnerable seeds and seedlings in a common habitat. Similarly, a bird or bat which scatters seeds singly or in small groups will have a different effect on seedling recruitment than will a large mammal which leaves thousands or even millions of seeds in a huge mass. Suffice it to say that the integration of frugivory with plant demography has only begun for relatively well studied birds and bats and their food plants. For equally important monkeys, arboreal carnivores, and terrestrial mammals, the record is almost blank.

VI. COMMUNITY EFFECTS

Community effects of frugivory fall into two natural categories. On one hand it is useful to ask whether fruit-eating birds and mammals influence floral composition, and *vice versa*. On the other, one wants to know whether plants and dispersal agents co-evolve or simply co-occur in contemporary communities. Each topic deserves a review of its own; here I highlight critical points.

A. Community Composition

Available fruits might determine which animals can live in a community, and the dispersal agents available could determine which plants colonize and reproduce there. Implications differ for continents, remote habitat islands, and habitat patches within a given community.

1. Continents

Continents differ in floral and faunal composition. Orians (1969), Karr (1971), Crome (1978), Snow (1981) and others have noted that tropical forests in the Americas and Australasia harbour a high diversity and abundance of fruit-eating animals because fruits are available all year. In contrast, fruits are scarce or absent for much of the year in temperate forests, and consequently fruit specialists are unusual and are usually nomadic. Furthermore, African forests may have fewer berries, drupes, and compound fruits — and fewer avian fruit specialists — than other tropical forests (Snow, 1980, 1981). For instance, Snow (1980) notes that there are only 50 species of palms in Africa, compared with 1140 in America and 1150 in tropical Asia and Australia. On the scale of continental biota, frugivore diversity and abundance appear to be closely associated with the diversity and abundance of fleshy fruits.

Historically, fruit-eating animals extended the geographical ranges of many plant taxa. Recent geological history and an obviously mixed biota characterize some rich contemporary floral regions, such as south-east Asia, indicating both that angiosperms originated elsewhere, and that progenitors of the plants there had immigrated to south-east Asia from other places (Raven and Axelrod, 1974). More to the point, some genera of South American origin appeared in North America well before a land bridge connected the two continents (e.g. *Annona* [Annonaceae], *Eugenia* [Myrtaceae], and many North American plants that produce fleshy fruits spread from North America to South America after the Isthmus of Panama bridged the two continents two to three million years ago (Raven and Axelrod, 1974). Elephant-like gomphotheres and other large mammals may have carried some large-seeded legumes from Eurasia and North America to the plains, forests, and mountains of South America during this 'Great American Interchange' (Simpson, 1980; Janzen and Martin, 1982). Extremely widespread groups like the Rubiaceae are most parsimoniously viewed as resulting from dispersal by berry-eating birds (Raven and Axelrod, 1974). Historically animals must have profoundly influenced continental plant distributions.

Annual variation in animal ecologies may influence plant community composition. Many birds that breed in the temperate zone migrate south, where they feed heavily on fruits (e.g. Leck, 1972; Howe and De Steven, 1979; Howe and Vande Kerckhove, 1979; Greenberg, 1981). These birds might compete so successfully with resident tropical birds that they limit resident abundance, and influence the reproductive success of bird-dispersed trees fruiting at other times of the year (Howe, 1984b). So far, the evidence is not strong. Fruits are only occasionally limited when migrants are common (Leck, 1972; Foster, 1977; Fleming, 1979), and migrants at one South American site are a minuscule proportion of the total avifauna (Hilty, 1980b). Fleming (1979) is probably correct in discounting competition among frugivorous birds and bats during most months of the neotropical year, although competition need operate only during the worst month in a year (or a lifetime) to regulate animal community structure.

More likely, dynamics in plant reproduction influence frugivore numbers. Seasonality in flower and fruit production is well recognized in temperate climates (Thompson and Willson, 1979; Herrera, 1984a), but both also occur in tropical forest whether or not the climate is highly seasonal (Hilty, 1980; Howe and Smallwood, 1982). Strong seasonality in fruit production in some Asian forests limits the variety of highly dependent birds (Fogden, 1972); occasional fruit failure in less seasonal Central American forests can force emigration of fruit-eating parrots and toucans, and mass starvation

174   *Henry F. Howe*

of monkeys (*Alouatta*), opossums (*Didelphis*), peccaries (*Tayassu*) and other mammals (Foster, 1982b). Marked seasonality also restricts the reproductive activity of many frugivores. The breeding seasons of fruit-eating birds such as the manakins *Manacus* and *Pipra* (Snow, 1962b,c; Snow and Snow, 1964; Worthington, 1982), the oilbird and fruit pigeons of several genera (Crome, 1975, 1978) are tied to peaks of fruit production in neotropical and Australian forests. Similar congruence of breeding and fruit abundance is well known in several gerera of Central American bats (*Artibeus, Carollia, Glossophaga, Uroderma, Vampyrops, Vampyrodes,* and *Vampyressa;* see Fleming *et al.,* 1972; Morrison, 1978a; Bonaccorso, 1979), and certainly should be expected in other animals dependent on fruit for food. Frugivores probably do not compete for food when fruits are seasonally abundant (Fleming, 1979), but seasonal scarcity of fruit in Asian, African, Austral-asian, and American forest communities probably restricts faunal richness by forcing competition among obligate fruit specialists and by restricting breeding seasons (Howe, 1984b). Data are not yet good enough to tell whether frugivore richness is negatively correlated with fruiting seasonality in tropical forests with similar faunal histories.

2. Habitat Islands

The very existence of terrestrial animals and plants on remote islands of recent origin implies successful dispersal. Plant colonists of distant oceanic islands are usually highly vagile forms with fruits adapted for dispersal by water (Chapter 2), adhesion to bird feathers, or consumption by birds (Carlquist, 1974, 1981). A preponderance of highly dispersable plants, many of them with fruits consumed by birds, characterizes both real islands surrounded by ocean (Chapter 2. IV) and habitat islands surrounded by very different terrestrial habitats. Two examples illustrate the phenomenon of bird-mediated colonization of remote islands.

Geologically recent oceanic islands, which may be thousands of kilome-ters from continental floras, often harbour rich floras of highly dispersable plants. For instance, Carlquist (1966) found that many plant genera of the Hawaiian Islands bear fleshy drupes, arillate seeds, or compound fruits like those eaten by frugivorous birds in mainland forests. Most fascinating, island forms (e.g. *Tetraplasandra* [Araliaceae], *Elaeocarpus* [Elaeocarpa-ceae], *Cryptocarya*[Lauraceae], *Rubus*[Rosaceae], *Alectryon*[Sapindaceae] and many others) often exhibit gigantism, in comparison with close relatives in mainland floras. Carlquist (1966) suggests that adaptation to competitive conditions in the interior forests of tropical islands has selected for large seed size. Coincidently, these derivative plants have lost the original mode of dispersability. Other Hawaiian taxa, perhaps more recently arrived or adapted to secondary succession in island forests (e.g. *Bobea* [Rubiaceae]), retain fleshy fruits that are small and still dispersable. In general, oceanic islands bear witness to both the astonishing dispersability of fruit eaten or otherwise carried by birds, and to the adaptability of fruit and seed size of plants derived from vagile colonists.

Distinctive habitats may be ecological islands. A recent report by Sugden (1983) concerns isolated low-elevation cloud forests in the Macuira, Santa Ana, and Margarita Mountains on the Caribbean coast of Colombia and Venezuela. Each forest is an ecological island surrounded by arid lowlands, which are forbidding habitats for forest plants. Sugden finds that these forests are floristically different, but share the general features of secondary successional plants bearing fruits dispersed by birds. One would expect many bird-dispersed fruits in a neotropical cloud forest (Wheelwright, 1983, 1984, 1985a,b), but it is remarkable that these coastal mountains share some species (*Clusia rosea* [Clusiaceae], *Croton hircinus* [Euphorbiaceae], *Miconia laevigata* [Melastomataceae], *Rapanea guianensis* [Myrsinaceae], and *Guapira fragrans* [Nyctaginaceae] dispersed by birds, and found in continuous forests hundreds of kilometers away. Macedo and Prance (1978) have similarly reported that 59% of the shrubs and trees in white sand 'campinas' surrounded by vast Brazilian rain forests are dispersed by birds, although it is not clear that this incidence is higher than that which might be expected in the surrounding rainforest. As many as 90% of the trees and even more of the shrubs bear animal-dispersed fruits in other neotropical rainforests (Frankie *et al.*, 1974; Opler *et al.*, 1980; Howe and Smallwood, 1982). Certainly, the reproductive characteristics of campina and rainforest plants in the Amazon deserve more attention.

3. Local Communities

Frugivores may influence floral composition within local communities. On a general level, they may mediate successional changes that occur with either large or small disturbances. On another, frugivores may influence spatial patterns of adult and juvenile plants.

Naturalists have long recognized that successional forest habitats, even as small as light gaps caused by treefalls, are dominated by highly dispersable plants which are often disseminated by birds or bats in either temperate or tropical forests (Ridley, 1930; Richards, 1952; Snow, 1971; van der Pijl, 1972; Grubb 1977; Denslow 1980a,b). The propagules involved may arrive during and shortly after gap formation (McDonnell and Stiles, 1983), or be represented by seeds in a soil bank (Marks, 1974; see Cook, 1980) or by repressed seedlings in the understorey (Hartshorn, 1978; Ng, 1978; Brokaw, 1982). On a gross scale, almost any forest is a mosaic of plant associations, with vagile bird- or bat-dispersed species well represented in at least the early stages of local succession (see Denslow, 1980a,b)

The influence of animal frugivory on early forest succession has been acknowledged for decades, but its potential for leaving a lasting imprint on forest composition has been underrated. Residual effects may be impressive. Some colonists, such as the bird- and bat-dispersed tree *Muntingia calabura* (Elaeocarpaceae) in western Costa Rica, perform the classic early successional role of wide dissemination, dense aggregations in early successional communities, rapid reproductive maturity (< 2 years), and early demise (Fleming *et al.*, 1985). The species virtually disappears within 20 years after a colonizing event. Other colonists may last for several decades. The practised eye can distinguish old landslides from surrounding vegetation by the species composition of the forest at least 50–60 years after the event (Garwood *et al.*, 1979), and long-lived pioneers may give the aspect of forest maturity to communities with known histories of human disturbance (e.g. Foster and Brokaw, 1982). Such persistence can confound identification of 'natural' climax communities on a short human time scale.

One example of recovery from a documented human disturbance illustrates the complexity of regeneration. In Central Panama, half of Barro Colorado Island was clear-cut during the building of the Panama Canal 80 years ago, while the remainder was only selectively logged (Knight, 1975). Today, the 'young forest' has a 20 to 30 m (sometimes 40 m) canopy, while the 'old forest' (with some trees as old as 450 years) averages 10 m higher (Foster and Brokaw, 1982). As might be expected, this 'young forest' has an increasingly mature aspect, although the species composition differs substantially in the two halves of the island. Today, however, the 'young forest' contains large stands of enormous fig trees (*Ficus insipida* [Moraceae]) and other large individuals of species that are not replacing themselves (Foster and Brokaw, 1982), as well as substantial numbers of 'old forest' trees (e.g. *Virola sebifera* and *V. surinamensis* [Myristicaceae]) which were advantaged by human activity, and are reproducing both in the 'young' (Howe, 1981, 1983) and 'old' forests (Hubbell and Foster, 1983).

The point is that some 'successional' species invade and recede in short order, others persist for many decades. Persistent trees, which recruit only in large gaps, take space, and remain as potential colonists of newly disturbed areas. Other 'colonists' such as the *Virola* species, seemingly adapted to mature forest (Croat, 1978), occupy the successional habitat by virtue of vagility, ability to take advantage of either well lighted or shaded conditions (Howe and Richter, 1982; Howe *et al.*, 1985), and simple proximity. The imprint of disturbance will remain obvious on animal-dispersed 'successional' and primary forest tree distributions for well over a century.

Frugivores can influence the spatial distribution of the species whose fruits they eat, whatever the successional status of the community. Most seeds fall in leptokurtic distributions, with the mode near the parent plant

(Levin and Kerster, 1974). Animals alter the shape of the seed shadow by carrying seeds beyond the modal seed density, which is almost always directly under the parental crown (Howe and Primack, 1975; Burbidge and Whelan, 1982; Howe *et al.*, 1985). Perhaps most importantly, animals extend the tails of the seed shadow and in so doing determine the density of seeds on those tails (Fig. 6). Either distance or density may determine the likelihood that a juvenile occupies a favorable environment without competing with several or many siblings (Harper, 1977, and above).

The interplay of dispersal and seed and seedling mortality influences the spatial patterns of saplings and adult trees in several ways. A sibling cohort may occupy a gap, leading to an aggregation of trees of the same general age (see above, and Augspurger, 1983a, b). Juvenile and adult tree distributions could also reflect seed 'shadows' if mortality of seeds and seedlings is random with respect to distance from the parent. The degree of aggregation would then be entirely dependent on patterns of seed dispersal (Fig. 6). More likely, some degree of disproportionate mortality is likely to kill more seeds and seedlings near adults than farther away (Janzen, 1970). Unless mortality is extreme, however, more juveniles are likely to be found under or near parents than farther away, simply because so many more seeds fall near the parent (Hubbell, 1979, 1980). In the hypothetical example mentioned in the previous section, a seed is ten times more likely to survive 30 m from a fruiting tree than under the crown. If 10 000 seeds fall under the crown and only 100 are dispersed 30 m, there will be ten adults near the parent and one 30 m away. Over many generations, a large aggregation of saplings will accumulate near the parent in the face of 'disproportionate' seed mortality.

Although little is known of the interplay between dispersal and seed and seedling mortality, some generalizations are warranted. First, tropical tree species are usually aggregated, for whatever reason, in those Asian (Ashton, 1969) or Central American (Fleming and Heithaus, 1981; Hubbell, 1979; Hubbell and Foster, 1983) forests which have been thoroughly mapped. Secondly a high degree of aggregation of saplings and adults does not imply that the species lacks disproportionate seed or seedling mortality under and near parent plants (Howe and Smallwood, 1982; Clark and Clark, 1984; Howe *et al.*, 1985), but it does suggest the extent to which modes of dispersal influence the tail of the seed shadow, and consequently the tail of the seedling shadow. In one excellent study, Hubbell (1979) discovered that the slope of the density of conspecifics as one moved away from an adult tree was far steeper for Costa Rican trees with gravity-dispersed, and even wind-dispersed seeds, than for species with bird- or bat-dispersed seeds (Fig. 7). In other words, the 'sapling' and 'adult shadows' were less peaked and had longer tails for bird- and bat-dispersed plants than for

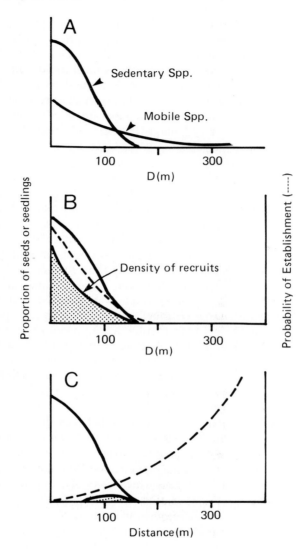

Fig. 6. Hypothetical relationships between seed deposition and distance (D) from fruiting trees for 'seed shadows' created by vertebrates that scatter seeds rather than leave them in massive clumps. (A) 'seed shadows', representing the density of seeds as a function of distance; (B) seedling or sapling density where there is little mortality near the parent; (C) seedling or sapling density where mortality is disproportionately high near the parent (from Fleming and Heithaus, 1981; see also Janzen, 1970).

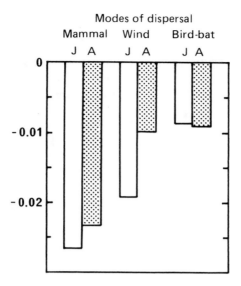

Fig. 7. Slopes of adult (A, closed bars) and juvenile (J, open bars) density as a function of distance from putative parent trees in nine mammal-dispersed, twelve wind-dispersed, and nine bird- or bat-dispersed species in Costa Rican dry forest. The analysis shows that tree species dispersed by terrestrial mammals are far more clumped than those dispersed by wind, which are in turn more clumped (at least as saplings) than those dispersed by the more mobile birds and bats (from Hubbell, 1979).

others in a complex tropical forest. Finally, the general results that seeds actually do suffer disproportionate mortality under many forest trees (Table V) and that there is more competition between conspecific seedlings than seedlings of different species (Connell *et al.*, 1984) suggest that the mixing of seed and seedling cohorts, which results from seed dispersal by animals, actually does increase plant species diversity. At the very least, there is now a far stronger rationale for examining the relationships between dispersal process and seed and seedling demography than there was when Janzen (1970) and Connell (1971) suggested that the link was important.

B. Co-evolution or Co-occurrence?

Fruits may evolve attractive or defensive characteristics in response to selection from one particularly important dispersal agent, or in response to a group of dispersal agents that affect the plant in more or less the same way. Conversely, frugivore traits may conceivably adapt to one fruit, or to many of similar size, color, and nutritional quality. A range of possibilities exists between co-evolution of species pairs and 'diffuse co-evolution' of guilds of plant and animal species (Janzen, 1980). Several lines of evidence indicate that co-evolution between plants and their dispersal agents are of the 'diffuse' sort in all but the most unusual circumstances (Howe and Smallwood, 1982; Howe, 1984a).

First, most fruits are eaten and dispersed by many animals, not one or a few species. The sections on foraging for fruits (Section III) and fruit syndromes (Section IV) indicate that frugivores do select among fruits on a number of criteria, but that many fruits (e.g. green figs, small red *Muntingia* berries) are eaten by a variety of birds and mammals (Morrison, 1978a; Fleming *et al.*, 1985). Even apparent dependence by a tree on one or a few dispersal agents is likely to be a local phenomenon (Howe and Estabrook, 1977). For instance, *Casearia corymbosa* (Flacourtiaceae) is primarily dispersed by a cotinga (*Tityra semifasciata*, thought to be a flycatcher by some) in Costa Rican rainforest (Howe, 1977), and by a vireo (*Vireo flavoviridis*) in dry forest (Howe and Vande Kerckhove, 1979). Within a single Panamanian rainforest, monkeys visited some *Tetragastris panamensis* trees and birds visited others (Howe, 1980) Extensive observations at the same site showed that most fruiting *Virola sebifera* trees are visited by toucans (*Ramphastos swainsonii* and *R. sulfuratus*) and tityras (*Tityra semifasciata*), but that motmots (*Baryphthengus martii*), trogons (*Trogon massena*), and aracaris (*Pteroglossus torquatus*) visit some trees and not others (Howe, 1981). There are no obvious differences in the influence of frugivory by these birds on *V. sebifera* seed and seedling survival, although subtle differences might eventually show up (Howe *et al.*, 1985). The point is that seed dispersal in both cases depends on several or many animals of very different taxonomic affinities. A similar prevalence of 'diffuse' interactions between temperate fruiting plants and birds has been discussed in depth by Herrera (1982a, 1984a, b), and is at least implicit in interactions between tropical trees and shrubs and bats (Fleming *et al.*, 1977; Morrison, 1978a) between neotropical trees and arboreal frugivores (Hladik and Hladik, 1969; Howe, 1980; Charles-Dominique *et al.*, 1981) and between west African trees and ruminants, rodents, and primates (Gautier-Hion *et al.*, 1980, 1985).

Secondly, ecological variation in the distribution and abundance of fruiting plants and animals, and dramatic spatial and seasonal differences in fruit production, make it virtually impossible for pairwise co-evolution to occur among plants and dispersal agents (Howe, 1984a). Sometimes either a plant or its animal mutualist may be specialized to use another organism which relies on many other species. Such asymmetries occur in dispersal ecology (Howe, 1977), and are common in pollination biology (Schemske, 1983), but 'diffuse' relationships seem to be the rule in dispersal ecology (Wheelwright and Orians, 1982, Janzen, 1983a).

Thirdly, the time frame of evolutionary change suggests that co-evolution between pairs of plant and disperser species is unlikely. Snow (1981) rightly points out that some bird-dispersed African trees occur in fossil strata that predate the animal taxa that now eat their fruits. Herrera (1985a) notes that plant taxa last 27 to 38 million years in the fossil record, while bird and mammal taxa last 0.5 to 4 million years. The reproductive characteristics of plants are likely to be conservative (Chaper 7, V, B); frugivores adjust to whatever plants are available, rather than the other way around. Similar asymmetries occur in most pollination systems in which animals play a role (Schemske, 1983).

In short, fruiting plants probably evolve traits acceptable to some array of dispersal agents (e.g. birds, bats, other mammals), but this array is seldom if ever as narrow as a species (Howe and Smallwood, 1982; Howe, 1984a; Herrera, 1985a). Pairwise co-evolution may occur in depauperate habitats, such as deserts or isolated islands, where interactions are forced and intense (Howe and Vande Kerckhove, 1979), but will almost certainly be the exception rather than the rule.

VII. SUMMARY AND CONCLUSION

This review of seed dispersal by fruit-eating birds and mammals has documented a small fraction of the enormous variety of relevant observations, has sketched the conceptual frameworks used to study dispersal phenomena, and has discussed in some detail the difficulties of teasing apart the causes and consequences of dispersal process and mode. The hope is that such a critical review will complement more intuitive reviews of the subject (Janzen, 1983b) by extending a pragmatic observational and experimental approach in ecology to the study of seed dispersal by vertebrates. In that light, some closing comments are in order.

The organization of the chapter reflects inevitable biases of perspective. Relevant studies usually concern foraging animals, fruiting plants, or are necessarily diffuse investigations of entire communities of animals and plants. Caveats go beyond these inevitable biases of perspective, however. Studies of frugivory and seed dispersal mix anecdote, conjecture, quantitative biology, and theory to an extent that is tolerated only in an embryonic science, still groping for plausible generalizations and testable predictions (Lakatos, 1970). Almost none of the empirical papers discussed here meets the standards of sample size, controlled experimentation, or replication that are possible — and even necessary — in more developed arenas in ecology (Hurlburt, 1984). Yet the work reviewed here is among the most quantitative available. The fact is that many seminal papers rely on little more than casual anecdotes of bird or monkey activity, often at one or two individual trees. Imagine a science of chemistry in which a few observations of a single proton in one medium are used to describe the properties of hydrogen! A chemist, used to interpreting the results of interactions between millions of atoms, might not even be impressed by data on the activity of 25 protons in each of ten media. Yet this is a large number of samples and replicates in dispersal ecology. Limitations of studying large shrubs or trees and furtive animals notwithstanding, the practical aspects of studying frugivory and seed dispersal have at least as far to go as they have come in the past decade.

Finally, some fundamental biological issues have yet to be addressed. This review has emphasized the interplay of frugivory and local plant demography in order to understand the whys and wherefores of animal activity and plant traits. This is justifiable both on the grounds of completeness and because so many animal-dispersed plants are endemics or have restricted distributions (Snow, 1981; Prance, 1982; Steyermark, 1982). But other animal-dispersed plants are astonishingly widespread (Ridley, 1930; Raven and Axelrod, 1974; Carlquist, 1981; Sugden, 1983). It is intriguing to wonder whether those berries that colonize a distant island, mountaintop, or even continent are substantially different in quality or presentation from conspecific berries ignored by the errant bird which mediates the colonization event. For instance, individual trees within a population vary enormously in the size of the fruit crop (Howe and Vande Kerckhove, 1979; Howe, 1982, 1983; Bullock, 1982); a very large crop might satiate local birds but attract a few critical transients. If so, the processes that enhance seed dispersal in a local community may or may not be associated with those that enhance the chance dispersal to distant islands or continents. Now that the integration of frugivore activity and plant demography is underway, the next productive synthesis may be between consequences of local and long-distance dispersal.

ACKNOWLEDGEMENTS

I am grateful to L. Emmons, R. Black, J. Garcia, S. Hendrix, A. Herre, A. Pascual-Diaz, L. Westley and M. Turner for comments on the manuscript. This research was supported by the National Science Foundation (U.S.A.) and the University of Iowa.

REFERENCES

Alexandre, D. Y. (1978). *Terre et Vie* **32**, 47–71.

Ali, S. and Ripley, S. D. (1983) 'Birds of India and Palestine'. Oxford University Press, Delhi.

Ashton, P. (1969). *Biol. J. Linn. Soc.* **1**, 155–196.

Augspurger, C. K. (1981). *Ecology* **62**, 775–788.

Augspurger, C. K. (1983a). *Oikos* **40**, 189–196.

Augspurger, C. K. (1983b). *J. Ecol.* **71**, 759–772.

August, P. V. (1981). *Biotropica (Supplement)* **13**, 70–76.

Ayensu, E. S. (1974). *Annals of the Missouri Botanical Garden* **61**, 702–727.

Becker, P. (1983). *In* 'Tropical Rain Forest: Ecology and Management' (S. L. Sutton, T. C. Whitmore and A. C. Chadwick, eds), pp. 241–252. Blackwell Scientific, Oxford.

Becker, P. and Wong, M. (1985). *Biotropica* **17**, 230–237.

Beehler, B. (1983). *Auk* **100**, 1–12.

Bonaccorso, F. J. (1979). *Bulletin of the Florida State Museum* **24**, 359–408.

Bonaccorso, F. J., Glanz, W. E. and Sanford C. M. (1980). *Revista de Biologia Tropical* **28**, 61–72.

Brokaw, N. V. L. (1982). *In* 'The Ecology of a Tropical Forest: Seasonal Rhythms and Long-term Changes' (E. G. Leigh, Jr., A. S. Rand and D. M. Windsor, eds), pp. 101–108. Smithsonian Press, Washington.

Bullock, S. H. (1982). *Oecologia* **55**, 238–242.

Burbidge, A. H. and Whelan, R. J. (1982). *Aust. J. Ecol.* **7**, 63–67.

Carlquist, S. (1966). *Brittonia* **18**, 310–335.

Carlquist, S. (1974). 'Island Biology'. Columbia University Press, New York.

Carlquist, S. (1981). *American Scientist* **69**, 509–516.

Charles-Dominique, P., Atramentowicz, M., Charles-Dominique, M., Gerard, H., Hladik, A., Hladik, C. M. and Prevost, M. F. (1981). *Terre et Vie* **35**, 341–435.

Clark, D. A. and Clark, D. B. (1981). *Oecologia* **49**, 73–75.

Clark, D. A. and Clark, D. B. (1984). *Am. Nat.* **124**, 769–788.

Connell, J. H. (1971). *In* 'Dynamics of Populations' (P. J. Den Boer and G. R. Gradwell, eds), pp. 298–312. PUDOC, Wageningen, Netherlands.

Connell, J. H., Tracey, J. G. and Webb, L. J. (1984). *Ecological Monographs* **54**, 141–164.

Cook, R. (1980). *In* 'Demography and Evolution in Plant Populations' (O. Solbrig, ed.), pp. 107–130. University of California Press, Berkeley.

Corner, E. J. H. (1949). *Ann. Bot.* **13**, 367–414.

Croat, T. B. (1978). 'The Flora of Barro Colorado Island'. Stanford University Press, Stanford, California.

Crome, F. H. J. (1975). *Aust. Wildlife Res.* **2**, 155–185.

Crome, F. H. J. (1976). *Emu* **76**, 8–14.

Crome, F. H. J. (1978). *Aust. J. Ecol.* **3**, 195–212.

Culver, D. C. and Beattie, A. J. (1980). *Am. J. Bot.* **67**, 710–714.

Davidar, P. (1983a). *Biotropica* **15**, 32–37.

Davidar, P. (1983b). *Oecologia* **60**, 271–273.

Davidson, D. W. and Morton, S. R. (1981). *Science* **213**, 1259–1261.

Denslow, J. S. (1980a). *Biotropica* (Supplement) **12**, 47–55.

Denslow, J. S. (1980b). *Oecologia* **46**, 18–21.

Denslow, J. S. and Moermond, T.C. (1982). *Oecologia* **54**, 170–174.

Diamond, J. M. (1973). *Science* **179**, 759–769.

Docters van Leeuwen, W. M. (1954). *Beaufortia* **41**, 105–206.

Douglas-Hamilton, I., and Douglas-Hamilton, O. (1975). 'Among the Elephants'. Collins & Harvill Press, London.

Ellner, S. and Shmida, A. (1981). *Oecologia* **51**, 133–144.

Estrada, A. and Coates-Estrada, R. (1984). *Am. J. Primatol.* **6**, 77–91.

Fleming, T. H. (1979). *Am. Zool.* **19**, 1157–1172.

Fleming, T. H. (1981). *Oecologia* **51**, 42–46.

Fleming, T. H. (1982). *In* 'Ecology of Bats' (T. H. Kunz, ed.), pp. 287–325. Plenum Press, New York.

Fleming, T. H. and Heithaus, E. R. (1981). *Biotropica* (Supplement) **13**, 45–53.

Fleming, T. H., Heithaus, E. R. and Sawyer, W. B. (1977). *Ecology* **58**, 619–627.

Fleming, T. H., Hooper, E. T. and Wilson, D. E. (1972). *Ecology* **53**, 555–569.

Fleming, T. H., Williams, C. F., Bonaccorso, F. J. and Herbst, L. H. (1985). *Am. J. Bot.*

Fogden, M. P. L (1972). *Ibis* **114**, 307–343.

Foster, M. S. (1977). *Ecology* **58**, 73–85.

Foster, M. S. (1978). *Tropical Ecol.* **19**, 131–154.

Foster, R. B. (1982a). *In* 'The Ecology of a Tropical Forest: Seasonal Rhythms and Long-term Changes' (E. G. Leigh, A. S. Rand and D. S. Windsor, eds), pp. 151–172. Smithsonian Press, Washington.

Foster, R. B. (1982b). *In* 'The Ecology of a Tropical Forest: Seasonal Rhythms and Long-term Changes' (E. G. Leigh, A. S. Rand and D. S. Windsor, eds), pp. 201–212. Smithsonian Press, Washington.

Foster, R. B. and Brokaw, N. V. L. (1982). *In* 'The Ecology of a Tropical Forest: Seasonal Rhythms and Long-term Changes' (E. G. Leigh, Jr., A. S. Rand and D. M. Windsor, eds), pp. 67–82. Smithsonian Press, Washington.

Frankie, G. W., Baker, H. G. and Opler, P. A.. (1974). *J. Ecol.* **62**, 881–919.

Frith, H. J., Crome, F. H. J. and Wolfe, T. O. (1976). *Emu* **76**, 49–58.

Frost, P. G. H. (1980). *In* 'Fruit-frugivore Interactions in a South African Coastal Dune Forest' (R. Nohring, ed.), pp. 1179–1184. Acta XVII Congressus Internationalis Ornithologici, Vol. II. Berlin.

Garwood, N. C. (1982). *In* 'The Ecology of a Tropical Forest: Seasonal Rhythms and Long-term Changes' (E. G. Leigh, A. S. Rand, D. M. Windsor, eds), pp. 173–185. Smithsonian Press, Washington.

Garwood, N. C., Janos, D. P. and Brokaw, N. (1979). *Science* **205**, 997–999.

Gautier-Hion, A., Emmons, L. H. and Dubost, G. (1980). *Oecologia* **45**, 182–189.

Gautier-Hion, A., Duplantier, J. M., Quris, R., Feer, F., Sourd, C., Decoux, J.-P., Dubost, G., Emmons, L. H., Erard, C., Hecketsweiler, P., Moungazi, A., Roussilhon, C. and Thiollay, J.-M. (1985). *Oecologia* **65**, 324–337.

Glanz, W. (1982). *In* 'The Ecology of a Tropical Forest: Seasonal Rhythms and Long-term Changes' (E. G. Leigh, A. S. Rand and D. M. Windsor, eds), pp. 455–468. Smithsonian Press, Washington.

Gleason, S. K. (1981). *Oecologia* **51**, 294–295.

Glyphis, J. P., Milton, S. J. and Siegfried, W. R. (1981). *Oecologia* **48**, 138–141.

Gould, J. (1967). 'Birds of Australia'. Methuen, London.

Goulding, M. (1980). 'The Fishes and the Forest'. University of California Press, Berkeley.

Grant, P. R., Smith, J. N. M., Grant, B., Abbott, R. and Abbott, L. K. (1975). *Oecologia* **19**, 239–259.

Greenberg, R. (1981). *Biotropica* **13**, 215-223.

Gross, K. L. and Werner, P. A. (1982). *Ecology* **63**, 921–921.

Grubb, P. J. (1977). *Biol. Rev.* **52**, 107–145.

Harper, J. L. (1977). 'The Population Biology of Plants'. Academic Press, New York.

Hartshorn, G. S. (1978). *In* 'Tropical Trees as Living Systems' (R. B. Tomlinson and M. H. Zimmermann, eds), pp. 617–638. Cambridge University Press, Cambridge.

Heithaus, E. R. (1982). *In* 'Ecology of Bats' (T. H. Kunz, ed.), pp. 327–367. Plenum Press, New York.

Heithaus, E. R. and Fleming, T. H. (1978). *Ecological Monographs* **48**, 127–143.

Heithaus, E. R., Fleming, T. H. and Opler, P. A. (1975). *Ecology* **56**, 841–854.

Herrera, C. M. (1981a). *Am. Nat.* **118**, 132–144.

Herrera, C. M. (1981b). *Oikos* **36**, 51–58.

Herrera, C. M. (1982a). *Ecology* **63**, 773–785.

Herrera, C. M. (1982b). *Am. Nat.* **120**, 819–822.

Herrera, C. M. (1982c). *Am. Nat.* **120**, 218–241.

Herrera, C. M. (1984a). *Ecological Monographs* **54**, 1–23.

Herrera, C. M. (1984b). *Ecology* **65**, 609–617.

Herrera, C. M. (1984c). *Oecologia* **63**, 386–393.

Herrera, C. M. (1985a). *Oikos* **44**, 132–141.

Herrera, C. M. (1985b). *In* 'Habitat Selection in Birds' (M. L. Coor, ed.), pp. 000–000, Academic Press, New York.

Herrera, C. M. and Jordano, P. (1981). *Ecological Monographs* **51**, 203–218.

Hilty, S. L. (1980a). *Biotropica* **12**, 292–306.

Hilty, S. L. (1980b). *In* 'Migrant Birds in the Neotropics' (A. Keast and E. S. Morton, eds.), pp. 265–272. Smithsonian Press, Washington.

Hladik, A. and Hladik, C. M. (1969). *Terre et Vie* **23**, 25–117.

Hladik, C. M. (1967). *Mammalia* **31**, 120–147.

Hladik, C. M., Haldik, A., Bousset, J., Valdebouze, P., Viroben, G. and Delort-Laval, J. (1971). *Folia Primatologica* **16**, 85–122.

Howe, H. F. (1977). *Ecology* **58**, 539–550.

Howe, H. F. (1979). *Am. Nat.* **114**, 925–931.

Howe, H. F. (1980). *Ecology* **61**, 944–959.

Howe, H. F. (1981). *Auk* **98**, 88–98.

Howe, H. F. (1982). *In* 'The Ecology of a Tropical Forest: Seasonal Rhythms and Long-term Changes' (E. G. Leigh, A. S. Rand and D. Windsor, eds.), pp. 189–200. Smithsonian Press, Washington.

Howe, H. F. (1983). *In* 'Tropical Rain Forest: Ecology and Management' (S. L. Sutton, T. C. Whitmore and A. C. Chadwick, eds.), pp. 211–227. Blackwell Scientific, Oxford.

Howe, H. F. (1984a). *Am. Nat.* **123**, 764–777.

Howe, H. F. (1984b). *Biological Conservation* **30**, 261–281.

Howe, H. F. (1985). *Am. Nat.* **125**, 853–865.

Howe, H. F. and DeSteven, D. (1979). *Oecologia* **39**, 185–196.

Howe, H. F. and Estabrook, G. F. (1977). *Am. Nat.* **111**, 817–832.

Howe, H. F. and Primack, R. B. (1975). *Biotropica* **7**, 278–283.

Howe, H. F. and Richter, W. (1982). *Oecologia* **53**, 347–351.

Howe, H. F. and Smallwood, J. (1982). *Annu. Rev. Ecol. Syst.* **13**, 201–228.

Howe, H. F. and Vande Kerckhove, G. A. (1979). *Ecology* **60**, 180–189.

Howe, H. F. and Vande Kerckhove, G. A. (1981). *Ecology* **62**, 1093–1106.

Howe, H. F., Schupp, E. W. and Westley, L. C. (1985). *Ecology* **66**, 781–791.

Hubbell, S. P. (1979). *Science* **203**, 1299–1309.

Hubbell, S. P. (1980). *Oikos* **35**, 214–229.

Hubbell, S. P. and Foster, R. (1983). *In* 'The Tropical Rain Forest: Ecology and Management' (S. L. Sutton, T. C. Whitmore and A. C. Chadwick, eds.), pp. 25–41. Blackwell Scientific, Oxford.

Hurlbert, S. H. (1984). *Ecological Monographs* **54**, 187–211.
Jackson, G. and Gartlan, J. S. (1965). *J. Ecol.* **53**, 573–597.
Janson, C. (1983). *Science* **219**, 187–189.
Janzen, D. H. (1969). *Evolution* **23**, 1–27.
Janzen, D. H. (1970). *Am. Nat.* **104**, 501–528.
Janzen, D. H. (1971). *Annu. Rev. Ecol. Syst.* **2**, 465–492.
Janzen, D. H. (1980). *Evolution* **34**, 611–612.
Janzen, D. H. (1981a). *Auk* **98**, 841–844.
Janzen, D. H. (1981b). *Biotropica (Supplement)* **13**, 59–63.
Janzen, D. H. (1981c). *Ecology* **62**, 587–592.
Janzen, D. H. (1981d). *Ecology* **62**, 593–601.
Janzen, D. H. (1982a). *Oikos* **38**, 150–156.
Janzen, D. H. (1982b). *Ecology* **63**, 1887–1900.
Janzen, D. H. (1983a). *Biol. J. Linn. Soc.* **20**, 103–113.
Janzen, D. H. (1983b). *In* 'Coevolution' (D. Futuyma and M. Slatkin, eds.), pp. 232–262. Sinauer Associates, Sunderland, Massachusetts.
Janzen, D. H. and Martin, P. (1982). *Science* **215**, 19–27.
Janzen, D. H., Miller, G. A., Hackforth-Jones, J., Pond, C. M., Hooper, K. and Janos, D. P. (1976). *Ecology* **57**, 1068–1075.
Jarman, P. J. and Sinclair, A. R. E. (1979). *In* 'Serengeti: Dynamics of an Ecosystem' (A. R. E. Sinclair and M. Norton-Griffiths, eds.), pp. 130–163. University of Chicago Press, Chicago.
Jenkins, R. (1969). 'Ecology of Three Species of Saltators in Costs Rica with Special Reference to Their Frugivorous Diet'. Ph.D. Dissertation, Harvard University, Cambridge, Massachusetts.
Jordano, P. (1984). *Oikos* **43**, 149–153.
Jordano, P. and Herrera, C. M. (1981). *Ibis* **123**, 502–507.
Karr, J. R. (1971). *Ecological Monographs* **41**, 207–233.
Keast, A. (1958). *Emu* **58**, 195–206.
Kendeigh, S. C., Dol'nik, V. R. and Gavrilov, V. M. (1977). Avian energetics. *In* 'Granivorous Birds in Ecosystems' (J. Pinkowski and S. C. Kendeigh, eds.), pp. 127–204. IBP Programme 12. Cambridge University Press, Cambridge.
Kiltie, R. A. (1981). *Biotropica* **13**, 234–236.
Kiltie, R. A. (1982). *Biotropica* **14**, 188–195.
Klein, L. L. and Klein, D. B. (1977). *In* 'Primate Ecology' (T. H. Clutton-Brock, ed.), pp. 153–182. Academic Press, London.
Knight, D. H. (1975). *Ecological Monographs* **45**, 259–284.
Knight, R. S. and Siegfried, W. R. (1983). *Oecologia* **56**, 405–412.
Lakatos, I. (1970). *In* 'Criticism and the Growth of Knowledge' (I. Lakatos and A. Musgrave, eds.), 91–196. Cambridge University Press, Cambridge.
Lamprey, H. F., Halevy, G. and Makacha, S. (1974). *East Afr. Wildlife J.* **12**, 81–85.
Leck, C. G. (1972). *Condor* **74**, 54–60.
Leigh, E. G., and Windsor, D. M. (1982). *In* 'The Ecology of a Tropical Forest: Seasonal Rhythms and Long-term Changes' (E. G. Leigh, Jr., A. S. Rand and D. M. Windsor, eds), pp. 111–122. Smithsonian Press, Washington, D.C.
Leighton, M. and Leighton, D. R. (1982). *Biotropica* **14**, 81–90.
Leighton, M. and Leighton, D. R. (1983). *In* 'Tropical Rain Forest: Ecology and Management' (S. L. Sutton, T. C. Whitmore and A. C. Chadwick, eds), pp. 181–196. Blackwell Scientific, Oxford.
Levey, D. J., Moermond, T. C. and Denslow, J. S. (1984). *Ecology* **65**, 844–850.

Levin, D. A. and Kerster, H. W. (1974). *Evolutionary Biology* 7, 139-220.

Lieberman, D., Hall, J. B., Swaine, M. D. and Lieberman, M. (1979). *Ecology* 60, 65-75.

Livingston, R. B. (1972). *Ecology* 53, 1141-1147.

MacArthur, R. H. and Wilson, E. O. (1967). 'The Theory of Island Biogeography'. Princeton University Press, Princeton.

Macedo, M. and Prance, G. T. (1978). *Brittonia* 30, 203-215.

Marks, P. L. (1974). *Ecological Monographs* 44, 73-88

Martin, T. E. (1985). *Oecologia* 66, 563-573.

McDiarmid, R., Ricklefs, R. E. and Foster, M. S. (1977). *Biotropica* 9, 9-25.

McDonnell, M. J. and Stiles, E. W. (1983). *Oecologia* 56, 109-116.

McKey, D. (1975). *In* 'Coevolution of Animals and Plants' (L. E. Gilbert and P. Raven, eds), pp. 159-191. University of Texas Press, Austin.

McKey, D. and Waterman, P. G. (1982). *Folia Primatologica* 39, 264-304.

McKey, D., Gartlan, J. S., Waterman, P. G. and Choo, G. M. (1981). *Biol. J. Linn. Soc.* 16, 115-146.

Milkman, R. (1982). *In* 'Perspectives on Evolution' (R. Milkman, ed.), pp. 105-118. Sinauer Associates, Sunderland, Massachusetts.

Milton, K. (1980). 'The Foraging Strategy of Howler Monkeys: A Study in Primate Economics'. Columbia University Press, New York.

Milton, K. (1981). *Am. Nat.* 117, 496-505.

Moermond, T. C. and Denslow, J. S. (1983). *J. Animal Ecol.* 52, 407-421.

Moermond, T. C. and Denslow, J. S. (1985). *In* 'Neotropical Ornithology' (P. A. Buckley, M. S. Foster, E. S. Morton, R. S. Ridgely and N. G. Smith, eds), pp. 865-897. A.O.U. Monographs.

Morrison, D. (1978a). *Ecology* 59, 716-723.

Morrison, D. (1978b). *Animal Behaviour* 26, 852-855.

Morton, E. S. (1973). *Am. Nat.* 107, 8-22.

Morton, E. S. (1977). *Auk* 94, 97-106.

Ng, F. S. P. (1978). *In* 'Tropical Trees as Living Systems' (P. B. Tomlinson and M. H. Zimmermann, eds), pp. 187-208. Cambridge University Press, Cambridge.

Nobel, J. C. (1975). *J. Ecol.* 63, 979-984.

Opler, P. A., Frankie, G. W. and Baker, H. G. (1980). *J. Ecol.* 68, 167-188.

Oppenheimer, J. R. (1982). *In* 'The Ecology of a Tropical Forest: Seasonal Rhythms and Long-term Changes' (E. G. Leigh, Jr., A. S. Rand and D. M. Windsor, eds), pp. 253-272. Smithsonian Press, Washington.

Orians, G. H. (1969). *Ecology* 50, 783-801.

Owadally, A. W. (1979). *Science* 203, 1363-1364.

Platt, W. J. (1976). *Oecolgia* 22, 399-409.

Poole, R. W. and Rathcke, B. J. (1979). *Science* 203, 470-471.

Prance, G. T. (1982). *In* 'Biological Diversification in the Tropics' (G. T. Prance, ed.), pp. 137-157. Columbia University Press, New York.

Prance, G. T. and Mori, S. A. (1978). *Brittonia* 30, 21-33.

Pratt, T. K. (1983a). *Emu* 82, 283-285.

Pratt, T. K. (1983b). *Am. Nat.* 122, 797-805.

Pratt, T. K. (1984). *Condor* 86, 123-129.

Pullainen, E., Helle, P. and Tunkkari, P. (1981). *Ornis Fennica* 58, 21-28.

Pyke, G. H., Pulliam, H. R. and Charnov, E. L. (1977). *Quart. Rev. Biol.* 52, 137-154.

Ramirez, W. (1970). *Evolution* 24, 680-691.

Raven, P. H. and Axelrod, D. I. (1974). *Annals of the Missouri Botanical Garden* 61, 539-673.

Regal, P. J. (1977). *Science* **196**, 622–629.

Richards, P. W. (1952). 'The Tropical Rain Forest'. Cambridge University Press, Cambridge.

Ridley, H. N. (1930). 'The Dispersal of Plants Throughout the World'. Reeve, Ashford.

Russell, J. K. (1982). *In* 'The Ecology of a Tropical Forest: Seasonal Rhythms and Long-term Changes' (E. G. Leigh, Jr., A. S. Rand and D. M. Windsor, eds), pp. 413–431. Smithsonian Press, Washington.

Salmonson, M. G. (1978). *Oecologia* **32**, 333–339.

Salmonson, M. G. and Balda, R. P. (1977). *Condor* **79**, 148–181.

Sarukhan, J., Martinez-Ramos, M. and Pinero, D. (1984). *In* 'Perspectives on Plant Population Ecology' (R. Dirzo and J. Sarukhan, eds), pp. 83–106. Sinauer Associates, Sunderland, Massachusetts.

Schemske, D. W. (1983). *In* 'Coevolution' (M. H., Nitecki, ed.), pp. 67–111. University of Chicago Press, Chicago.

Schoener, T. W. (1971). *Annu. Rev. Ecol. Syst.* **2**, 369–404.

Simpson, G. G. (1980). 'Splendid Isolation: The Curious History of South American Mammals'. Yale University Press, New Haven.

Snow, B. K. and Snow, D. W. (1984). *Ibis* **126**, 39–49.

Snow, D. W. (1962a). *Zoologica* **47**, 199–221.

Snow, D. W. (1962b). *Zoologica* **47**, 65–104.

Snow, D. W. (1962c). *Zoologica* **47**, 183–198.

Snow, D. W. (1965). *Oikos* **15**, 274–281.

Snow, D. W. (1971). *Ibis* **113**, 194–202.

Snow, D. W. (1980). *In* 'Regional differences between tropical floras and the evolution of frugivory' (R. Nohring, ed.), pp. 1193–1198. Acta XVII Congressus Internationalis Ornithologici, Berlin.

Snow, D. W. (1981). *Biotropica* **13**, 1–14.

Snow, D. W. (1982). 'The Cotingas' Cornell University Press, Ithaca, New York.

Snow, D. W. and Snow, B. K. (1964). *Zoologica* **49**, 1–39.

Sorensen, A. E. (1981). *Oecologia* **50**, 242–249.

Sorensen, A. E. (1983). *Oecologia* **56**, 117–120.

Sorensen, A. E. (1984). *J. Animal Ecol.* **53**, 545–557.

Sork, V. L. (1985). *Biotropica* **17**, 130–136.

Sporne, K. R. (1965). 'The Morphology of Gymosperms'. Hutchinson University Library, London.

Stebbins, G. L. (1974). 'Flowering Plants: Evolution Above the Species Level'. Harvard University Press, Cambridge, Massachusetts.

Steenbergh, W. F. and Lowe, C. H. (1969). *Ecology* **50**, 825–834.

Steenbergh, W. F. and Lowe, C. H. (1977). 'Ecology of the Saguaro: II. Reproduction, Germination, Establishment, Growth and Survival of the Young Plant'. National Park Service Scientific Monograph Series, No. 8. Washington D.C.

Stiles, E. W. (1980). *American Naturalist* **116**, 670–688.

Steyermark, J. A. (1982). *In* 'Biological Diversification in the Tropics' (G. T. Prance, ed.), pp. 182–220. Columbia University Press, New York.

Stocker, G. C. and Irvine, A. K. (1983). *Biotropica* **15**, 170–176.

Sugden, A. M. (1983). *In* 'Tropical Rain Forest: Ecology and Management' (S. L. Sutton, T. C. Whitmore and A. C. Chadwick, eds), pp. 43–56. Blackwell Scientific, London.

Temple, S. A. (1977). *Science* **197**, 885–886.

Temple, S. A. (1979). *Science* **203**, 1364.

Thompson, J. N. and Willson, M. F. (1979). *Evolution* **33**, 973–982.

Turcek, F. J. (1963). *Proceedings XIII International Ornithological Congress:* 285–292

Turner, R. M., Alcorn, S. M. and Olin, G. (1969). *Ecology* **50**, 835–844.

van der Pijl, L. (1972). 'Principles of Dispersal in Higher Plants', Second Edition. Springer-Verlag, Berlin.

Walker, E. P. (1975). 'Mammals of the World, Volumes I and II'. Third Edition. Johns Hopkins University Press, Baltimore.

Walsberg, G. E. (1975). *Condor* **77**, 169–174.

Walsberg, G. E. (1977). *University of California Publications in Zoology* **108**, 1–63.

Werner, P. A. (1976). *Systematic Botany* **1**, 246–268.

Wheelwright, N. T. (1983). *Auk* **100**, 286–301.

Wheelwright, N. T. (1985a). *Ecology* **66**, 808–818.

Wheelwright, N. T. (1985b). *Oikos* **44**, 465–477

Wheelwright, N. T. and Orians, G. (1982). *Am. Nat.* **119**, 402–413

Wheelwright, N. T., Haber, W. A., Murray, K. G. and Guindon, C. (1984). *Biotropica* **16**, 173–192.

White, S. C. (1974). 'Ecological Aspects of Growth and Nutrition in Tropical Fruit-eating Birds'. Ph.D. Dissertation, University of Pennsylvania, Philadelphia.

Williams, G. (1966). 'Adaptation and Natural Selection: A Critique of Some Current Evolutionary Thought'. Princeton University Press, Princeton.

Williams, G. (1975). 'Sex and Evolution'. Princeton University Press, Princeton.

Willson, M. F. and Melampy, M. N. (1983). *Oikos* **41**, 27–31.

Willson, M. F. and Thompson, J. N. (1982). *Can. J. Bot.* **60**, 701–713.

Worthington, A. (1982). *In* 'The Ecology of a Tropical Forest: Seasonal Rhythms and Long-term Changes" (E. G. Leigh, Jr., A. S. Rand and D. M. Windsor, eds), pp. 343–430. Smithsonian Press, Washington.

Wright, S. J. (1983). *Ecology* **64**, 1016–1021.

CHAPTER 5

Rodents as Seed Consumers and Dispersers

M. V. PRICE AND S. H. JENKINS

I.	Introduction	191
II.	Granivorous Rodents and the Fates of Seeds	192
	A. Foraging Behavior	192
	B. The Seed Fate Diagram	193
	C. Utility of the Fate Diagram	196
III.	Determinants of the Fate Path	197
	A. Probability of Being Harvested	197
	B. Probability of Being Cached	205
	C. Fate of Seed Caches	211
	D. Probability of Becoming Established	215
VI.	Net Effect of Granivorous Rodents	221
	A. Consequences for Populations	221
	B. Evolutionary Consequences	226
V.	Directions for Further Research	230
	Acknowledgements	231
	References	231

I. INTRODUCTION

When rodents radiated during the Eocene they displaced several similar groups dating from the Paleocene and soon became the most diverse group of mammals. Today the order Rodentia contains 1687 named species, one-third of all mammals, grouped into approximately 45 families and 354 genera (Kurten, 1971; Eisenburg, 1981).

All rodents are equipped with chisel-like, continually growing incisors, followed by a gap and a series of grinding cheek teeth, and a jaw that can move laterally as well as forward and backward. This feeding apparatus

SEED DISPERSAL
ISBN 0 12 511900 3

allows them to process seeds, fruits, and herbaceous material efficiently, and few species are specialized carnivores. Being herbivorous, abundant, and ubiquitous, rodents have great potential for influencing plant reproductive success.

In this chapter we consider the consquences to plants of their interactions with granivorous (seed-eating) rodents. We will focus on two central questions: (1) What is the net effect of these rodents on plant populations? (2) What is their net effect on the evolution of plant reproductive traits?

A thorough analysis of plant-granivorous rodent interactions requires that one understand the individual and population biology of the animals and the plants, as well as the ecological and evolutionary aspects of interspecific interactions, and we cannot do full justice to all these topics. We refer the reader to reviews on foraging behavior by Schoener (1971), Pyke *et al.* (1977), Krebs (1978), and Pyke (1984); on behavior of seed-eating rodents by Reichman (1981, 1983) and Smith and Reichman (1984); on plant population biology by Harper (1977) and Silvertown (1982); on granivorous rodent population biology by Terman (1968), Thompson (1978), Munger *et al.* (1983), and Heaney (1984); on seed 'predation' by Janzen (1971); and on seed dispersal by Howe and Smallwood (1982) and Janzen (1983a,b).

The empirical literature relating granivorous rodent activities to plants is large. We have concentrated on North American desert-dwelling heteromyid rodents and north-temperate forest-dwelling sciurids, systems we know best, rather than attempting an exhaustive review. For further information about deserts and temperate forests, see Noy-Meir (1973, 1974), Brown *et al.* (1979b), and Reichle (1981).

We begin by developing a 'Fate Diagram' that summarizes ways in which rodents can influence the fate of a seed (Section II,B), and discuss how such a diagram is useful for studying the net consequences of rodents for plant population dynamics and evolution. We then outline factors that determine the probabilities that a seed follows different fate paths (Section III). Finally, we discuss alternative approaches to the two questions posed above and illustrate what can be learned by applying them (Sections IV, V). Hereafter when we refer to 'rodents' we mean 'granivorous rodents'.

II. GRANIVOROUS RODENTS AND THE FATES OF SEEDS

A. Foraging Behavior

Rodent-plant interactions are complex, for rodents husk, move, and bury seeds in addition to eating them; hence they affect seed spatial distributions and establishment probabilities as well as seed numbers. To analyse the potential effects of rodents, one must understand their foraging behavior.

A foraging animal first must decide where to begin searching, and whether to search for seeds not yet shed or for those dispersed on the ground. In this context the term 'decide' does not imply conscious choice (Krebs, 1978). While searching, the animal encounters seeds or seed patches. At each encounter it can harvest or continue searching. If a seed is harvested, the animal can eat all or part of it, or store it either husked or intact in a cache for later consumption. Seed stores can be buried under soil or litter, or placed in crevices, burrows or nests, and can consist of a few large caches ('larderhoards') or many small scattered caches ('scatterhoards').

This scheme applies generally to granivores; the major variation lies in where and how they search for seeds, and in the probabilities of alternative behaviors thereafter. For example, birds generally locate seeds by sight, whereas rodents generally rely on olfaction. This causes birds and rodents to encounter buried and surface seeds with different probabilites. Differences in digestive physiology and morphology also affect foraging behavior. A seed toxic to some animals may be edible and palatable to others; seeds that are too large or small for some animals can be utilized by others. These differences affect the probability that an animal will harvest a seed once this has been encountered. Other aspects of seed morphology determine the probability that a granivore consumes seeds rather than caching them; for example, birds and rodents that readily cache seeds also have specialized structures for transporting them (Smith and Reichman, 1984).

B. The Seed Fate Diagram

We can use this behavioral schema to illustrate ways in which rodents can affect seed fates (Fig. 1). Numbered boxes indicate possible events in the life of a seed, arrows indicate alternative event sequences, and values beside arrows indicate the probabilities that a seed will move along particular paths.

Since rodents can collect seeds directly from plants, it is useful to start with the unshed mature seed (Box 1). This seed has two possible immediate fates: it can be shed (Box 2) or encountered and harvested (Box 3). The probabilities of being harvested ($1-a$) and of escaping harvest (a) must sum to one. Once shed (Box 2), a seed may be harvested from the ground (Box 5) with probability ($1-b$) or may continue to escape harvest (Box 4) with probability b, and then either die (Box 8) or become established, perhaps after a period of dormancy (Box 9). The paths from Box 1 to

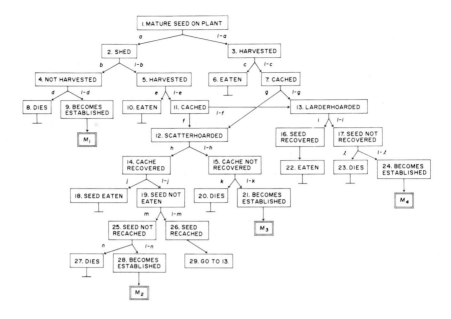

Fig. 1. A fate diagram for seeds subject to harvest by granivorous rodents. Numbered boxes indicate possible events in the life of a seed, and arrows connecting boxes indicate possible transitions between events. Values beside arrows indicate the probability a transition between two boxes will occur. M_1 through M_4 indicate expected reproductive outputs of seeds that traverse particular fate paths; other paths, ending in horizontal bars, have zero reproductive output.

2 leading to Box 4, and then to Boxes 8 or 9, which occur with total probability *ab*, are the only ones in which a seed avoids direct interaction with rodents.

A seed that is encountered and harvested (Boxes 3, 5) can either be cached (Boxes 7, 11) or eaten (Boxes 6, 10). If eaten, the seed dies, since granivorous rodents generally chew what they eat. Cached seeds, in contrast, have some chance of becoming established so long as they are not subsequently recovered and eaten.

As Figure 1 indicates, several paths lead to establishment, further seed production, and recruitment to subsequent generations. A seed can be shed (Box 2), escape harvest (Box 4), and become established (Box 9). If so, it produces some average number of further seeds, M_1. Alternatively, a seed can be harvested (Box 3) and cached (Box 7) in a scatterhoard (Box 12) that is not recovered (Box 15). If so, it has a probability $(1-k)$ of becoming established and then producing M_3 seeds. There are nine paths to establishment in Figure 1, out of a total of 26, which are summarized in Table I along with their reproductive consequences and probabilities of occurrence.

Table 1. Synopsis of Figure 1 Fate Paths. Each path is indicated by a sequence of numbers corresponding to the boxes in Figure 1. 'Probability of occurrence' and 'expected reproductive output' refer, respectively, to the probability that a seed traverses a particular path and its reproductive output if it does. Probabilities and reproductive outputs correspond to values in Figure 1. See text for further discussion.

Path Number	Path Description	Probability of Occurrence	Expected Reproductive Output
1	1-2-4-8	abd	0
2	1-2-4-9	$ab(1-d)$	M_1
3	1-2-5-10	$a(1-b)e$	0
4	1-2-5-11-12-14-18	$a(1-b)(1-e)fhj$	0
5	1-2-5-11-12-14-19-25-27	$a(1-b)(1-e)fh(1-j)mn$	0
6	1-2-5-11-12-14-19-25-28	$a(1-b)(1-e)fh(1-j)m(1-n)$	M_2
7	1-2-5-11-12-14-19-26-29-13-16-22	$a(1-b)(1-e)fh(1-j)(1-m)i$	0
8	1-2-5-11-12-14-19-26-29-13-17-23	$a(1-b)(1-e)fh(1-j)(1-m)(1-i)l$	0
9	1-2-5-11-12-14-19-26-29-13-17-24	$a(1-b)(1-e)fh(1-j)(1-m)(1-i)(1-l)$	M_4
10	1-2-5-11-12-15-20	$a(1-b)(1-e)f(1-h)k$	0
11	1-2-5-11-12-15-21	$a(1-b)(1-e)f(1-h)(1-k)$	M_3
12	1-2-5-11-13-16-22	$a(1-b)(1-e)(1-f)i$	0
13	1-2-5-11-13-17-23	$a(1-b)(1-e)(1-f)(1-i)l$	0
14	1-2-5-11-13-17-24	$a(1-b)(1-e)(1-f)(1-i)(1-l)$	M_4
15	1-3-6	$(1-a)c$	0
16	1-3-7-12-14-18	$(1-a)(1-c)ghj$	0
17	1-3-7-12-14-19-25-27	$(1-a)(1-c)gh(1-j)mn$	0
18	1-3-7-12-14-19-25-28	$(1-a)(1-c)gh(1-j)m(1-n)$	M_2
19	1-3-7-12-14-19-26-29-13-16-22	$(1-a)(1-c)gh(1-j)(1-m)i$	0
20	1-3-7-12-14-19-26-29-13-17-23	$(1-a)(1-c)gh(1-j)(1-m)(1-i)l$	0
21	1-3-7-12-14-19-26-29-13-17-24	$(1-a)(1-c)gh(1-j)(1-m)(1-i)(1-l)$	M_4
22	1-3-7-12-15-20	$(1-a)(1-c)g(1-h)k$	0
23	1-3-7-12-15-21	$(1-a)(1-c)g(1-h)(1-k)$	M_3
24	1-3-7-13-16-22	$(1-a)(1-c)(1-g)i$	0
25	1-3-7-13-17-23	$(1-a)(1-c)(1-g)(1-i)l$	0
26	1-3-7-13-17-24	$(1-a)(1-c)(1-g)(1-i)(1-l)$	M_4

The latter are the products of probabilities of each path segment, and the contribution of each path to total expected reproductive output is the product of its probability and the expected fecundity of a seed that has followed it. In turn, total expected reproductive output, or fitness, of a seed is the sum of contributions through all paths, and the fitness of its maternal parent is the per-seed fitness multiplied by seed output.

We have truncated the fate diagram in several places to eliminate infinite loops. For example, we assume that a scatterhoarded seed that is recovered and recached (Box 26) is placed in a larderhoard (Box 13) rather than in another scatterhoard, and that a recovered larderhoarded seed (Box 16) is eaten (Box 22) rather than recached. A more important assumption is that seeds do not have prolonged dormancy. If they do, their reproductive output must be discounted by any delay, using methods detailed by Hubbell and Werner (1979). It is also important to realize that all the variables in Figure 1 may depend on the total number of seeds that follow particular fate paths. For example M_1, the reproductive output of a seed that escapes rodents altogether, will depend on the number of seeds of like fate, for seed density affects competition and the chance of being consumed by other granivores or by pathogens.

C. Utility of the Fate Diagram

The fate diagram in Figure 1 is a conceptually simple and useful tool for analysing rodent-seed interactions. An analogous approach to complex plant life histories was proposed by Hubbell and Werner (1979). Our approach is an extension of theirs, and is related also to techniques of path analysis used in such diverse areas as decision theory (Raiffa, 1968), community ecology (Levins, 1975), population genetics (Wright, 1921, 1968), and ecosystems ecology (Watt, 1968).

Fate diagrams have a number of uses. By outlining avenues of interaction, they expose perhaps unappreciated net consequences of an interaction and facilitate the design of experiments. A formal diagram also simplifies the task of tracing multiple consequences of changes in one component of an interaction, and is the basic procedure used to develop predictions about its populational and evolutionary effects.

Consider the question of whether the consumption of seeds by rodents harms or benefits plants. A common casual assumption is that rodents are detrimental, since they reduce the survival of seeds. However, it is clear from Figure 1 and Table I that rodents might have a net positive effect, depending on the probabilities that a seed travels along various paths and its fitness in doing so. If caching seeds sufficiently enhances their subsequent establishment and fecundity, and if the probabilities of their escaping being eaten once harvested are great enough, then the net effect of rodents will be positive. Sufficient conditions can be inferred from Figure 1 and Table I: the expected fitness must be small in the absence of rodents (i.e., fitness gained through paths 1 and 2, when $ab = 1$) and must increase as the proportion of seeds processed by rodents increases (i.e., fitness gained through paths 3-26 combined, as a and b decrease). This suggests that a critical comparison is between the fitness of unharvested and unmanipulated seeds, and the fitness of seeds that are processed by rodents. Very few studies have attempted to make this comparison.

There are two basic methods of assessing the net effects of rodents. Short-term populational effects can be explored by observing the consequences for plant numbers of fluctuations in rodent densities. This approach has the virtue that plant responses integrate all steps in the rodent-seed interaction, so that one can quickly evaluate short-term populational consequences of rodents for a particular system; the disadvantage is that it is difficult to use results from such a study to predict results of another, unless the mechanisms producing the initial results are understood. Alternatively, one can estimate parameters of a fate diagram and calculate the expected reproductive output of a seed as a function of rodent density. While this approach has the virtue of treating interaction mechanisms explicitly, completely specifying a fate diagram is not a trivial task; any final estimate of the net effect of rodents compounds errors in estimates of the many component parameters, and extrapolation to other systems is feasible only if functional relationships between rodent densities and fate diagram variables are equivalent.

Although dissecting fate diagrams is less efficient than a direct experimental manipulation for some purposes, it is in most cases the only way to determine how rodents should influence the evolution of traits affecting seed morphology or reproductive phenology. It may also be the only feasible way to analyse possible geographic variation in the effects of rodents; rather than repeating a difficult manipulation in different places, one can consider how changes in the ecological context of the interaction should alter the fate diagram. We will review a number of studies that illustrate how direct experimentation and analyses of fate diagrams have been used to study rodent-seed interactions, but first we discuss in more detail how rodents can affect the fates of seeds.

III. DETERMINANTS OF THE FATE PATH

A. Probability of Being Harvested

The probability that a seed is harvested before germinating is a function of the probability of encounter and of harvest, given that encounter occurs. Encounter probability depends on the time to germination, rodent density, where and how intensively individual rodents search, and the likelihood of an individual detecting seeds in its search path. In turn, the probability that an encountered seed is harvested is a function of its quality as a food source, the ease with which it can be harvested, and the abundance of alternative foods.

Harvest rates are straightforward to measure, and much is known about what influences them. It is more difficult to determine whether encounter or harvest phases of foraging are involved, because situations in which a seed is encountered but ignored are indistinguishable from those in which it is not detected; in either case the animal may show no outward response to the seed. The only unambiguous indication of encounter is a harvest attempt. Depending on the question being asked, however, it may or may not be important to distinguish encounter from harvest.

1. Probability of Encounter

This depends on how rodents forage. Although diurnal rodents (e.g., sciurids) often use sight, most rodents probably locate seeds by olfaction, and this faculty certainly must be used to find buried seeds (Cahalane, 1942; Reymolds, 1958; Howard and Cole, 1967; Lockard and Lockard, 1971; Reichman and Oberstein, 1977; Barnett, 1977; Hutchins and Lanner, 1982; Evans *et al.*, 1983). A seed's olfactory conspicuousness affects encounter rates by determining how close a rodent must come before the seed is detected. All else being equal, olfactorily inconspicuous seeds will escape harvest more frequently.

Little is known about what determines olfactory conspicuousness. Reichman (1981) has assumed that seeds give off volatile chemicals at some rate dependent on their size and surface properties. As these molecules diffuse, their concentration decreases as the square of the distance from the seed. Total odor concentration at any point on the soil surface is assumed to be the sum of contributions from all sources, so that odor concentrations provide cues to the locations, densities, and identities of buried seeds.

Reichman's model suggests a number of largely unexplored possibilities. Soil characteristics could affect odor diffusion, and hence seed detectability (c.f. Reichman, 1981, 1983; Johnson and Jorgensen, 1981). There may be extensive odor mimicry systems involving unpalatable species. Husk and fruit characteristics may act to reduce odor escape and hence make seeds olfactorily cryptic. Finally, under certain conditions of seed depth and spacing, odor concentrations may be highest between seeds rather than directly over them (Reichman and Oberstein, 1977; Reichman, 1981), thus misleading rodents about the locations of seeds.

Rodents probably vary in their olfactory acuity, but we know of no studies that have directly measured this ability. Johnson and Jorgensen (1981) presented individuals of several desert rodent species with a choice of buried and surface seed clumps. They observed interspecific differences in the relative numbers of surface and buried seeds taken. Reichman (1981,

Reichman and Oberstein, 1977) obtained similar results with two heteromyid species. However, such observations cannot distinguish interspecific differences in olfactory acuity from other factors influencing the probability of harvest once a clump of seeds has been detected.

Encounter rates are also functions of where and how carefully individuals search. Search paths are not random, for animals concentrate their activity in particular patches, habitats, or microhabitats, and pass quickly through others (c.f. Thompson, 1982; Bowers, 1982). This means that the probability that a seed is encountered depends on its spatial location.

Rodents preferentially harvest seeds from dense patches, although individual rodent species differ in their degree of preference for such patches (Hutto, 1978; Price, 1978a; Reichman, 1981, 1983). It is not known to what extent this is a function of clumped seeds being more conspicuous, or more likely to be harvested once encountered. Whatever the cause(s), seed removal rates and total amounts harvested from artificial or naturally broadcast seed baits usually increase with seed density (Brown *et al.*, 1975; Price, 1978a; Reichman, 1979; Lewis, 1980; Heithaus, 1981; Sork *et al.*, 1983; Stapanian and Smith, 1984; Price and Waser, 1985). By selectively cropping clumped seeds, rodents tend to reduce the spatial patchiness of seeds in the soil (Reichman, 1979).

On a larger spatial scale, rodents show pronounced species-specific microhabitat or habitat preferences. Hence the vulnerability of seeds in a given habitat can change depending on the species composition of the rodent fauna. For example, heteromyids exhibit pronounced microhabitat preferences, with some species foraging primarily in open spaces between shrubs and others in areas under shrub or tree canopies (Price and Brown, 1983; Price, 1984). The relative vulnerability of seeds in open and closed microhabitats seems to vary depending on relative densities of open and shrub foragers. O'Dowd and Hay (1980) and Hay and Fuller (1981) found that seeds were removed more rapidly from under shrubs than from open spaces, a difference that corresponded with the relative densities of rodents in the two microhabitats. In another experiment, however, heteromyids removed seeds more rapidly from open spaces (Brown *et al.*, 1975). Such differences in removal rates could occur if sites differ in relative abundances of open and shrub-foraging species, or if microhabitat preferences change geographically. Both are probable. The species composition of rodent communities does vary geographically (Brown, 1973, 1975; Price and Brown, 1983), and microhabitat preferences can vary seasonally (Price 1978b), between sites (Price and Waser, 1985), and in response to temporal changes in moonlight intensity (Price *et al.*, 1984).

Habitat affects a seed's vulnerability in non-desert habitats as well. Red squirrels (*Tamiasciurus hudsonicus*) rarely forage far from trees, and so are less likely to harvest seeds from meadows or open woodlands than from forests (Hutchins and Lanner, 1982; Benkman *et al.*, 1984). More guanacaste tree seeds (*Enterolobium cyclocarpum*) are removed from horse dung placed in Costa Rican deciduous forest than in adjacent grassland (Janzen, 1982). In California grasslands, rodents are most abundant near shrubs, and consquently seeds are more likely to be removed if placed near (86% removed) than far (12% removed) from shrubs (Bartholomew, 1970). In old fields, seeds are removed more rapidly from vegetated than disturbed (plowed) habitats (Mittelbach and Gross, 1985).

2. Probabiliy of Harvest Given Encounter

Having encountered a seed, a rodent can either harvest it or ignore it. This has been treated extensively in the literature on optimal foraging theory (e.g., Pyke *et al.*, 1977; Krebs, 1978; Pyke, 1984), which assumes animals forage so as to achieve the highest possible net rate of food intake. Empirical studies have found that animals generally do preferentially harvest food types that yield high intake rates, but that other factors, such as toxin content or trace nutrient balance, also can influence food preferences. Because intake rate appears to be an important factor in food choice, we can identify factors likely to dictate the probability that seeds are harvested once encountered if we know what seed attributes influence the net food reward a rodent gains per seed unit time while harvesting and eating them.

A seed's value is dictated by its nutrient content and the time and energy costs of finding, harvesting, and consuming it. The first property is determined by intrinsic seed characteristics, like size and chemistry, and by rodent digestive physiology, while harvest and consumption costs are functions of intrinsic seed properties (such as size, shape, and husk thickness) and extrinsic factors (such as seed density, soil parameters, and rodent morphology). Rodent seed preferences appear to be affected both by intrinsic and extrinsic factors.

Preferences can be measured by presenting individual rodents with equally-accessible seeds from several species, and determining whether seeds are taken in equal numbers. Rodents almost always prefer certain seeds in such situations, and relative harvest rates in the field often parallel preferences shown in the laboratory. Hay and Fuller (1981) found that sacred datura (*Datura meteloides*) seeds were avoided in the laboratory by two heteromyid species and were removed at low rates in the field. Borchert and Jain (1978) noted that mice (*Microtus californicus* and *Mus musculus*) preferred wild oat (*Avena fatua*) seeds over those of three other grasses, and that the decrease in *Avena* seed densities in plots exposed to rodents was approximately twice that of the others. Price (1983) found that millet (*Panicum miliaceum*) seeds were highly preferred by heteromyids, and Lockard and Lockard (1971) observed that heteromyids removed millet more rapidly than other seeds in the field. Stapanian and Smith (1984) found that fox squirrels (*Sciurus niger*) preferred black walnuts (*Juglans nigra*) over acorns of two oak (*Quercus*) species and harvested more walnuts, carried them further to caches, and collected them from lower density patches than acorns.

Seed preferences often correspond to the highest energetic value. Smith and Follmer (1972; see also Stapanian and Smith 1984) found that fox and gray (*S. carolinensis*) squirrels obtain higher net energetic return from bur oak (*Quercus macrocarpa*) acorns than from the smaller shumard (*Q. shumardii*) or white oak (*Q. alba*) acorns, and they preferred bur oak acorns. This example underscores the importance of a seed's size in determining its energy value. Large seeds generally contain more energy than small ones (Smith and Follmer, 1972; Smith, 1975; Reichman, 1976), and most rodents prefer large over small seeds, so long as they are not too large to handle (Reynolds, 1958; Abbott, 1962; Reichman, 1975, 1979; Price, 1983).

A seed's energy value can be influenced not only by its size, but also by the costs of processing and consuming it. A major component of processing costs is the time required to remove the husk (c.f. Rosenzweig and Sterner, 1970). Rodents are sensitive to this cost and prefer husked over unhusked seeds (e.g., Kaufman and Collier, 1981). Smith and Follmer (1972) noted that although bur oak and black walnut kernels were of similar caloric value, it took gray squirrels 900 s to gnaw through the husk of a walnut and eat the kernel, compared to only 404 s for the thinner-husked acorn. Squirrels therefore achieved higher net rates of energy intake when eating acorns, and indeed preferred bur oak acorns over whole walnuts. This preference was reversed when the walnuts were husked.

When several seeds are packaged together, the value of the entire package may be more relevant than the values of individual seeds. Elliott (1974), for example, noted that red squirrels did not selectively forage from lodgepole pine (*Pinus contorta*) trees bearing cones with large seeds, but instead preferred trees whose cones had a high ratio of seed to total cone weight, and were easily detached from the branch.

In some cases seed chemistry may also affect seed value. Heteromyid rodents, which often must subsist largely on water produced by the oxidation of their food, prefer seeds high in carbohydrate (Lockard and Lockard, 1971; Price, 1983), presumably because more water is formed per unit of energy produced by oxidizing carbohydrate than either lipid or protein, and more water is required to eliminate waste products of protein catabolism (Schmidt-Nielsen, 1964).

Toxins also affect rodent dietary preferences (c.f. Janzen, 1971; Freeland

and Janzen, 1974). Seeds of *Datura meteloides*, for example, contain alkaloids known to be highly toxic to mammals (p. 247 of Bell, 1984), and heteromyids avoid these seeds (Hay and Fuller, 1981). Sherbrooke (1976) documented another case of toxicity. The seeds of jojoba (*Simmondsia chinensis*), a Sonoran Desert shrub, contain appetite-suppressing compounds. Most heteromyids in his study did not eat jojoba and when given a pure diet of jojoba seeds lost weight and died. However, one species, *Perognathus baileyi*, flourished on jojoba seeds. It is intriguing to note that this pocket mouse is the only heteromyid species common on the rocky slopes where jojoba is a dominant plant. Abbott (1962) speculated that toxins were responsible for the low ranking of balsam fir (*Abies balsamea*) seeds by three forest-dwelling rodent species: white-footed mice (*Peromyscus leucopus*), red-backed voles (*Clethrionomys gapperi*), and *Microtus pennsylvanicus*.

The extent to which intrinsic seed attributes actually affect harvest probability is not known. If all else is equal, high-quality seeds should be harvested more often than low-quality seeds, but several extrinsic conditions also are important. One is the availability of better items; rodents generally avoid low-quality food unless better food is scarce. For example, the success of reseeding programs can be increased by mixing target seeds with highly preferred seeds such as sunflower (*Helianthus annuus* ; Sullivan and Sullivan, 1982). Similarly, heteromyid rodents prefer the seeds of Indian rice grass (*Oryzopsis hymenoides*) to those of Russian thistle (*Salsola paulsenii*), and will switch from an almost pure diet of *Salsola* seeds to a pure diet of *Oryzopsis* seeds when the latter are produced (McAdoo *et al.*, 1983). Everett *et al.* (1978) also found that seed availability had a pronounced effect on diets in the laboratory.

The probability of harvest may also vary seasonally. Gray and fox squirrels consume mostly hickory (*Carya ovata*) and black walnut seeds in the fall and spring, and rely on acorns (*Quercus spp.*) during the winter. Smith and Follmer (1972) speculated that this occurred because the weight of food in the gut was an important constraint on foraging in fall and spring, but time spent out of the nest was a more important constraint in winter. Acorns provide more energy per unit of handling time, but less energy per edible gram of seed than seeds of hickory or black walnut (see also Smith, 1975).

Spatial distribution also modifies seed value. Price (1983, Price and Heinz, 1984) demonstrated that heteromyid rodents can harvest seeds at a faster rate from dense seed patches than from sparse seed patches, and Shaffer (1980) observed that eastern chipmunks (*Tamias striatus*) required less time to fill their cheek pouches when foraging from caches than when first harvesting dispersed seed. This ease of harvesting may explain why clumped seeds are more desirable (Hutto, 1978; Price, 1978a, 1983; Trombulak and Kenagy, 1980: Janzen, 1982; Price and Waser, 1985).

How deeply a seed is buried, and in what type of soil, also affects its value. All rodents will avoid deeply buried seeds if given an option, presumably because digging is costly (Cahalane, 1942; Lockard and Lockard, 1971; Barnett, 1977; Reichman and Oberstein, 1977; Reichman, 1979; Heithaus, 1981; Johnson and Jorgensen, 1981; Abramsky, 1983; Evans *et al.*, 1983), but some species dig more readily than others. Non-burrowers (e.g., *Peromyscus*) are less likely to dig than burrowers (e.g. heteromyids), presumably because it costs them more in energy to dig (Johnson and Jorgensen, 1981). Similarly, smaller heteromyid species are less likely than larger ones to dig for seeds (Reichman and Oberstein, 1977).

Soil attributes also affect the costs of harvesting buried seeds. Price and Heinz (1984) observed that the rate at which heteromyid rodents extracted buried seeds varied with the diameter of soil particles relative to the size of seeds. Harvest rates were high when seeds were larger than soil particles and decreased as soil particle size increased. Heteromyids appeared to be reluctant to forage in patches of coarse mineral soil, although there were differences between species in the degree of reluctance (Price, 1978b; Price and Waser, 1985).

How much protection a seed gains by being unpalatable, small, surrounded by a hard husk or unwieldy cone, buried, mixed with coarse soil, or in a low density area, depends on the capabilities of rodents and the availabilities of more profitable seeds. Although unprofitable seeds will tend to be overlooked, they will still be harvested under some circumstances.

3. Total Probability of Harvest

A variety of methods has been used to estimate the probability that a seed is harvested by a rodent. The most indirect method is to infer harvest rates by calculating the amounts of seed necessary to maintain a granivore population from metabolic rates, population densities, diets, and energy contents of available seeds. Consumption estimates can then be divided by seed production estimates to determine the proportion of seeds consumed. By this method, Pearson (1964) estimated that rodents consumed 93% of seeds produced in a California annual grassland, and Soholt (1973) estimated that a population of the Merriam kangaroo rat (*Dipodomys merriami*) consumed 95% of the *Erodium cicutarium* (filaree) seeds produced at a Mojave desert site. Chew and Chew (1970) estimated that rodents consumed 87% of seeds produced at a Chihuahuan desert site. In contrast to these high estimates from deserts, Pulliam and Brand (1975) estimated that rodents took less than 1% of forb seeds produced in an Arizona grassland.

In some cases a more direct estimate of seed consumption can be obtained from the remains of seeds. Heithaus (1981) noted that rodents (mainly *Peromyscus leucopus*) discarded the elaiosome when consuming seeds of four ant-dispersed plants. If the numbers of discarded elaiosomes accurately reflect the numbers of seeds consumed, then the initial seed number and final elaiosome number can be used to estimate the proportion of seeds eaten by rodents. By this method Heithaus estimated that rodents consumed 42% of seed capsules before seeds were shed, and a further 67% of shed seeds on the ground. This is probably an underestimate, because rodents sometimes carried seeds away before consuming them, and ants sometimes collected discarded elaiosomes. Boyd and Brum (1983) used a similar technique to estimate the impact of desert rodents on creosote bush (*Larrea tridentata*). They observed that rodents husked seeds in a characteristic way before eating them, and so they used husk numbers to estimate that rodents destroyed 28% of shed seeds. The estimate may be inaccurate, because the rodents common to this desert site often husk seeds before placing them in cheek pouches and carrying them to a cache, and sometime pouch seeds without husking them at all.

If rodents harvest and cache more seeds than they consume, then actual values for probabilities of harvest will be larger than consumption-based estimates. Harvest probabilities can be estimated more directly by observing seed removal from artificial baits. The relative contribution of rodents can be assessed by comparing disappearance rates of seeds from plots that vary in rodent density or activity. The difficulty with this technique is that the types, densities, and spatial arrangements of seeds used in an experiment may be unrealistic, despite attempts at realism, and hence baits may be harvested at unnaturally high or low rates. The technique is therefore most useful in determining what factors affect harvest probabilities, rather than in estimating actual harvest probabilities in nature.

Seed removal experiments indicate that rodents are important seed harvesters in most habitats, often far more important than other granivores. Abramsky (1983) set out replicate seed trays in the Israeli desert, some covered to exclude ants but not rodents, others covered to exclude rodents but not ants. More seeds disappeared, and at a higher rate, from trays accessible to rodents, and rodents often removed 97% of available seeds overnight. Using similar methods, Brown *et al.* (1975), Reichman (1979) and Evans *et al.* (1983) obtained similar results in various North American desert and montane habitats, but Mares and Rosenzweig (1978) found that rodents removed very little seed in the Monte Desert of north-western Argentina, where granivorous rodents are much rarer than in North America.

One can avoid some difficulties associated with artificial seed baits by following the disappearance of seeds of native plants. For example, Sork *et al.* (1983) observed the disappearance of red oak (*Quercus rubra*) acorns in plots, and estimated that vertebrates had removed 50% of the acorns over winter. In many cases, however, the relative importance of rodents can be inferred only from the difference in disappearance rates for seeds made inaccessible and seeds made accessible to them. The commonest way to exclude rodents is to cover seeds with wire cages. Cages generally enhance seed survival, an observation of practical importance in reseeding programs (Smith and Aldous, 1947). Shaw (1968b) noted that the apparent percent germination of sessile oak (*Quercus petraea*) acorns in North Wales could be increased from 2% to 46% by caging the acorns to exclude rodents selectively. Christy and Mack (1984) increased western hemlock (*Tsuga heterophylla*) germination from 7% to 28% by caging. In the Sonoran Desert, Reichman (1979) found that soil in plots containing rodents but not granivorous ants contained approximately one-quarter to one-sixth the seed biomass of soil in plots lacking both rodents and ants. Natural spatial or temporal variation in rodent densities can also be used to assess the importance of rodents (c.f. Vincent, 1977; Sullivan and Sullivan, 1982). By correlating rates of natural decrease of soil seed densities with changes in rodent density, Nelson and Chew (1977) estimated that rodents accounted for 30–80% of the seasonal decrease in soil seed pools in Nevada. By selectively poisoning rodents with treated grain, Nelson *et al.* (1970) enhanced the survival of seeds in a Nevada range reseeding program from 48% to 79%. Batzli and Pitelka (1970) found that the density of *Avena* spp. seeds outside a rodent-proof fence was about half that inside the fence. Using similar techniques, Borchert and Jain (1978) estimated that mice consumed from 30–65% of the seeds of four grass species.

Although these estimates may be inaccurate individually, the conclusion is inescapable that rodents harvest many seeds. What they do with them subsequently is therefore important.

B. Probability of Being Cached

A major distinction between rodents and many other seed consumers is the tendency of rodents to cache harvested seeds. Some of these seeds are recovered and eaten, but some escape detection, germinate, and become established. What factors determine the probability that a seed is cached, the characteristics of the cache, and the probability of recovery?

1. Initial Probability of being Cached or Consumed
Caching is an expensive activity. Not only must seeds be harvested in the first place, but they must then be transported, stored, and reharvested before any benefit is gained from them. In addition, the effort of caching may

be wasted if stored seeds spoil, germinate, or are stolen by competitors before being recovered. Animals should therefore cache only so long as the expected benefits outweigh these costs.

Caches presumably are insurance against food shortages (Smith and Reichman, 1984). Fluctuations in food availability may be caused by variation in either competitor abundance or seed production; the latter can be caused by seasonal or unpredictable year-to-year variation in weather, or by intrinsic variation in the output of seed producers, as occurs in mast-fruiting trees and shrubs (Silvertown, 1980). The net value of caching should be greatest in inconstant environments. This value should also depend on the likelihood that cachers will recover their caches, and on the costs of storage and recovery. These should be functions of animal morphology, rates of seed spoilage or germination, and competitor abundance.

Caching is most conspicuous in rodents inhabiting seasonal or unpredictable environments such as temperate forests and deserts (Smith and Reichman, 1984). Seed-eaters are more prone to cache than fruit-eaters, presumably because fruits spoil quickly, and caching appears uncommon in the moist tropics, where decomposition is rapid (Smith and Reichman, 1984.) Rodent species differ in caching tendency and in the types of caches made under controlled conditions (Smith and Reichman, 1984, and references therein; S. H. Jenkins 1984, unpubl. data). For example, large burrow caches are more prevalent in very small and very large heteromyids (Lawhon and Hafner, 1981; Smith and Reichman, 1984), perhaps because animals of intermediate size cannot effectively use narrow burrow entrances or aggression to exclude thieves. Heteromyids, which have cheek pouches, cache more than sympatric cricetids, which lack this efficient device for seed transport.

There are remarkably few quantitative estimates of the probabilities of seeds being cached rather than consumed. An impression of the extent of caching can be derived from observations of food stores recovered from nests or burrows (Vorhies and Taylor, 1922; Shaw, 1934; Hawbecker, 1940; Tappe, 1941; Broadbooks, 1958; Lanner, 1981; Poche *et al.*, 1982), but one needs additional information to estimate what fraction of the total harvest is initially stored, and what fraction is eventually eaten. Elliott (1978) attempted to determine how much eastern chipmunks store in excess of their seasonal food requirements by examining the amounts remaining in burrow caches in spring. He estimated these 'leftovers' would last burrow occupants for 30 to 330 days at expected summer resting metabolic rates, and noted that the excess was positively correlated with fall seed production.

It should be possible to derive better estimates of caching probabilities from detailed field observations of foraging animals. This method has not been applied to nocturnal rodents, but has been successfully used to study caching by a bird, Clark's nutcracker (*Nucifraga columbiana*). From observations of foraging and caching rates, Tomback (1982) estimated that the average nutcracker cached 35 000 whitebark pine (*Pinus albicaulis*) seeds in the fall; this cache size could supply 45% of the metabolic needs of an adult and its young through winter and spring. Vander Wall and Balda (1977) gave similar estimates for nutcrackers harvesting pinyon pine seeds (*Pinus edulis*).

An alternative method for estimating caching probability is to measure total harvest rates and subtract estimates of seed numbers corresponding to the consumer's immediate metabolic requirements. Hutchins and Lanner (1982) estimated that red squirrels (*Tamiasciurus hudsonicus*) harvested about seven times as many seeds per individual as Clark's nutcrackers; since squirrels are only about twice as heavy as these birds and have similar diets, this suggests that squirrels cache at least five times as much seed as they consume during winter.

Abbott and Quink (1970) placed seeds in feeders accessible only to small rodents, primarily red-backed voles (*Clethrionomys gapperi*) and white-footed mice (*Peromyscus leucopus*). Rodents removed about three times as much seed as they ate in place, judging from husk numbers compared to total seed removal. From laboratory studies of nightly food consumption rates and estimates of rodent population densities, Abbott and Quink argued that most of the removed seed was cached. Shaw (1968a) used similar reasoning to argue that bank voles (*Clethrionomys glareolus*) and wood mice (*Apodemus sylvaticus*) cached most of the sessile oak (*Quercus petraea*) acorns they harvested at a site in North Wales. Sullivan (1978), on the other hand, suggested that *P. maniculatus* ate about 90% of broadcast Douglas fir seed (*Pseudotsuga menziesii*) within two weeks.

Although these observations suggest that rodents often cache food far in excess of their needs, they do not indicate whether other factors regulate the amounts cached below the maximum set by food availability. Because caching is expensive, rodents should not accumulate substantially larger seed stores than they can use before the seeds rot or germinate, or before the next flush of seeds is produced. There is some laboratory evidence that rodents indeed regulate cache size. Lockner (1972) and Wrazen and Wrazen (1982) noted that chipmunks (*Eutamias ruficaudus* and *Tamias striatus*) decrease caching rates as cache sizes increase, and Vandermeer (1979) found that the rate at which spiny pocket mice (*Heteromys desmarestianus*) removed seeds was positively affected by seed availability up to a point, and negatively affected by the amounts already cached. Lawhon and Hafner (1981) observed that pocket mice (*Perognathus* spp.), which subsist entirely on stored food over winter, cached more seeds than kangaroo rats (*Dipodomys* spp.), which forage year-round, and that all heteromyids cached less during winter, a time of naturally low seed availability, than during fall or spring.

Even though caching rodents cannot be 'satiated' in the same sense as non-caching seed eaters, their tendency to stop caching when adequate food has been stored would cause the proportion of seeds harvested and cached to be inversely related to seed production above some threshold production level. This is effectively the 'predator satiation' phenomenon Janzen offered to account for the evolution of mast fruiting in trees subject to intense seed 'predation' (Janzen, 1971; Vandermeer, 1979).

2. Cache Characteristics

More can be said about how seeds are cached. Zoologists traditionally distinguish two primary cache types, 'scatter-' and 'larderhoards', and sometimes distinguish burrow larderhoards from surface larderhoards, or 'middens'. Archetypal scatterhoarders are gray and fox squirrels (*Sciurus* spp.) of North American eastern deciduous forests, which bury individual nuts throughout their home ranges (Stapanian and Smith, 1978; Thompson and Thompson, 1980); archetypal larderhoarders are red and Douglas squirrels (*Tamiasciurus* spp.) of Pacific Northwest conifer forests, which accumulate large middens of whole cones near the centers of their territories (Smith, 1968). Some rodents make both scatterhoards and larderhoards. What determines cache type?

Smith and Reichman (1984) proposed that scatterhoarding is adaptive behavior in situations where competitors can freely steal from larderhoards; small, dispersed scatterhoards are less easily exploited. The probability of scatterhoarding thus depends on the nature of the food source and on competitive relationships within the granivore community.

Heteromyid rodents place larderhoards in burrows at depths of 2 to 57 cm (Vorhies and Taylor, 1922, for *Dipodomys spectabilis*). Scatterhoards are groups of seeds (as many as 250 *Oryzopsis hymenoides* seeds per cache, McAdoo *et al.*, 1983) placed either throughout an individual's home range (Reynolds and Glendening, 1949) or near the burrow (Shaw, 1934). Not all heteromyids scatterhoard: Lawhon and Hafner (1981) reported that in the laboratory, *Perognathus formosus* and *P. longimembris* cache greater proportions of seeds in larderhoards than do *Dipodomys merriami* and *D. panamintinus*. This may be related to the fact that most *Perognathus* hibernate, so they cannot use surface scatterhoards during winter. Perhaps most interesting for our purposes are heteromyids that make both larderhoards and scatterhoards: these include *D. venustus* (Hawbecker, 1940) and *D. ingens* (Shaw, 1934) in the field, and *D. merriami* and *D. panamintinus* in the laboratory (S. H. Jenkins 1984, unpubl. data). Shaw (1934) reported that *D. ingens* store seeds in surface caches temporarily and later transfer the seeds to burrows. He speculated that these kangaroo rats make scatterhoards for drying food, so that it will last longer in the burrow before rotting. Hawbecker (1940) suggested instead that *D. venustus* fill larderhoards in burrows first, then scatterhoard less desirable or later maturing seeds in surface caches. If plant establishment is more likely from scatterhoards than from larderhoards, then these two explanations imply different selection pressures on plant reproductive phenology. Temporary scatterhoarding may select for rapid germination before surface caches are recovered. On the other hand, permanent scatterhoarding of non-preferred or late-maturing seeds may favor delayed seed set, or characteristics that make seeds acceptable but not highly preferred.

Chipmunks (*Tamias striatus*) also make both scatterhoards and larderhoards (Elliott, 1978). Scatterhoarding predominates in spring and larderhoarding in fall, although juveniles scatterhoard more than adults during fall, perhaps because young animals have not yet secured permanent burrows. Most scatterhoards are temporary; only three of 16 lasted more than twelve days (Elliott, 1978).

In terms of the ultimate fates of cached seeds, the distinction between larderhoards and scatterhoards may not be as clear cut as is sometimes assumed. For example, Shellhammer (1966) observed a Douglas squirrel (*Tamiasciurus douglasi*) store sequoia cones (*Sequoiadendron giganteum*) in groups of three or four in shallow depressions around the base of its nest tree. It may be important for interpreting the adaptive value of caching to determine whether this represents a concentration of scatterhoards or a diffuse midden, but from the seed's perspective it probably is a larderhoard. Glanz (1984) reported that red squirrels in Maine store acorns individually, pine cones individually or in large larderhoards, and hemlock (*Tsuga canadensis*) cones usually in larderhoards of 10–80 cones, each type of cache scattered throughout their territories. Large red and Douglas squirrel (*Tamiasciurus hudsonicus* and *T. douglasi*) caches are usually assumed to be unfavorable germination sites (Tomback 1982, Benkman *et al.* 1984), but more subalpine fir seedlings (*Abies lasiocarpa*) are found on red squirrel middens than elsewhere (Hutchins and Lanner, 1982). This may occur because subalpine fir seeds are less preferred than whitebark pine seeds, and so are more likely to survive the winter, or because the microenvironment of the cache is more favorable for germination of subalpine fir than whitebark pine. Even seeds stored in burrows may become successfully established if larderhoards are shallow enough. For example, Tappe (1941) observed *Dipodomys heermanni* caching only in burrows, but some of these burrow caches were quite near the surface, and filaree (*Erodium cicutarium*) appeared to be more abundant near burrows than in other locations. Perhaps *Erodium* seeds, a major constituent of larderhoards, germinated and successfully established themselves from those caches near the surface.

These examples indicate that the zoologists' classification of caches as 'larder-' or 'scatterhoards' is not particularly relevant from the plant's perspective. Instead, caches may be more fruitfully classified according to microhabitat, depth, size, and distance from the parental plant. The most favorable situation for some plants is likely to be a series of small caches just below the soil surface, but even here variation among species in germination requirements may blur the picture. Large seeds may germinate best when planted deeper than others, and cache microsite may be critical. Little is known about details of cache position, but it may be important whether they are placed under or near large objects, as agoutis (*Dasyprocta punctata*) do with hard nuts of tropical trees (Smythe, 1970), or in open spaces, as kangaroo rats appear to do (S. H. Jenkins, pers. obs.).

3. Dispersal Distances

Studies using a variety of techniques have yielded estimates of how far rodents move seeds before caching them. Abbott and Quink (1970) followed radioactively labelled white pine (*Pinus strobus*) seeds and found that virtually all those harvested by white-footed mice and red-backed voles from feeders were moved less than 30 m, and 60% were moved less than 15 m. Radvanyi (1970) reported that radiotagged white spruce (*Picea glauca*) seeds were moved at most 9 m. Sork (1984) found essentially no dispersal of metal-tagged acorns beyond 20 m from feeders. On the other hand, Smythe (1970) observed agoutis carrying seeds up to 50 m before caching them, and Olmsted (1937) found some acorn caches 60 m from the nearest mature black oak (*Quercus velutina*), though most were within 30 m of a potential source. Glanz (1984) and Stapanian and Smith (1984) showed that red and fox squirrels, respectively, carried preferred foods greater distances than less-preferred foods before caching them. Reynolds (1954) reported that velvet mesquite (*Prosopis juliflora*) seeds were dispersed an average of 14 m and within a range of 1–32 m from feeders. He calculated that Merriam kangaroo rats (*Dipodomys merriami*) could thus spread mesquite 0.16 km in 50 years. From this he concluded that cattle (*Bos taurus*) were probably responsible for the rapid recent spread of mesquite in Arizona rangelands.

If rodents tend to concentrate scatterhoards near their burrows, as kangaroo rats (Shaw, 1934; Hawbecker, 1940) and eastern chipmunks may (Shaffer, 1980), then maximum dispersal distances should equal home range radii, and would thus depend on rodent species, food abundance, and other environmental factors that affect home range size. However, this does not seem to be a general characteristic of caching behavior. Elliott (1978) observed eastern chipmunks travelling more than 100 m to a seed-producing red maple (*Acer rubrum*) tree when food within their home ranges was scarce, but they cached these seeds near the source tree rather than returning to cache them near their burrows. There may be selection operating on squirrels to optimize scatterhoarded nut density; although high density would increase the rate at which an individual could recover its own caches, it might also increase cache thievery (Stapanian and Smith, 1978). Stapanian and Smith (1978) argued that optimal cache density for fox squirrels feeding on black walnuts (*Juglans nigra*) is about 1 nut per 100 m^2, which is close to that observed (but see Kraus, 1983). If it is advantageous to spread out caches, as argued by Stapanian and Smith (1978, 1984; see also Smith and Reichman, 1984), then mean dispersal distances would probably be less than if caches were concentrated near nest trees or burrow entrances. Stapanian and Smith (1978) made the suggestion that one advantage of mast-seeding for trees, assuming constant cache spacing, may be increased dispersal distances in mast years.

There are numerous theoretical treatments of seed dispersal in recent ecological literature (e.g., Bullock, 1976; Hamilton and May, 1977; Ellner and Shmida, 1981; Howe and Smallwood, 1982). These papers suggest that dispersal need not be over long distances to be advantageous. Hence even the short distance dispersal effected by rodents may be beneficial to plants, and may sometimes complement dispersal by other means. Vander Wall and Balda (1977) emphasized the effects of long distance dispersal by birds on geographical distributions of plants, and argued that Clark's nutcrackers may be responsible for the spread of pinyon pine (*Pinus edulis*) from one mountain range to another, perhaps several kilometers away. However, R. L. Everett (pers. comm., 1983) has suggested that rodent caching may cause the slow but steady spread of pinyon up or down slopes, which is often observed on mountain ranges of the North American Great Basin.

C. Fate of Seed Caches

Because plants can benefit from rodent hoarding activities only if cache recovery is incomplete, factors that affect recovery are critical. Recovery should be dictated by rodent densities and social organization, and by the mechanisms used to relocate caches. For example, if individuals have overlapping, undefended home ranges, and if olfaction is the primary means of locating caches, then recovery parallels exactly the process of initial seed harvest (Section III, A). In this case the death of the cacher will have only a minor effect on recovery of its caches. However, if exclusive territories are maintained and vacancies are not filled immediately, or if caches are relocated primarily from memory, as is true for some birds (Shettleworth and Krebs, 1982; Vander Wall, 1982), then the death of a cacher is likely to minimize the probability that caches are recovered.

The latter scenario is unlikely for rodents because they can locate buried seeds by smell (Section III, A), although cache recovery is likely to be more efficient if individuals can remember where to begin searching. As is true of initial harvest probabilities, cache recovery depends on the species of rodent, clump size and depth of buried seeds, and soil conditions.

Studies of cache recovery have usually relied on measurements of the disappearance rates of artificially or naturally cached seeds. Cahalane (1942) found that 99% of hickory nuts which he covered with leaf litter were removed by fox squirrels, but only 56% of those buried 1–3 cm deep were recovered. Whether the incomplete recovery of buried nuts is biologically significant depends on the depths of natural caches. Cahalane (1942) reported that most fox squirrel caches were covered just by leaf litter, and Thompson and Thompson (1980) stated that only 5% of gray squirrel caches were covered by more than 0.5 cm of soil. These results imply that fox and probably gray squirrels should be able to recover their own (and each other's) caches easily. Heteromyid scatterhoards are not much deeper (e.g., 1–5 cm; Shaw, 1934; Reynolds and Glendening, 1949).

Evans *et al.* (1983) planted bitterbrush (*Purshia tridentata*) seeds at depths of 1 cm in clump sizes ranging from 1 to 100 seeds. Single seeds were never harvested, but 75% of two-seed clumps and higher percentages of larger clumps were harvested. The surprising result here is the great difference between singletons and pairs. However, kangaroo rats scatterhoard these seeds in clumps larger than five seeds, at least in the laboratory, so failure to harvest single planted seeds does not necessarily imply that animals would be unlikely to locate each other's caches.

Barnett (1977) reported significant differences in recovery of white oak acorns (*Quercus alba*) and pignut hickory nuts (*Carya glabra*) as functions of depth. Hickory nuts were more often recovered than acorns at all depths (surface to 2.5 cm). Acorns were more often recovered from the surface than from 2.5 cm underground, but hickories were recovered equally well from any depth (to 2.5 cm). These patterns could result from differences in detectability (hickory nuts are larger and more aromatic, at least to humans), or preference (captive gray squirrels strongly prefer hickories over acorns; Lewis, 1982), or both. Barnett's experimental design did not exclude the possible effects of natural abundance of nuts and acorns. He planted acorns under oaks and hickory nuts under hickories, and the year of his study was a mast year for oaks, but not for hickories. Therefore, background acorn densities were high where acorns were planted, but background hickory nut densities were low where hickory nuts were planted. Perhaps squirrels became satiated more quickly by acorns than hickory nuts, because of the greater natural availability of acorns at the time of the study, and thus took fewer of the acorns provided in the experimental arrays. As we discussed earlier (Section III, A), it is difficult to distinguish situations in which an animal fails to detect a seed from situations in which an animal detects it but 'chooses' not to harvest it. This problem is as important for studies of mechanisms of cache recovery as it is for studies of initial harvest of naturally available or artificially provided seeds. For example, one might be tempted to argue from Barnett's (1977) study that the ability of gray squirrels to locate scatterhoards is severely circumscribed — they can do so if caches are not too deep or if they contain sufficiently aromatic seeds. However, as described above, there are at least two other plausible explanations for Barnett's results, neither of which suggests limitation of olfactory sensitivity.

These experiments with artificial seed caches could not detect any effects of memory on cache recovery. McQuade *et al.* (1986) have provided indirect evidence that memory might be important in cache recovery by gray squirrels. They found that squirrels attended to extrinsic visual cues in preference to olfactory cues or spatial positions in collecting food items from hidden locations in an array, and argued that squirrels might relocate caches in the field by 'memorizing' their locations in relation to visual landmarks. Even if memory plays a significant role in cache relocation by some species of rodents, the widespread abilities of rodents to locate buried seed by smell and the shallow depths at which scatterhoards are usually made indicate that plants cannot rely on memory failure or the death of cachers to escape substantial seed loss.

The extent to which rodents recover their own scatterhoards has been measured in two excellent studies. Abbott and Quink (1970) used radioactively labelled white pine seeds to locate caches made by white-footed mice (*Peromyscus leucopus*) and red-backed voles (*Clethrionomys gapperi*). In three experiments, they found a total of 192 caches by this method. Of 63 caches made in summer, all were completely retrieved within 19 days. Of 129 made in fall 1967 and fall 1968, 44 (34%) still contained some intact seeds the following spring, although many caches had been partially retrieved. Fall caches were 5–10 cm deep; most temporary summer caches were less than 2.5 cm deep. Fourteen of the 44 partially intact fall caches produced seedlings, but only four contained live seedlings at the end of the first growing season. Out of 3280 seeds cached in fall, 27 seedlings existed at the end of the next summer (0.8%). Radioactive labelling is a technique that deserves more use, although a serious problem can be the use of some isotopes that depress germination (for example, the scandium-46 used by Abbot and Quink, 1970).

Thompson and Thompson (1980) observed hoarding and cache recovery by gray squirrels in a Toronto cemetery. These authors planted horse chestnuts (*Aesculus hippocastanum*) in fall, and found that 85% of these were recovered during the winter. Since the rates of recovery of these artificially planted nuts over the course of the winter were highly correlated with the rates of recovery of squirrel-made caches, they argued that about 85% of squirrel caches were recovered. Of the 77 unrecovered horse chestnuts planted by the experimenters, 8% germinated (6 of the original 500 planted).

These studies show that cache recovery may not be complete, and some germination may occur from unrecovered caches. By contrast, Sork (1983) stated that no hickory seedlings (*Carya glabra*) occurred at her study site in southern Michigan, and argued that most of the mortality of hickory seeds, in both mast and non-mast years, was due to rodents. There are numerous studies, however, which document highly clumped distributions of tree and shrub seedlings that could easily have arisen from caches (Olmsted, 1937; Paulsen, 1950; West, 1968; La Tourrette *et al.*, 1971; Sherman and Chilcote, 1972; McAdoo *et al.*, 1983; Evans *et al.*, 1983). Olmsted (1937) found clumps of 23–84 acorns in grassland vegetation on sandy soil in Connecticut, up to 60 m from the nearest mature oak. About half of the mature oaks invading the grassland at this site occurred in clumps of two or more stems; furthermore, stems within a clump were the same age, indicating that they had not arisen from basal sprouting. Olmsted believed these clumps arose from squirrel caches. West (1968) found that bitterbrush and ponderosa pine (*Pinus ponderosa*) seedlings often occurred in clumps, while seedlings of other woody species at his study site almost always occurred as singletons. Sherman and Chilcote (1972) showed that a large majority of seemingly individual bitterbrush shrubs were actually clumps of individuals, when their partly buried stems were excavated and carefully examined. These clumps may also have arisen from rodent caches.

La Tourrette *et al.* (1971) and McAdoo *et al.* (1983) measured densities of clumps of cheatgrass (*Bromus tectorum*) and Indian rice grass, presumably representing heteromyid caches. Densities were 3 clumps m^{-2} for cheatgrass and 1.37 clumps m^{-2} for Indian rice grass. Many of these caches contained both seedlings and ungerminated seeds. The inference that clumps of seeds or seedlings always represent rodent caches must be treated cautiously, because Reichman (1984) showed that large numbers of seeds may accumulate in natural depressions in open areas of the Sonoran desert. However, cached seeds may occur in distinctive patterns when cached, permitting unambiguous cache recognition (J. A. Young, pers. comm, 1985).

One of the most complete studies of successful establishment from rodent caches is that of Reynolds and his colleagues, on *Dipodomys merriami* and velvet mesquite in southern Arizona (Reynolds and Glendening, 1949; Paulsen, 1950; Reynolds, 1954, 1958). They studied rodent caching by putting out mesquite seeds mixed with a 'tracer' seed that germinated after the first rain and enabled them to locate mixed caches. They also found many clumps of 4–13 velvet mesquite seedlings which may have germinated from rodent caches. Only 16% of the seedlings were under crowns of mature mesquite plants. Since this is a fairly large seed, wind dispersal is insignificant, and movement of seeds into open areas by animals seems more likely. However, some animal dispersal was attributed not to rodents, but to cattle, native ungulates (deer, *Odocoileus* spp., and peccaries, *Tayassu tajacu*), and coyotes (*Canis latrans*), which eat mesquite pods and defecate viable seeds.

D. Probability of Becoming Established

If the only effect of rodents were to remove seeds from the pool of viable recruits, one would expect them to have a negative effect on plant populations, and seeds would be expected to evolve characteristics that reduce their conspicuousness or value to rodents. As we will discuss below, neither of these expectations seems to be met in all cases, perhaps because by reducing densities of unharvested seeds and by manipulating, moving, and burying seeds, rodents may so enhance establishment probabilities and expected seed outputs of survivors that their net effect on plant fitness is positive. In this section we discuss the probable consequences of rodent activities for components of seed fitness.

1. Establishment of Unharvested Seeds

One of the primary effects of any seed harvester is to lower average soil seed densities. This occurs if seeds are removed at random, but is more pronounced if harvesters selectively forage from dense seed aggregations, as rodents do. Harvesters thus decrease the crowding of unharvested seeds. It may be a different story for the harvested seeds, however, because they may be cached in dense clumps. Let us ignore the fate of cached seeds for now and consider the consequences for unharvested seeds of a reduction in density, i.e. in number per unit area.

In the simplest population models, which assume constant probabilities of survival and reproduction, a reduction in seed number will cause a proportional reduction in recruitment to the next generation. These simple models rarely apply to plants, however, because seedling survival and reproductive success are highly density-dependent, and seed germination may be as well. Seed germination can either be positively or negatively density-dependent in the laboratory, but the situation in nature has not been investigated thoroughly (Harper, 1977; Antonovics and Levin, 1980). The only study of which we are aware is that of Inouye (1980), who found that the rate of appearance of new desert annual seedlings was enhanced if established plants were thinned.

Density also affects seedling growth and survival (Chapter 6 of Harper, 1977). The course of growth and mortality in a pure-species, even-aged stand of plants is often like that shown in Figure 2. When seedlings first emerge, they increase in size for a period of time with no mortality, until they become large enough to compete with neighbors. The point at which this begins to occur depends on the initial density. For example, in Figure 2 competition begins at Time 1 in the highest density stand ($D4$) when plants are still relatively small, and at much later times and with larger plant sizes in lower-density stands. As neighbors compete, some individuals die in a process called 'self-thinning' and the survivors continue to grow (arrows pointing upwards and to the left, Fig. 2). The relationship between survivor density and biomass per survivor is remarkably constant across plant species and follows a power law of the form:

$$w = c\,D^{-3/2}$$

where w is average plant biomass, c is a constant, and D is the density of survivors (Harper, 1977). Self-thinning continues along a trajectory described by this relationship until the maximum supportable biomass per unit area has been reached, at which time the exponent changes from $-3/2$ to -1 (Fig. 2). Harper (1977) and Silvertown (1982) discuss this phenomenon further.

Since the reproductive output of an individual is directly proportional to its biomass, it becomes apparent from Figure 2 that by reducing seed densities a seed consumer may either increase, not affect, or decrease the expected fitness of seeds. The net result depends on how much the densities are reduced, and how great is the density-dependence. For example, consider a hypothetical population of desert annuals in which individuals continue to survive but cease to grow after a cohort of seedlings has bumped up against the self-thinning curve (Went, 1949; Beatley, 1967; Klikoff, 1966). If the expected reproductive output of each seedling is proportional to its biomass, then it would be described well by a $-3/2$ power function of initial seed density. Suppose seed density is D in the absence of rodents; reproductive output per individual would equal

$$cD^{-3/2}.$$

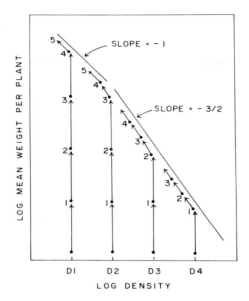

Fig. 2. Relationship between plant density and mean plant weight during growth in an even-aged stand of plants: the 'self-thinning' law. Arrows indicate the trajectory followed by a stand planted at a particular seed density (D_1 to D_4), and numbers indicate successive points in time. For a while after germination, seedlings increase in weight and there is no mortality (vertical arrows). After plants become large enough to compete with their neighbours, increases in size occur only at the expense of mortality, and arrows point upwards and to the left. The slope of the trajectory during this phase of growth is -3/2 until the canopy closes, at which point it changes to -1. After Harper (1977, Figure 6/31) and Silvertown (1982, Figure 5.7).

This value is the expected fitness per individual, in the absence of rodents. Now suppose that rodents consume a fraction ($1-p$) of seeds, thereby reducing seed densities by an equivalent proportion. Reproductive output of the survivors would be $c(pD)^{-3/2}$, and weighted mean seed fitness (the probability of surviving multiplied by the expected reproductive output of survivors) would be

$$(1-p)\,0 + pc\,(pD)^{-3/2} = pc\,(pD)^{-3/2}.$$

As long as p is a positive fraction between 0 and 1, this value is greater than $cD^{-3/2}$, which means the expected fitness is actually enhanced in the presence of seed consumers!

For this to occur the density-dependence must be extreme to compensate for the probability of seeds being eaten, and competing plant species must be affected to the same extent. Even if seed consumption does not enhance expected fitness of individual seeds, however, it still will have little effect on plant populations so long as seed densities are not reduced below replacement level (see Platt, 1976; p 460 of Harper, 1977).

The result that seed consumers can enhance expected seed fitness is similar to results of models of the evolution of disperal in general (c.f. Bullock, 1976; Hamilton and May, 1977). In these models, the frequencies of traits which cause dispersal should increase even though there is a risk to dispersal, because if density-dependence is strong enough, parents that disperse seeds away from the crowded local patch are at a selective advantage. However, this does not necessarily mean that plants will evolve to attract seed consumers! The situations are different in a number of respects, such as the fact that fitness of consumed seeds is zero while that of dispersers is not; when there is spatial mixing of offspring from different parents, any mutant whose seeds are more attractive to consumers will suffer a fecundity disadvantage at the same time as it is 'helping' neighboring non-mutants by lowering the level of competition their seeds experience. Conditions for the spread of such a mutant would be restrictive (c.f. Wilson, 1980; Nunney, 1985).

2. Establishment of Harvested Seeds

a. Effect of handling by rodents. — Rodents may kill seeds before caching them. Fox (1982), for example, reported that gray squirrels often excise the embryos of white oak (but not red oak) acorns before caching them. Unlike red oak acorns, white oak acorns typically germinate in fall within a few weeks of dropping, so by excising the embryos, squirrels prevented early germination of their seed stores.

For other plant species, however, germination may be unlikely *unless* seeds are handled by rodents. McAdoo *et al.* (1983) showed that kangaroo rats husk Indian rice grass seeds before caching them, a procedure which markedly increased percent germination in standard laboratory tests. Soil samples analysed by Young *et al.* (1983) contained more empty husks and fewer germinable seeds as winter progressed, suggesting that rodents probably also cache husked seeds in the field. Another intriguing aspect of this study is that Indian rice grass seeds are strikingly polymorphic, and rodents select the most germinable seed morphotypes.

In addition to husking seeds, rodents sometimes bite them, presumably to ascertain whether they are acceptable. Reynolds and Glendening (1949) reported that Merriam's kangaroo rat tests mesquite seeds in this way, and that nicking seed coats of mesquite seeds in the laboratory increases germination form 6% to more than 90%. It is not known whether or not these rodent-mediated increases in germination are beneficial to the plant; conceivably, increasing the proportion of seeds that germinate immediately could be detrimental in the long term in a fluctuating environment.

Rodents may also serve as vectors for microorganisms. Rebar and Reichman (1983) found spores of seed-decomposing fungi in heteromyid cheek pouches. Presumably these spores inoculate cached seeds and hasten their decomposition in a moist burrow environment. It is conceivable that cheek pouches could also harbor spores of mycorrhizal fungi that might enhance seedling establishment.

b. Effect of caching. — One of the primary ways rodents can affect the establishment of seedlings is by burying seeds. Some seeds, such as acorns (Griffin, 1971) and horse chestnuts (Thompson and Thompson, 1980) may fail to germinate if they are not buried; caching is probably the major process by which these large nuts are buried.

The depth of burial is probably critical. Olmsted (1937) observed heavy mortality of black oak seedlings (*Quercus velutina*) emerging from squirrel caches. Barnett (1977) reported equal germination rates for white oak acorns and pignut hickory nuts whether they were buried 2.5 cm deep, or covered only by leaf litter. Griffin (1971) showed that acorns (*Q. douglasii, Q. lobata,* and *Q. agrifolia*) buried at 7 cm had much higher germination success than acorns on the surface, but Shaw (1968b) reached the opposite conclusion for *Q. petraea.* These divergent results may have been due to differences in soil moisture and temperature between Griffin's California study site and Shaw's in North Wales, or to the different species of oaks involved. In arid sites, burial may enhance germination by minimizing desiccation; at a humid site, burial may stimulate decomposition before germination can occur. Wood (1938) investigated relationships between physical conditions and germination of chestnut oak (*Q. prinus*) acorns in the field. The optimum depth of burial appeared to be 2 to 5 cm.

Because cache location is likely to affect the rate of rotting or germination, rodents should be selective about the location of caches. Douglas squirrels may store sequoia cones in moist areas, where spontaneous opening of the cones and release of seeds is inhibited (Shaw, 1936). By contrast, Shellhammer (1966) observed a Douglas squirrel storing cones in dry microsites even though water was present nearby. The squirrel observed by Shellhammer stored most cones at the bases of trees and near fallen logs, which is also characteristic of agoutis (Smythe, 1970). Storing seeds under or near objects could be important in relocating caches if memory is a primary mechanism of cache recovery.

Cache microsite undoubtedly has a large effect on seedling establishment. Harper (1977) has reviewed the pronounced effects on seed germination of substrate type, microtopography, and even the presence of objects on the soil surface. On a large scale, the microhabitat in which the cache is placed is also important. For example, in tropical forests, seedlings have much higher chances of establishing themselves in light gaps (Augspurger, 1984; Howe *et al.,* 1985); in deserts many species can become established only in the shade of a 'nurse plant' (Steenbergh and Lowe, 1969; Yeaton, 1978; D. Samson, unpubl. data).

c. Mortality of seeds and seedlings — The rodents that cache seeds may also eat seedlings that emerge from caches. Paulsen (1950) showed that 53% of mesquite seedlings protected from small mammals by wire cones survived for one year, but only 6% of seedlings protected from cattle but not small mammals, and 4% exposed to both cattle and small mammals, survived as long. Wood (1938) reported that 77 of 78 planted oak seedlings were harvested within five days and 1218 of 1224 were taken by mice which dug under a protective screen. Watt (1919), Evans *et al.*, (1983), and Mills (1983) also suggested that rodents kill many seedlings, although Boyd and Brum (1983) showed that survivorship of creosote bush seedlings for 90 days was less than 5% whether they were protected from mammals or not (seedlings exposed to mammals died earlier in the 90 day period). The results of caging experiments must be interpreted carefully, however, because cages may create more favorable microclimatic conditions for seedlings in addition to excluding rodents (Shaw, 1968b). For heteromyid rodents, and perhaps other species, we expect substantial consumption of seedlings because of the importance of green vegetation in their diet for successful reproduction in desert environments (Munger *et al.*, 1983). Abbott and Quink (1970) reported evidence of repeated visits of mice to caches of white pine seeds and the uprooting of seedlings as the mice harvested additional seeds from these caches.

One hypothesis for the adaptive value of scatterhoarding as opposed to larderhoarding is that seeds last longer before rotting in scatterhoards (Elliot, 1978; Stapanian and Smith, 1978). From the plant's perspective, however, the key question is whether fungal attack occurs sooner for seeds not harvested by rodents than for scatterhoarded seeds. Augspurger (1984) found that seeds of several tropical tree species survived better in light gaps than under canopies of parent plants; the difference was mainly due to reduced rates of pathogen attack in light gaps. If rodents move seeds to open areas, then they may enhance the probabilities of survival of these seeds.

There is conflicting evidence on the extent of mortality caused by insects for seedlings emerging from presumed rodent caches. Evans *et al.* 1983) found insects to be a major cause of mortality in bitterbrush seedlings, but Mills (1983) found no significant effect of insects on chamise (*Adenostoma fasiculatum*) or ceanothus (*Ceanothus greggii*) seedling survival. In theory, by moving seeds away from adult plants, rodents could reduce the losses of seeds and seedlings caused by animals that respond mainly to distance in searching for food, but to the extent that they store seeds in clumps, rodents might increase mortality caused by density-responsive seed consumers (Clark and Clark, 1984).

IV. NET EFFECT OF GRANIVOROUS RODENTS

It should be abundantly clear by now how complex the interaction between seed-eating rodents and plants is. Not only do fate diagrams of the interaction contain many branching paths that feed into one another, but all their variables are functions of seed and rodent characteristics, as well as of other biotic and abiotic features of the environment in which the interaction occurs. These variables can change as populations fluctuate or as climatic conditions vary or as the interactants evolve. It is therefore not surprising that so little is known about the net effects of seed-eating rodents on plant populations and on the evolution of seed traits.

In this section we review several experimental field studies. Our goal is twofold: to draw tentative conclusions about the probable effects of granivorous rodents; and to discuss critically methods of approach.

A. Consequences for Populations

The most straightforward method for assessing the impact of rodents on plant populations is to control rodents with fences and observe the effects of these treatments on plants. Several such studies have been conducted in North American desert and grassland sites.

Reynolds (1950, 1958) reported the effects of excluding rodents from a desert scrub-grassland transition zone site in southern Arizona. After exclosures were established, a six-year drought ensued during which large-seeded shrub and perennial grass populations decreased more rapidly on rodent-grazed than on rodent-protected sites. The drought eased in 1936, and in the next six years of favorable conditions the trend was reversed, and large-seeded species increased much more rapidly on rodent-grazed than rodent-protected sites. Because these large-seeded species are long-lived perennials, the pattern persisted once it had become established. Reynolds explained this result as follows. Merriam's kangaroo rat, the most common rodent on the site, selectively harvests and caches large seeds (greater than 4 mg in weight). In drought years individuals consume most of what they cache and thereby reduce the abundance of these species relative to those with small, non-preferred seeds. When seed production is high, rodent populations increase, but not fast enough to keep pace with the increased seed production. Rodents cache more seed than they consume and, being in a favorable microsite, cached seeds become established more frequently than do those on the soil surface. Hence during favorable years recruitment of large-seeded species is higher when rodents are present. Expressed in terms of the fate diagram, the change in climatic conditions after 1936 decreased the probability that a cached seed was recovered and consumed.

This example underscores the potential importance of recognizing year-to-year variation in path probabilities when assessing the net effect of rodents. For the first six years, rodents had a small net negative effect on populations of large-seeded plants, but for the next six years they had a large positive effect. The net effect of rodents over the reproductive life of a large-seeded plant depends on the frequencies of good and bad years.

In this light it is interesting to consider mast-seeding, which has the same effect on seed-to-rodent ratios as climate-induced fluctuations in deserts. Mast-seeding is usually interpreted as an adaptation for ensuring seed survival by periodically satiating seed consumers with huge seed crops. Most animal populations cannot respond rapidly enough to track strong yearly swings in food availability, so more of the seeds produced during a mast year are likely to escape being eaten. Indeed, the probability of a seed being removed appears generally to be lower in mast years (e.g. Vincent, 1977; Jensen, 1982; O'Dowd and Gill, 1984). Mast-seeding also can affect the benefit realized from seed-planting animals, by determining the average probability that seeds are cached and then not recovered. The optimal masting periodicity and amplitude should vary depending on whether or not the effect of seed consumers is completely negative. It may be of interest, therefore, to compare mast year frequencies of plant species subject to caching and non-caching seed consumers.

J. H. Brown and his co-workers have conducted experimental studies of interactions among seed-eating rodents, ants, and Sonoran and Chihuahuan desert annuals (Brown *et al.*, 1979a,b, 1986; Reichman, 1979; Inouye *et al.*, 1980; Davidson *et al.*, 1985). They used fences and poisoning to remove various combinations of ants and rodents from experimental plots. One caveat is in order when interpreting the results of these studies, and perhaps also those of Reynolds (1958): the fences probably lowered the abundance of mammalian herbivores (rabbits, *Lepus* spp. and *Sylvilagus* spp.) as well as granivores. It is possible that herbivores were responsible for the effects of the experimental treatments, but for purposes of discussion let us assume that rabbit exclusion had no effect.

At the Sonoran Desert site, the removal of rodents produced significant increases in soil seed densities and somewhat smaller, but still significant, increases in vegetative plant densities after three years (Inouye *et al.*, 1980; Inouye *et al.*, 1981; Brown *et al.*, 1986). Rodents had a greater effect on large-seeded than on small-seeded plants (large seeds weighed more than 1 mg); in rodent-free plots, the densities of large-seeded annuals were 1.5-8.2 times greater than on control plots. Similar increases in densities of large-seeded species following the removal of rodents have been observed over a 12-year period at another Sonoran Desert site, and over five years at a Chihuahuan desert site (Brown *et al.*, 1986; Davidson *et al.*, 1985).

Presumably removal of rodents had a negative effect on the abundance of small-seeded annuals because rodents do not consume these seeds as avidly as large seeds, and small-seeded species were thus released from interspecific competition (Inouye *et al.*, 1980; Brown *et al.*, 1986; Davidson *et al.*, 1985). Such indirect effects make it extremely difficult to predict the net effect of a change in rodent abundance on plant populations; direct effects on recruitment of preferred plant species produce a cascade of indirect effects mediated through the responses of other granivore species, pathogens (c.f. Inouye 1981), and competing plant species. This means that it is not advisable to infer the net effect of rodents from the analysis of only one component of a fate diagram (e.g., from the total amount of seed consumed).

The difference between these results and those of Reynolds is intriguing, for in one case heteromyid rodents enhanced abundance of large-seeded species in good years, while in the other case they had the opposite effect. One reason for the discrepancy might be the set of plant species considered. Brown and his co-workers were primarily concerned with winter-germinating annuals, whereas Reynolds was concerned with summer-germinating perennial grasses and shrubs. Winter annuals set seed in late spring, when rodent densities are at their highest, and the seeds are exposed to rodents throughout the summer and fall until they germinate after the onset of winter rains. The seeds of perennial shrubs can germinate during the summer rains, not long after they are shed, and the seeds of many perennial grasses are shed in October after some heteromyids have become inactive above ground. It is conceivable that the species studied by Reynolds, by virtue of their flowering and seeding phenologies, enjoy the benefits of being cached without paying high costs in being recovered and eaten. Davidson *et al.* (1985, compare their Tables 4 and 5) also noted that rodents affected summer annuals less than winter annuals. Summer annuals may be less preferred, but it is also possible they experience less consumption by being available when fewer granivores are active. Another possibility is that shrubs benefit disproportionately from their seeds being cached, because their large seed size enables seedlings to emerge from depths of 2–5 cm in the soil, typical depths for rodent seed caches, whereas smaller seeds lack sufficient reserves for seedlings to emerge.

Fitch and Bentley (1949) compared the effects of various rodent species on the composition of California annual grasslands. In 1936 four enclosures were stocked, one with herbivorous ground squirrels (*Spermophilus beecheyi*), another with largely granivorous kangaroo rats (*Dipodomys heermanni*), a third with herbivorous pocket gophers (*Thomomys bottae*), and the fourth with no rodents at all. Enclosed rodent populations were allowed to fluctuate somewhat, but were monitored to ensure that they did not go below a minimum stocking density or become contaminated with other species. The enclosures were observed until 1943, at which time the animals were removed, and the plant community was recensused in 1946 and 1947 to assess intrinsic differences between enclosures. The most striking result was that, of the three stocked enclosures, the vegetation in the one with kangaroo rats was consistently most similar to the control, in terms of both aboveground biomass and species composition. Even highly-preferred large-seeded species, such as filaree (*Erodium botrys*), were more affected by ground squirrel grazing than by kangaroo rat seed removal. The effect could not be accounted for by intrinsic differences between enclosures.

Borchert and Jain (1978) coupled exclosure studies with more detailed analyses of interaction mechanisms. From laboratory feeding trials, they determined that *Mus musculus* and *Microtus californicus* prefer wild oat (*Avena fatua*) seeds to those of three other grasses. Rates of seed disappearance from areas accessible to these mice were much higher than from fenced control plots (where loss was negligible), and wild oat seeds were selectively removed. However, reductions in vegetative densities and dry weights did not correspond directly to changes in seed numbers. For example, when *Hordeum leporinum* seeds were placed out with those of wild oat, rodents removed 62% of the latter and only 30% of the former. At the end of the growing season *Hordeum leporinum* dry mass was the same, and wild oat dry mass was reduced by only 41%, on rodent-grazed compared to rodent-protected plots. Similar results were obtained in pairwise plantings of wild oat with the other species. Interpretation of these results is complicated by the fact that *Microtus californicus* will eat substantial amounts of vegetative tissue as well as seeds. However, the relationship between relative seed numbers and relative biomass is clearly not a simple one. By reducing overall seed abundance, rodents increased the per-capita biomass of established plants. Each plant species appeared to respond more to a reduction in its own density than to reductions in those of other species, but non-preferred species such as *Bromus diandrus* and *Lolium multiflorum* clearly benefited as well from a reduction in the abundance of *Avena*. The net effects of rodents were to increase the probability that a surviving seed produced a mature plant, and to change the proportional representation of plant species in the community by reducing the abundance of wild oat relative to each of the other three species.

All of the studies discussed so far used a direct experimental approach. It is also possible, at least in principle, to infer the net effect of rodents on plant populations by dissecting the fate diagram. However, this is so difficult that there are few studies that present an analysis complete enough to permit firm conclusions to be drawn. The best examples of this approach that we have found are those of Shaw (1968a,b) and McAdoo et al. (1983), and even here enough steps remained unexplored to render conclusions tentative at best. Shaw focused on the effect of rodents (especially wood mice and bank voles) on establishment of sessile oak. His approach was to compare the success of acorns placed on the surface of the ground, where they would fall if naturally shed, with that of acorns planted either under a shallow layer of oak litter or 1–2.5 cm deep in the soil, corresponding to the depth where rodents would cache them. All planting treatments were either caged to exclude various classes of granivores, or made accessible to granivores, with suitable controls for cage effects. Rodents were the most important granivores. No rodent-accessible seeds on the surface produced seedlings, which indicated that rodents removed practically all acorns. The fate of these harvested seeds was not determined. Forty-four percent of caged surface seeds germinated, whereas 58% germinated under litter, and 36% under soil. Approximately 1% of uncaged seeds under litter germinated, and 4% of uncaged seeds under soil germinated. If one had an estimate of the proportion of harvested seeds that were cached in various locations, and the probability that they were recovered before they could germinate, then it would be possible to calculate the net effect of rodents on germination probability. In this particular case, since such a high percentage of surface acorns germinated, the net effect surely was negative. However, Shaw remarked that this high success of surface acorns was probably a consequence of the unusually high humidity in his North Wales study area, and not representative of conditions elsewhere in the species' range.

McAdoo *et al.* (1983) attempted to develop a detailed accounting of Indian rice grass seed fates in Nevada, based particularly on observations of rodent behavior, seed germinability under various conditions, and densities of seed caches. As yet, the accounting is not sufficient to allow them to estimate how rodents affect recruitment, but their observations suggest recruitment from seed caches could be large enough to offset the number of seed consumed. Only 72% of the seeds of this plant are viable, the rest being empty, and heteromyid rodents selectively harvest 43% of viable seeds, judging from the densities of seed husks in the soil. They cache approximately 8% of these, judging from cache densities. By removing the husk before caching seeds, rodents increase germinability to about 33%, and burying the seeds greatly enhances the chances of seedling establishment. About 0.02% of seeds produced in a good year become seedlings, and rodents appeared responsible for much of this recruitment. Exactly how much will not be clear until quantitative information is available about the probability that an unharvested seed on the soil surface becomes established.

The only general conclusion that can be drawn from these studies is that one would be foolish to attempt a general statement about the effect of granivorous rodents on plant populations. Depending on competitor densities, climatic and edaphic conditions, rodent densities, seed preferences, and caching behavior, a plant's abundance may be increased, decreased, or not affected by rodents.

B. Evolutionary Consequences

Little is understood about the evolutionary effects of rodents on plants. For all practical purposes it is impossible to use a straightforward experimental approach to answer evolutionary questions; variation in most interesting characters is limited, and changes in morph frequencies within populations are often too slow to measure. Hence, attempts to compare the course of change in experimental populations with and without rodents would be successful only in exceptional circumstances. To approach evolutionary problems one must instead use one of two indirect methods. The effects of rodents can be inferred by comparing plant traits in regions that vary systematically in their rodent densities (c.f. Cook *et al.*, 1971). The difficulty here is that one can rarely exclude the possibility that the effect is due to some factor, other than rodents, which also varies between regions. Alternatively, one can extrapolate from relative phenotype success over the short term to what should happen to phenotype frequencies over the long term. Such extrapolation is always risky but can usually be tested by analysing patterns in the distribution of plant traits. A combination of these two approaches is especially effective, in our opinion. Pattern analysis, the first approach, is an efficient way to identify traits that are likely to be significant in shaping seed fates. It is also a means of testing predictions derived from fate diagram analysis, the second approach, which is an effective way to verify that a trait does affect fitness components, and to determine whether the postulated mechanism is correct. Because the analysis of fate diagrams is time-consuming, it is best used to study traits that pattern analysis indicates are likely to be significant.

The interactions that dictate the relative success of phenotypes are so complex that developing an unambiguous prediction by the second approach is itself a difficult matter. While it is not too difficult to construct a generalized fate diagram, estimates of the effects of a phenotypic character on path probabilities and reproductive outputs must be derived experimentally. Not only is it difficult to devise an experiment that captures the important elements of the natural situation, but the short-term nature of most experiments may restrict their utility. If fitness is strongly affected by climatic patterns over long periods, then an experiment conducted under one climatic regime may be misleading. The value of predictions based on fate diagram analysis thus depends on how thorough the analysis has been. If estimates of variables are based on unrealistic experimental conditions, then the fitness estimates and the evolutionary predictions derived from them may be inaccurate. Similarly, if only a small portion of the fate diagram has been explicitly considered, then some subtle but important consequence of altering one portion of the diagram or some trade-off between fitness gained through one or another path may be overlooked. If so, the prediction may be worthless.

We have found few instances where an evolutionary hypothesis has arisen from a thorough fate diagram analysis. Instead, hypotheses are proposed *a posteriori* to account for an observed pattern, or one subset of a diagram is examined and expectations developed on the assumption that other elements of the diagram do not substantially change the picture.

The fate diagram component that has received most attention is the probability of harvest, and most evolutionary predictions are based on the assumption that a seed harvested by rodents has lower expected fitness that one which is overlooked entirely. Elliott (1974; see also Smith, 1975) examined how attributes of the cones of serotinous lodgepole pine affect squirrel feeding preference. Serotinous cones remain closed until triggered by fire; non-serotinous cones open immediately. Elliott found that the number of cones removed by red squirrels from individual trees was significantly higher in trees with many seeds per cone, and hence high ratios of seed-to-total-cone weight, and in trees whose cones were shaped so as to be easily detached from the branch. From these observations, he predicted that squirrel feeding preferences for profitable and easily harvested cones would tend to increase the relative investment in protective cone tissue, even though that would increase resource investment per seed and hence lower the number of seeds a tree could produce. Indeed, only 1% of the weight of a cone consisted of seed material. This proportion was higher — 2% — in non-serotinous lodgepole pines that inhabit moister areas and release seeds immediately upon cone maturation. Elliott argued that the shorter seed retention time in non-serotinous races caused the characteristics of the cone to be less important in dictating harvest probabilities than in serotinous races, hence shifting the balance towards production of cones with relatively less expensive protective tissue. Difficulties with this interpretation are: (1) that cone characteristics affecting squirrel feeding preferences also probably are important in affecting seed susceptibility to hot fires, so that observed patterns of allocation to protective cone tissue could be functions of fire rather than squirrels; and, (2) it is plausible, but not certain, that the net effect of harvest by squirrels is negative.

This example illustrates the importance of keeping in mind that the interaction between rodents and seeds does not take place in a vacuum, and that the evolutionary effects of rodents may depend on the strengths of conflicting selective pressures mediated through the physical environment, or by competitors. Smith (1975) has discussed the ways in which the effects of rodents may be superimposed upon physical constraints. For example, many forest or arid-zone plants may maintain large seeds, even though this increases their desirability to rodents, because otherwise their seedlings would not have the reserves to send roots deep enough to obtain moisture. Ellner and Shmida (1981) used a similar argument to account for the low incidence in desert plants of morphological characteristics that facilitate other means of dispersal, such as plumes or fleshy fruits. They argued that the cost of producing specialized dispersal structures or losing seeds to an animal dispersal agent is not outweighed by an associated benefit from being situated in post-dispersal microsites that are more favorable than pre-dispersal microsites, because suitable microsites are not rare in deserts.

Other seed consumers may also modify the evolutionary effects of rodents. Elliott (1974) argued that, depending on fire frequency, squirrels should select either for retention of seeds in tough, closed, few-seeded cones, or for rapid seed release from less expensive cones. Tomback (1982, 1983) and Hutchins and Lanner (1982) have argued that, despite heavy consumption of seeds by squirrels, whitebark pine has evolved large wingless seeds retained in open cones. The difference may occur because an alternative seed harvester, Clark's nutcracker, is common in the range of whitebark pine. Nutcrackers remove seeds from open cones on trees, cache them in small, widely-dispersed scatterhoards, and apparently use memory to relocate caches many months later. They are probably not as efficient as rodents at recovering caches. Red squirrels usually cache seeds in middens at the base of established trees. They harvest some cones directly from trees, but they also harvest seeds after they have been shed. Hutchins and Lanner (1982) reported that no seedlings emerged from squirrel stores. Seeds cached by nutcrackers have a good chance of being overlooked by squirrels, especially if they are in meadow rather than forest habitat, and of emerging from the 3 cm cache depth. Although nutcrackers consume seeds as well as dispersing them, whitebark pine appears to have developed traits that increase the chance that nutcrackers will harvest seeds before the more destructive rodents get to them. Unfortunately, these investigators have not compared establishment probabilities of buried and surface seeds protected from other consumers. Without this information, it is impossible to judge whether nutcrackers act as beneficial seed dispersers or are simply the least of several evils. If the latter, then whitebark pine should evolve traits to escape nutcrackers in the absence of rodents.

Another instance of avoiding rodents by selectively attracting a different dispersal agent has been described in Nuttall's violet (*Viola nuttallii*; Turnbull and Culver 1983). This myrmecochorous species sheds its seeds in mid-morning, when ants are most active and rodents are inactive. The common rodents in the montane habitats of this species usually consume seeds as soon as these are encountered. They probably have a net detrimental effect on seed fitness, so plants that can make their seeds selectively accessible to ants will enjoy a fitness advantage.

In some cases rodents appear to be the lesser evil. Bradford and Smith (1977; Smith, 1975), in one of the most thorough fate analyses we have encountered, measured the relative fitnesses of one, two and three-seeded fruits of *Scheelea rostrata* (Palmae) in two areas with different suites of seed consumers. In Costa Rica, bruchid beetles attacked 52% of shed fruits, whereas rodents attacked 10%. In Panama, rodents were more important than bruchids, attacking 63% of fruits compared to the bruchids' 18%. Seedlings from multiseeded fruits suffer a pronounced competitive disadvantage because they are small and close to siblings. However, multiseeded fruits are attacked slightly less often by bruchids, and bruchid larvae in the infested fruits usually kill only one seed.

Rodents affect single and multiseeded fruits equally. The fruits contain a pulpy mesocarp attractive to rodents, which eat the pulp and then scatterhoard the stony endocarp with its enclosed seeds. By carrying seeds away from the vicinity of the parent tree and then burying them, rodents presumably provide some degree of escape from bruchids. Bradford and Smith argued that *Scheelea rostrata* has evolved the nutritional mesocarp and stony endocarp to take full advantage of rodent dispersal abilities. If only rodents were present, single-seeded fruits should predominate, because they produce more vigorous seedlings. However, when bruchids are common, multi-seeded fruits should have higher success. Indeed, the frequency of multi-seeded fruits is much higher (27%) in Costa Rica, where bruchids are abundant, than in Panama (6%) where they are not. The primary ingredient missing from this account is information about the net effect of rodents by themselves. The geographical patterns in fruit frequencies should occur regardless of rodents, and Bradford and Smith have not documented any variation in traits that should be influenced only by the interaction with rodents.

Where rodents are relatively rare, seed characteristics seem to have responded primarily to other seed consumers. Pulliam and Brand's (1975) analysis of seed production and consumption in an Arizona grassland provides an example. Rodents are scarce in these grasslands and less important seed consumers than ants and sparrows. In this area there are two peaks of seed production. Seeds produced in response to winter rains are accessible primarily to ants and rodents. These seeds are smooth and round, which makes them difficult for ants to handle, but desirable for rodents.

Seeds produced after summer rains are accessible primarily to wintering sparrows and rodents. These seeds have conspicuous awns and husks, which make them difficult for sparrows to eat. Pulliam and Brand suggest from this pattern that seed morphologies have been molded more by the seasonal importance of sparrows and ants than by the rodents, which are active year-round.

Therefore, where rodents are common, plants should evolve seed characteristics to avoid or attract them, depending on their net effect. Desert habitats near the grasslands studied by Pulliam and Brand support high densities of heteromyid rodents, which rarely consume seeds weighing less than 0.3 mg (Reichman, 1975, 1976; Price, 1983). Twenty-two of 40 seed species collected from soil samples near Tucson, Arizona weighed less than 0.3 mg (Reichman and Price, unpubl. data), whereas Pulliam and Brand (1975, Table 1) reported only seven of 29 grassland species in this weight class. From this it would appear that most, but by no means all, desert plants have evolved small seed size, perhaps to avoid harvest by rodents.

The general impression one gets from these examples is that granivorous rodents often reduce seed fitness, and that many seed traits are adaptations to escape them. We do not know whether this is accurate, because the studies just reviewed did not make the critical comparison of unharvested and harvested seed fitness. Most discussions of the net effect of rodents therefore should be viewed as plausible hypotheses, rather than rigorously supported conclusions.

V. DIRECTIONS FOR FURTHER RESEARCH

We can briefly summarize by quoting what has already been said about seed dispersal in general:

"Seed dispersal has to be one of the greyest words in biology. Everybody knows what it is yet we know next to nothing about it." (p. 105 of Janzen, 1983a).

"The biology of seed dispersal by animals ... has focused on the animals that arrive at a fruit crop and ... on how they treat the crop. The seed shadow thereby generated and what recruitment it yields to the female parent is *terra incognita* in field studies." (p. 104 of Janzen, 1983a).

"Astonishingly little is known about the advantages to a parent plant that are actually conferred by ... dispersal ..." (p. 201 of Howe and Smallwood, 1982).

With regard to rodent-seed interactions, something is understood about animal foraging choices and about factors that affect plant growth and reproduction. Much less is known, however, about how seed-eating rodents modify seed fates and what net effects they have on plant reproduction in nature. Rectifying this situation is tantamount to undertaking a complete fate diagram analysis.

Ecologists interested in seed dispersal have already begun such analyses. For example, as a first step in understanding why so many plants have specific characters for seed dispersal, Augspurger (e.g., 1984), Clark and Clark (1981, 1984), and Howe (e.g., Howe *et al.*, 1985), among others, have initiated systematic studies of how seed morphology affects seed density, dispersal distance, and microsite, and how these in turn affect seed survival and seedling growth.

Critics may argue that such a mechanistic focus will produce nothing more than a compendium of special case histories. We disagree. Robust generalizations should emerge from a mechanistic perspective so long as the large questions are kept in mind, and so long as empirical investigations are coupled with and guided by theoretical ones (e.g., Bullock, 1976; Hamilton and May, 1977).

ACKNOWLEDGEMENTS

N. Waser provided critical input during the development of the ideas expressed here. D. Janzen, J. Young, and N. Waser kindly commented on the manuscript. This work was partially supported by National Science Foundation grant BSR 8407602.

REFERENCES

Abbott, H. G. (1962). *J. Forestry* **60**, 97–99.
Abbott, H. G., and Quink, T. F. (1970). *Ecology* **51**, 271–278.
Abramsky, Z. (1983). *Oecologia* **57**, 328–332.
Antonovics, J., and Levin, D. A. (1980). *Annu. Rev. Ecol. Syst.* **11**, 411–452.
Augspurger, C. K. (1984). *Ecology* **65**, 1705–1712.
Barnett, R. J. (1977). *Am. Midland Nat.* **98**, 319–330.
Bartholomew, B. (1970). *Science* **170**, 1210–1212.
Batzli, G. O., and Pitelka, F. A. (1970). *Ecology* **51**, 1027–1039.
Beatley, J. C. (1967). *Ecology* **48**, 745–750.
Bell, E. A. (1984). *In* 'Seed Physiology. Vol. 1. Development' (D. R. Murray, ed.), pp. 245–264. Academic Press, Sydney.
Benkman, C. W., Balda, R. P., and Smith, C. C. (1984). *Ecology* **65**, 632–642.
Borchert, M. I., and Jain, S. K. (1978). *Oecologia* **33**, 101–113.

232   *M. V. Price and S. H. Jenkins*

Bowers, M. A. (1982). *J. Mammalogy* **63**, 361–367.
Boyd, R. S., and Brum, G. D. (1983). *Am. Midland Nat.* **110**, 25–36.
Bradford, D. F., and Smith, C. C. (1977). *Ecology* **58**, 667–673.
Broadbooks, H. E. (1958). *Miscellaneous Publications of the Museum of Zoology of the University of Michigan* **103**, 5–42.
Brown, J. H. (1973). *Ecology* **54**, 775–787.
Brown, J. H. (1975). *In* 'Ecology and Evolution of Communities' (M. L. Cody and J. M. Diamonds, eds), pp. 315–341. Belknap Press of Harvard University Press, Cambridge, Massachusetts.
Brown, J. H., Davidson, D. W., and Reichman, O. J. (1979a). *Am. Zool.* **19**, 1129–1143.
Brown, J. H., Reichman, O. J., and Davidson, D. W. (1979b). *Annu. Rev. Ecol. Syst.* **10**, 201–227.
Brown, J. H., Davidson, D. W., Munger, J. C., and Inouye, R. S. (1986). *In* 'Community Ecology' (J. O. Murie and G. R. Michener, eds), pp. 41–61. Harper and Row, N.Y.
Brown, J. H., Grover, J. J., Davidson, D. W., and Lieberman, G. A. (1975). *Ecology* **56**, 987–992.
Bullock, S. H. (1976). *Oecologia* **24**, 247–256.
Cahalane, V. H. (1942). *J. Wildlife Management* **6**, 338–352.
Chew, R. M., and Chew, A. E. (1970). *Ecologial Monographs* **40**, 1–21.
Christy, E. J., and Mack, R. N. (1984). *J. Ecol.* **72**, 75–91.
Clark, D. A., and Clark, D. B. (1981). *Oecologia* **49**, 73–75.
Clark, D. A., and Clark, D. B. (1984). *Am. Nat.* **124**, 769–788.
Cook, A. D., Atsatt, P. R., and Simon, C. A. (1971). *Bioscience* **21**, 277–281.
Davidson, D. W., Samson, D. A. and Inouye, R. S. (1985). *Ecology* **66**, 486–502.
Eisenberg, J. F. (1981). 'The Mammalian Radiations'. University of Chicago Press, Chicago.
Elliott, L. (1978). *Smithsonian Contributions Zoology* **265**, 1–107.
Elliott, P. F. (1974). *Evolution* **28**, 221–231.
Ellner, S., and Shmida, A. (1981). *Oecologia* **51**, 133–144.
Evans, R. A., Young, J. A., Cluff, G. J., and McAdoo, J. K. (1983). *United States Department of Agriculture Forest Service General Technical Report* INT-**152**, 195–202.
Everett, R. L., Meeuwig, R. O., and Stevens, R. (1978). *J. Range Management* **31**, 70–73.
Fitch, H. S., and Bentley, J. R. (1949). *Ecology* **30**, 306–321.
Fox, J. F. (1982). *Evolution*, **36**, 800–809.
Freeland, W. J., and Janzen, D. H. (1974). *Am. Nat.* **108**, 269–289.
Glanz, W. E. (1984). *American Society of Mammalogists 64th Annual Meeting*, abstract No. 209.
Griffin, J. R. (1971). *Ecology* **52**, 862–868.
Hamilton, W. D., and May, R. M. (1977). *Nature* **269**, 578–581.
Harper, J. L. (1977). 'Population Biology of Plants'. Academic Press, London.
Hawbecker, A. C. (1940). *J. Mammalogy* **21**, 388–396.
Hay, M. E., and Fuller, P. J. (1981). *Ecology* **62**, 1395–1399.
Heaney, L. R. (1984). *In* 'The Biology of Ground-Dwelling Squirrels' (J. O. Murie and G. R. Michener, eds), pp. 43–78. University of Nebraska Press, Lincoln.
Heithaus, E. R. (1981). *Ecology* **62**, 136–145.
Howard, W. E., and Cole, R. E. (1967). *J. Mammalogy* **48**, 147–150.
Howe, H. F., and Smallwood, J. (1982). *Annu. Rev. Ecol. Syst.* **13**, 201–228.
Howe, H. F., Schupp, E. W., and Westley, L. C. (1985). *Ecology* **66**, 781–791.
Hubbell, S. P., and Werner, P. A. (1979). *Am. Nat.* **113**, 277–293.
Hutchins, H. E., and Lanner, R. M. (1982). *Oecologia* **55**, 192–201.
Hutto, R. L. (1978). *Oecologia* **33**, 115–126.

Inouye, R. S. (1980). *Oecologia* **46**, 235–238.

Inouye, R. S. (1981). *Oecologia* **49**, 425–427.

Inouye, R. S., Byers, G. S., and Brown, J. H. (1980). *Ecology* **61**, 1344–1351.

Janzen, D. H. (1969). *Evolution* **23**, 1–27.

Janzen, D. H. (1971). *Annu. Rev. Ecol. Syst.* **2**, 465–492.

Janzen, D. H. (1983a). *Biol. J. Linn. Soc.* **20**, 103–113.

Janzen, D. H. (1983b). *In*: 'Coevolution' (D. J. Futuyma and M. Slatkin, eds), pp.232–262. Sinauer Associates, Sunderland, Massachusetts.

Jensen, T. S. (1982). *Oecologia* **54**, 184–192.

Johnson, T. K., and Jorgensen, C. D. (1981). *J. Range Management* **34**, 312–314.

Kaufman, L. W., and Collier, G. (1981). *Am. Nat.* **118**, 46–60.

Kiltie, R. A. (1981). *Biotropica* **13**, 141–145.

Klikoff, L. G. (1966). *Am. Midland Nat.* **75**, 383–391.

Kraus, B. (1983). *Ecology* **64**, 608–610.

Krebs, J. R. (1978). *In* 'Behavioural Ecology' (J. R. Krebs and N. B. Davies, eds), pp. 23–63, First Edition. Sinauer Associates, Sunderland, Massachusetts.

Kurten, B. (1971). 'The Age of Mammals'. Columbia University Press, New York.

Landeen, D. S., Jorgensen, C. D., and Smith, H. D. (1979). *Great Basin Nat.* **39**, 267–273.

Lanner, R. M. (1981). 'The Piñon Pine'. University of Nevada Press, Reno.

La Tourrette, J. E., Young, J. A., and Evans, R. A. (1971). *J. Range Management* **24**, 118–120.

Lawhon, D. K., and Hafner, M. S. (1981). *Oecologia* **50**, 303–309.

Levins, R. (1975). *In* 'Ecology and Evolution of Communities' (M. L. Cody and J. M. Diamond, eds), pp. 16–50. Belknap Press of Harvard University Press, Cambridge, Massachusetts.

Lewis, A. R. (1980). *Ecology* **61**, 1371–1379.

Lewis, A. R. (1982). *Am. Midland Nat.* **107**, 250–257.

Lockard, R. B., and Lockard, J. S. (1971). *J. Mammalogy* **52**, 219–221.

Lockner, F. R. (1972). *Z. Tierpsychol.* **31**, 410–418.

McAdoo, J. K., Evans, C. C., Roundy, B. A., Young, J. A., and Evans, R. A. (1983). *J. Range Management* **36**, 61–64.

McQuade, D. B., Williams, E. H., and Eichenbaum, H. B. (1986). *Z. Tierpsychol.*

Mares, M. A., and Rosenzweig, M. L. (1978). *Ecology* **59**, 235–241.

Mills, J. N. (1983). *Oecologia* **60**, 267–270.

Mittelbach, G. G., and Gross, K. L. (1985). *Oecologia* **65**, 7–13.

Munger, J. C., Bowers, M. A., and Jones, W. T. (1983). *Great Basin Naturalist Memoirs* **7**, 91–116.

Nelson, J. F., and Chew, R. M. (1977). *Am. Midland Nat.* **97**, 300–320.

Nelson, J. R., Wilson, A. M., and Goebel, C. J. (1970). *J. Range Management* **23**, 163–170.

Noy-Meir, I. (1973). *Annu. Rev. Ecol. Syst.* **4**, 25–51.

Noy-Meir, I. (1974). *Annu. Rev. Ecol. Syst.* **5**, 195–214.

Nunney, L. (1985). *Am. Nat.* **126**, 212–230.

O'Dowd, D. J., and Gill, A. M. (1984). *Ecology* **65**, 1052–1066.

O'Dowd, D. J., and Hay, M. E. (1980). *Ecology* **61**, 531–540.

Olmsted, C. E. (1937). *Botanical Gazette* **99**, 209–300.

Paulsen, H. A. Jr. (1950). *J. Range Management* **3**, 281–286.

Pearson, O. P. (1964). *J. Mammalogy* **45**, 177–188.

Platt, W. J. (1976). *Oecologia* **22**, 399–409.

Poche, R. M., Mian, M. Y., Haque, M. E., and Sultana, P. (1982). *J. Wildlife Management* **26**, 139–147.

Price, M. V. (1978a). *J. Mammalogy* **59**, 624–626.
Price, M. V. (1978b). *Ecology* **59**, 910–921.
Price, M. V. (1983). *Oecologia* **60**, 259–263.
Price, M. V. (1984). *Netherlands J. Zool.* **34**, 63–80.
Price, M. V., and Brown, J. H. (1983). *Great Basin Nat. Memoirs* **7**, 117–134.
Price, M. V., and Heinz, K. M. (1984). *Oecologia* **61**, 420–425.
Price, M. V., and Waser, N. M. (1985). *Ecology* **66**, 211–219.
Price, M. V., Waser, N. M., and Bass, T. A. (1984). *J. Mammalogy* **65**, 353–356.
Pulliam, H. R., and Brand, M. R. (1975). *Ecology* **56**, 1158–1166.
Pyke, G. H. (1984). *Ann. Rev. Ecol. Syst.* **15**, 523–575.
Pyke, G. H., Pulliam, H. R., and Charnov, E. L. (1977). *Quart. Rev. Biol.* **52**, 137–154.
Radvanyi, A. (1970). *Ecology* **51**, 1102–1105.
Raiffa, H. (1968). 'Decision Analysis'. Addison-Wesley, Reaching, Massachusetts.
Rebar, C., and Reichman, O. J. (1983). *J. Mammalogy* **64**, 713–715.
Reichle, D. E., ed. (1981). 'Dynamic Properties of Forest Ecosystems'. Cambridge University Press, Cambridge.
Reichman, O. J. (1975). *J. Mammalogy* **56**, 731–751.
Reichman, O. J. (1976). *Southwestern Nat.* **20**, 573–586.
Reichman, O. J. (1979). *Ecology* **60**, 1085–1092.
Reichman, O. J. (1981). *In* 'Foraging Behavior' (A. C. Kamil and T. D. Sargent, eds), pp. 195–213. Garland STPM Press, New York.
Reichman, O. J. (1983). *Great Basin Nat. Memoirs* **7**, 77–90.
Reichman, O. J. (1984). *J. Biogeography* **11**, 1–11.
Reichman, O. J., and Oberstein, D. (1977). *Ecology* **53**, 636–643.
Reynolds, H. G. (1950). *Ecology* **31**, 456–463.
Reynolds, H. G. (1954). *J. Range Management* **7**, 176–180.
Reynolds, H. G. (1958). *Ecological Monographs* **28**, 111–127.
Reynolds, H. G., and Glendening, G. E. (1949). *J. Range Management* **2**, 193–197.
Rosenzweig, M. L., and Sterner, P. W. (1970). *Ecology* **51**, 217–224.
Schmidt-Nielsen, K. (1964). 'Desert Animals'. Dover, New York.
Schoener, T. W. (1971). *Ann. Rev. Ecol. Syst.* **2**, 369–404.
Shaffer, L. (1980). *J. Mammalogy* **61**, 733–734.
Shaw, M. W. (1968a). *J. Ecol.* **56**, 565–583.
Shaw, M. W. (1968b). *J. Ecol.* **56**, 647–660.
Shaw, W. T. (1934). *J. Mammalogy* **15**, 275–286.
Shaw, W. T. (1936). *J. Mammalogy* **17**, 337–349.
Shellhammer, H. S. (1966). *J. Mammalogy* **47**, 525–526.
Sherbrooke, W. C. (1976) *Ecology* **57**, 596–602.
Sherman, R. J., and Chilcote, W. W. (1972). *Ecology* **53**, 294–298.
Shettleworth, S. J., and Krebs, J. R. (1982). *J. Experimental Psychology* **8**, 354–375.
Silvertown, J. W. (1980). *Biol. J. Linn. Soc.* **14**, 235–250.
Silvertown, J. W. (1982). 'Introduction to Plant Population Ecology'. Longman, Essex, England.
Smith, C. C. (1968). *Ecological Monographs* **38**, 31–63.
Smith, C. C. (1975). *In* 'Coevolution of Animals and Plants'. (L. E. Gilbert and P. H. Raven, eds), pp. 53–77. Univ. of Texas Press, Austin.
Smith, C. C. and Follmer, D. (1972). *Ecology* **53**, 82–91.
Smith, C. C., and Reichman, O. J. (1984). *Ann. Rev. Ecol. Syst.* **15**, 329–351.
Smith, C. F., and Aldous, S. E. (1947). *J. Forestry* **45**, 361–369.
Smythe, N. (1970). *Am. Nat.* **104**, 25–35.

Soholt, L. F. (1973). *Ecological Monographs* **43**, 357–376.
Sork, V. L. (1983). *Ecology* **64**, 1049–1056.
Sork, V. L. (1984). *Ecology* **65**, 1020–1022.
Sork, V. L., Stacey, P., and Averett, J. E. (1983). *Oecologia* **59**, 49–53.
Stapanian, M. A., and Smith, C. C. (1978). *Ecology* **59**, 884–896.
Stapanian, M. A., and Smith, C. C. (1984). *Ecology* **65**, 1387–1396.
Steenbergh, W. F., and Lowe, C. H. (1969). *Ecology* **50**, 825–834.
Sullivan, T. P. (1978). *Can. J. Zool.* **56**, 1214–1216.
Sullivan, T. P., and Sullivan, D. S. (1982). *J. Appl. Ecol.* **19**, 33–45.
Tappe, D. T. (1941). *J. Mammalogy* **22**, 117–148.
Terman, C. R. (1968). *In* 'Biology of *Peromyscus* (Rodentia)' (J. A. King, ed.), pp. 412–450. Special Publication No. 2, American Society of Mammalogists.
Thompson, D. C. (1978). *Ecology* **59**, 708–715.
Thompson, D. C., and Thompson, P. S. (1980). *Can. J. Zool.* **58**, 701–710.
Thompson, S. D. (1982). *Ecology* **63**, 1303–1312.
Tomback, D. F. (1982). *J. Animal Ecol.* **51**, 451–467.
Tomback, D. F. (1983). *In* 'Coevolution' (M. H. Nitecki, ed.), pp. 179–223. University of Chicago Press, Chicago.
Trombulak, S. C., and Kenagy, G. J. (1980). *Oecologia* **44**, 342–346.
Turnbull, C. L., and Culver, D. C. (1983). *Oecologia* **59**, 360–365.
Vander Wall, S. B. (1982). *Animal Behavior* **30**, 84–94.
Vander Wall, S. B., and Balda, R. P. (1977). *Ecological Monographs* **47**, 89–111.
Vandermeer, J. H. (1979). *Brenesia* **16**, 107–116.
Vincent, J. P. (1977). *Annales des Sciences Forestières* **34**, 77–87.
Vorhies, C. T., and Taylor, W. P. (1922). *United States Department of Agriculture Bulletin* **1091**, 1–40.
Watt, A. S. (1919). *J. Ecol.* **7**, 173–203.
Watt, K. E. F. (1968). 'Ecology and Resource Management'. McGraw-Hill, New York.
Went, F. W. (1949). *Ecology* **30**, 1–13.
West, N. E. (1968). *Ecology* **49**, 1009–1011.
Wilson, D. S. (1980). 'The Natural Selection of Populations and Communities'. Benjamin/Cummings Publishing Company Inc., Menlo Park, CA.
Wood, O. M. (1938). *Ecology* **19**, 276–293.
Wrazen, J. A., and Wrazen, L. A. (1982). *J. Mammalogy* **63**, 63–72.
Wright, S. (1921). *J. Agric. Res.* **20**, 557–585.
Wright, S. (1968). 'Evolution and the Genetics of Populations. Vol 1. Genetic and Biometric Foundations'. University of Chicago Press, Chicago.
Yeaton, R. I. (1978). *J. Ecol.* **66**, 651–656.
Young, J. A., Evans, R. A., and Roundy, B. A. (1983). *J. Range Management* **36**, 82–86.

CHAPTER 6

Seed Dispersal in Relation to Fire

R. J. WHELAN

I.	Introduction	237
II.	Fire as a Dispersal Vector	238
III.	Post-Fire Conditions	239
	A. Nutrients, Temperature, Water and Light	239
	B. Competition	240
	C. Animal Populations	241
	D. Summary	243
IV.	Long-Distance Dispersal	244
V.	Local Dispersal	245
	A. Post-fire Reproduction and Seed Release	247
	B. Soil-stored Seeds	255
VI.	General Theories	258
	A. Wind-dispersed Pioneers	259
	B. Patterns of Local Dispersal	262
	C. Variations within a Species	262
VII.	Summary	264
	Acknowledgements	265
	References	265

I. INTRODUCTION

Fire is a common natural disturbance in many parts of the world. Several recent reviews have investigated the responses of floras and faunas to fire (e.g. Kozlowski and Ahlgren, 1974; Gill *et al.*, 1981; Wright and Bailey, 1982; Wein and Maclean, 1983; Booysen and Tainton, 1984). Although some of these have considered the reproductive characteristics of plants in relation to particular fire regimes, a review of research into seed dispersal in relation to fire is difficult because so few studies have addressed the topic directly. Many reports of post-fire regeneration in fire-prone regions have focused on the remarkable regenerative powers of established plants

237

SEED DISPERSAL
ISBN   0 12 511900 3

through resprouting from both epicormic meristems and lignotubers (e.g. Purdie, 1977a; Gill, 1981b), and also through germination from a soil- or plant-stored seed bank. However, attention has not generally been given to the dispersal of the propagules that appear after fire. This is surprising in view of the importance that has been placed on long-distance dispersal of seeds in secondary successions such as old fields (Odum, 1971), and also on effective local dispersal within a relatively undisturbed community (Howe and Smallwood, 1982).

In this chapter, I examine first the possibility that the passage of a fire front may itself provide a means of dispersal for seeds. Secondly I describe the direct and indirect effects of fire on the physical and biotic characteristics of an area, because these effects have implications for the successful establishment of seeds that germinate after fire. I then consider long-distance dispersal of seeds into recently burned areas, compared with local dispersal within a burned site. The main aims of reviewing these topics are: (1) to illustrate the importance of understanding the seed-dispersal phase of the life histories of plants in a community if we are to predict post-fire changes in a community; (2) to point to areas in which information is particularly scarce; and, (3) to stimulate direct tests of the ideas presented.

II. FIRE AS A DISPERSAL VECTOR

The updraught in a wildfire is remarkable. A fire lit on a still day provides such an updraught that there is a marked inflow of air at ground level (Vines, 1974, 1981). The phenomenon provides a most difficult problem for firefighters, namely the transport of bits of smouldering bark and leaf material well ahead of the front of the fire to produce 'spotting'. There have been reports of spotting up to 25 km ahead of the fire front and material may be carried 3000 m up in a convection column (Foster, 1976). Fires 30 km inland commonly result in both charred and intact leaves being washed onto the beaches on the east coast of Australia (pers. obs.).

The convection column of a fire could clearly promote long-distance dispersal for wind-dispersed seeds. To my knowledge, however, there are no reports of such an occurrence. Charring of the pappus or wing of a wind-dispersed seed, very likely in the event of a fire, would reduce its powers of dispersal! It is possible that seeds on plants adjacent to a fire could be sucked into the convection column without damage. This would depend on the fortuitous combination of flowering time and spatial positioning in relation to the time and location of a fire. There is a record of high-level wind transport of seeds between New Zealand and Australia (Close, 1978, cited in Kloot, 1984) and the convection column of a large fire could provide the means by which seeds might reach high-altitude air streams.

III. POST-FIRE CONDITIONS

A. Nutrients, Temperature, Water and Light

Although various recent studies and reviews have described the effects of fires on the physical characteristics of soils (e.g. Ahlgren and Ahlgren, 1960; Kozlowski and Ahlgren, 1974; Raison, 1979, 1980; Humphreys and Craig, 1981; Woodmansee and Wallach, 1981; Wright and Bailey, 1982; Kruger *et al.*, 1983; MacLean *et al.*, 1983; Cass *et al.*, 1984), a few generalizations that relate directly to the conditions facing seeds and seedlings in a post-fire site should be made.

1. Nutrients

Many elements are volatilized by burning (e.g. nitrogen, phosphorus, magnesium; see Harwood and Jackson, 1975), but fire may actually increase the *availability* to plants of the nutrients in the top few millimetres of soil. This effect may be particularly important in communities growing on poor soils, especially those low in nitrogen and phosphorus, because nutrients are normally bound up in plant biomass and are recycled within the established plants (Vitousek and Reiners, 1975; Woodmansee and Wallach, 1981). For example, O'Connell *et al.* (1979) showed that a four-fold increase in accession of phosphorus and nitrogen to the soil occurred after an intense fire in eucalypt forest in Western Australia. They attributed this to crown scorching, which prevented the usual withdrawal of these nutrients into the plant before leaf drop. Increased availability of nutrients after fire clearly has implications for the establishment, survival and growth rates of the seedlings that appear at that time.

2. Temperature

One effect of fire that has been little studied is the sterilization of the soil surface by the heat. A few workers have suggested that allelopathic chemicals in chaparral litter and soils inhibit germination (Muller *et al.*, 1968; Wilson and Rice, 1968) and that high soil temperatures during fires destroy these compounds (Christensen and Muller, 1975). This effect has also been suggested for eucalypts in Australia (Ashton, 1970) and for bracken fern in South Africa (Granger, 1984).

Fires can also have delayed effects on soil temperatures by altering the albedo of the soil surface, by removing the shading normally provided by the vegetation and by destroying the insulating effect of the litter (Ahlgren and Ahlgren, 1960; Daubenmire, 1968; Old, 1969; Viro, 1974; Cass *et al.*, 1984). An increase in the extremes of soil temperature may increase productivity (Rice and Parenti, 1978) and has been shown to stimulate germination in some plant species (Tothill, 1969; 1977). Van Cleve and Viereck

(1981) pointed out that the warming of burned sites in permafrost areas of Alaska, caused by the changed albedo, led to an increase in the thickness of the active layer of the soil profile.

3. Water

The water relations of a soil change in response to fire, and characteristics of the fire, such as it duration and intensity, might be expected to determine the amount of change. Factors such as the removal of both live vegetation and the litter layer are expected to alter interception of precipitation, infiltration rate, runoff and evapotranspiration, as summarized by Humphreys and Craig (1981). Their general conclusion was: " ... we cannot predict with any precision the effects of a given fire regime."

4. Light

Light conditions differ in burnt and unburnt vegetation. Vogl (1974), for example, highlighted the importance of light availability in post-fire regeneration of North American grasslands and Ashton (1981) stated: "the failure of eucalypts to regenerate within forests in the absence of fire is related to low light intensities." Both intensity and wavelength of light are likely to be important here, but I know of no studies that have attempted to compare these factors in burned and unburned vegetation.

B. Competition

One effect of fire that is commonly cited is the reduction of competition because of death or 'pruning' of biomass present before the fire (Ashton, 1970; Gill, 1981a; Mott and Groves, 1981). However, few studies have attempted to distinguish empirically between the host of forms such competition might take; for light, water, space, nutrients, etc. These factors will certainly differ in burned and unburned sites, as described above, and in general, competition for them might reasonably be expected to be at a minimum immediately after fire. The effects of competition will then develop as the vegetation recovers. The time-span before competition becomes an important interaction will depend both on the severity of the fire and on the rapidity of regeneration of the vegetation. For example, Zimmer (1940) described a mallee woodland in south-eastern Australia which featured many gaps between the woody, perennial shrubs. He attributed the lack of smaller plants in these gaps to the effective capture of water by the dense mat of roots in the surface layer of soil. It was only in the first few months after wildfire, when transpiration of the woody plants was arrested, that seedlings were able to become established.

C. Animal Populations

The responses of animals to fire have implications for the recovery of plant populations for two main reasons: (1) herbivores have been shown to affect both survival of adult plants and recruitment to the population through seedlings; and, (2) invading animals might be expected to be direct agents of seed dispersal.

There is a large body of literature on the responses of mammal populations to fire (e.g. Christensen, 1980; Catling and Newsome, 1981; Main, 1981b; Wright and Bailey, 1982; Fox, 1983). This literature reveals a wide variety of responses, with different taxa responding differently. For example, populations of small mammals decline either during or soon after fire and may take more than five years to become re-established (Christensen, 1980; Fox, 1983) while large mammals may increase in numbers on patches of recently burned vegetation (Shea *et al.*, 1979; Whelan and Main, 1979; Christensen, 1980). This effect is apparently influenced by the patchiness of the fire: mammals will be able to graze small patches of regenerating vegetation more readily than an extensive, burned landscape far from refuges (Whelan, 1977; Leigh and Holgate, 1979). I used the measure of herbivory on preferred food items as an estimate of activity in the swamp wallaby (*Wallabia bicolor*). This showed a low intensity of feeding in the unburned vegetation at the boundary of the fire, a marked increase in the burned vegetation at the edge, then a decline with distance into the burn from the unburned edge (Table 1).

Wright and Bailey (1982) summarized a variety of responses of bird populations to fire in North America. These ranged from elimination of species that require dense thickets of vegetation for cover and nesting to the congregation on newly burned sites by species such as mourning doves (*Zenaida macroura*) feeding on dead insects. Seed-eaters, such as the doves, exploit the abundance of seeds that sometimes follows fire (Lawrence, 1966). Ground Parrots (*Pezoporus wallicus*) inhabit fire-prone heathland in southeastern Australia and may invade within a year of fire, when sedges seed prolifically (R. Jordan, pers. comm., 1985). Moreover, it is at this time that other plants such as *Xanthorrhoea resinosa*, on which *P. wallicus* has been reported to feed, have a peak in reproduction.

Studies on invertebrates also reveal a variety of patterns. Some demonstrate an increase in abundance of surface-active arthropods very soon after fire (Whelan *et al.*, 1980; Tap and Whelan, 1984), while others indicate that population sizes are reduced for many years (Ahlgren, 1974; Springett, 1976). Whelan and Main (1979) demonstrated that herbivory by insects on the seedlings that germinated after fire in a Western Australian woodland was lower in a large burned area than in small patches of burned vegetation,

Table I. Intensity of grazing by swamp wallabies on *Sowerbaea juncea* plants in relation to distance from boundary of fire. Many clumps of this lily at each distance were scored for presence or absence of grazing. Swamp wallabies were resident only in the unburned vegetation.

Site	No. samples	Total no. plants examined	Mean % grazed (S.E.)	
(a) Unburned	6	190	4.8	(2.5)
(b) Burned				
edge	6	119	28.5	(11.3)
20 m	5	143	18.1	(4.3)
50 m	4	163	23.7	(6.3)
100 m	5	177	28.3	(11.5)
200 m	5	194	14.4	(5.4)
300 m	5	175	4.0	(2.8)

and that this was correlated with different rates of re-invasion into the two sorts of sites by the principal insect herbivores.

The general conclusion from these studies is that burned vegetation may lack some taxa of animals for some time after fire. The actual time will depend primarily on the spatial characteristics of the fire in relation to the taxa of animals. In addition, some groups of animals may congregate in recently burned vegetation, or may simply increase their activity within a burned area.

The implications for seed dispersal that result from the wide array of patterns described are manifold, and speculation is facilitated by the lack of detailed study. The mammals that are attracted into newly burned vegetation, as described above, may encourage introduction of seeds, which either cling to fur or are defecated. Bullock and Primack (1977) modelled the pick-up and loss of seeds on animal fur using cloth models. This approach could be used to examine the likely input of seeds to burned areas. Although there are no reports of seeds in Australian heathlands being transported in this way (Clifford and Drake, 1981), several animals, including both mammals and birds, in these fire-prone heathlands have been recorded eating seeds. It would not be difficult to examine seed input in faeces.

The implications of animals feeding on seeds and fruits for dispersal depend not only on the activity of the animal but also on the fates of the seeds. Many birds are known to be 'messy' feeders (Clifford and Drake, 1981) but the escape of some seeds during feeding is only inferred. Similarly, many mammal species bury intact seeds and escape of seeds is inferred because some caches must be lost. Possums are known to disperse seeds of cycads some distance from parent plants (Chapter 5, Burbidge and Whelan, 1982; Ballardie and Whelan, 1986) but 'effective dispersal' will depend on some of these seeds escaping the notice of seed consumers. Attention needs to be paid to the seed's point of view in studies of post-fire activity of putative dispersal agents.

D. Summary

A general conclusion that may be reached from the above discussion is that post-fire conditions are considerably different from those that prevail in an unburned community. The conspicuous differences include increased space and light at the soil surface, increased range in soil temperatures, and increased availabilities of some nutrients. Other differences that have been less clearly demonstrated include: (1) either temporary absence or increase of animal populations; and, (2) decreased competition through increased availability of water, light and nutrients to both seedlings and resprouting adults.

Many studies have reported improved growth and survival of seedlings in burned areas compared with unburned areas. For example, Purdie (1977b) and Whelan and Main (1979) have shown, for dry sclerophyll forest in eastern and western Australia respectively, that survivorship of seedlings was higher for those that germinated in the first year after fire than for individuals of the same species that germinated in unburned samples. This difference between the burned and unburned sites had disappeared by the second year after burning (Whelan, 1977).

A few studies have implicated herbivores in this differential survival. Whelan and Main (1979), for example, showed that herbivory by insects was greater on the few seedlings that germinated in unburned sites than on the many that germinated in burned plots in Western Australian woodland. Bond (1984) showed a similar effect for seedlings of *Protea repens* in South African fynbos. Other factors, such as water availability (e.g. Zimmer, 1940, described above), nutrients and light may eventually be demonstrated to be as important.

Thus the post-fire environment is likely to promote safe sites for germination of seeds and establishment of seedlings. It may, in fact, provide the only opportunity for recruitment to populations of some species (i.e. a 'regeneration niche'; Grubb, 1977). Plants in fire-prone regions may therefore be expected to have evolved characteristics that allow them to take advantage of the predictable availability of favourable conditions after each fire.

IV. LONG-DISTANCE DISPERSAL

One characteristic that allows a plant species to exploit newly disturbed sites is the production of large numbers of seeds which may be dispersed long distances. A temptation in viewing post-fire vegetation changes is to treat fire as a disturbance which initiates a secondary succession, such as occurs after abandonment of farmland (old-field succession). With this model, one would expect a pioneer seral stage made up of short-lived, well-dispersed annuals. The timing of appearance of plants of later seral stages would be influenced in part by their respective dispersal abilities. Unfortunately, few studies have detailed the means and effectiveness of dispersal and fewer still have considered this in relation to disturbance by fire. The following discussion reviews a series of observations of apparent dispersal of seeds of pioneers into recently burned sites.

Apparent adaptations that would allow long-distance dispersal, i.e. small, light seeds with structures such as a pappus (Chapter 1), are certainly prominent among pioneers in newly burned sites. The application of the common name 'fireweed' to well dispersed, weedy plants such as *Epilobium angustifolium* (Fig. 21F of Chapter 1) is indicative of their early appearance and abundance in burned areas. Wildfire in the drier sites of Alaskan taiga results in a succession dominated, in its early stages, by light-seeded species such as *Epilobium angustifolium, Salix scouleriana* and *S. bebbiana* (Viereck, 1973, Rowe and Scotter, 1973). Many references are made in Kozlowski and Ahlgren (1974) to the well dispersed seeds that germinate soon after the fire. For example, species of *Taraxacum, Hieracium* and *Chamaenerion* (*Epilobium*) are the first pioneers after fire in Finland (Viro, 1974); aspen and birch (*Betula* spp.) are wind-dispersed pioneer trees in the north-eastern US (Little, 1974); burrow-weed (*Haplopappus tenuisectus*) and snakeweed (*Gutierrezia sarothrae*) are post-fire pioneer shrubs that are disseminted by wind in the Sonoran Desert. Jarratt and Petrie (1939) stated that seeds germinating from the surface layers of ash in eastern Australian eucalypt forest "were regarded as migrants from unburnt adjacent areas." In the first year after fire in the Mediterranean 'maquis', vegetation is dominated by annuals that are 'opportunistic invaders' not requiring fire to stimulate germination (Cody and Mooney, 1978). Finally, Granger (1984) reported that *Phillippia evansii* (Ericaceae) in South African forests appeared in recently burned sites because of widespread dispersal of its fine seeds by the summer 'bergwinds'.

The ability of later successional species to appear in post-fire succession may also be determined by their dispersal abilities. For example, Viereck (1973) commented that tree species such as white spruce (*Picea glauca*) attributed the failure of shrubs to invade open desert grasslands after fire to poor dispersal of their seeds.

As stated above, information on dispersal of seeds into burned areas by animals is extremely scanty. Ahlgren (1974) commented that after fire in northern Minnesota, the pioneer herbs almost all appeared from seeds, but only about one-third of the species were dispersed into the site by wind. A third of the species were dispersed by birds and small mammals and others may have been dispersed by large mammals such as deer and bear. Humphrey (1974), discussing the lack of invasion of shrubs into desert grasslands, considered that those cases of successful invasion could be explained by domestic stock bringing seed in from adjacent areas. Glyphis *et al.* (1981) found that *Acacia cyclops*, a species introduced from Australia, was dispersed selectively by birds to sites beneath tall shrubs on Cape Peninsula, South Africa.

Australian heathlands and eucalypt forests have fewer genera that are considered to be dispersed by birds than is the case for the adjacent temperate rainforests. Several studies indicate that rainforest species invade the surrounding *Eucalyptus* forest in the years between fires and this dispersal is effected primarily by birds. For example, Gleadow and Ashton (1981) reported that *Pittosporum undulatum*, a wet-forest species, was germinating under existing bushes and trees in central Victoria, Australia. I have made similar observations on both *P. undulatum* and also the 'cabbage palm', *Livistona australis*, in New South Wales. The effectiveness of birds as agents of dispersal could therefore determine the rate of rainforest invasion of eucalypt forest between fires (Ashton, 1981; Gilbert, 1959). K. Mills (pers. comm., 1985) reported that about 86% of tree shrub species in the Illawarra district of New South Wales, Australia, are dispersed by birds (Table II). Ashton (1981) suggested that slow re-invasion of rainforest species into large tracts of burned forest in south-eastern Australia might be partially attributable to their poor dispersal over long distances. The degree to which fruit-eating birds in rainforests are confined to the patches of rainforest dotted along the eastern coast of Australia is worthy of further study, particularly in relation to local, within-patch dispersal versus invasion of surrounding vegetation.

V. LOCAL DISPERSAL

Notwithstanding the above examples of apparent long-distance dispersal of seeds into recently burned sites, remarkably few studies have demonstrated, or even suggested, that such a process is common or widespread. In fact, a seral stage comprising mostly well dispersed fugitives appears to be lacking in many fire-prone communities. Kruger (1983), for example, summarized data from other sources which indicated that 'immigrant post-fire annuals' are not a conspicuous component of the post-fire vegetation

Table II. Dispersal mechanisms of seeds from rainforest trees and shrubs in the Illawarra region, New South Wales, Australia. Data are expressed as % of species (S) and of individual plants (I) dispersed by each vector.

Site	Rainforest type	Wind		Birds		Mammals[a]		N	
		S	I	S	I	S	I	S	I
Bulli Pass 34°18'S 150°53'E	Sub-tropical/ warm- temperate	7	25	90	75	3	0.1	40	3610
Goondarrin Ck. 34°23'S 150°48'E	Warm temperate	22	28	78	72	0	0	23	3490
Barren Grounds 34°40'S 150°43'E	Cool/warm temperate	24	49	76	51	0	0	21	2215

[a] Mammalian dispersal inferred because seed size > 2.5 cm

Data supplied by K. Mills.

in most Mediterranean-type vegetation. Moreover, 'fire-ephemerals' were lacking in all regions surveyed except Californian chaparral. Even in these areas, many of the fire ephemerals may, in fact, appear from dormant seeds stored in the soil between fires (Sweeney, 1968; Hanes, 1977). Sweeney (1956) suggested that dispersal of seeds from adjacent unburned sites does not contribute significantly to the herbaceous cover in burned chaparral. This was confimed by Keeley (1981), who concluded that the post-fire annual floras in certain regions with Mediterranean-type climate exhibit a specific adaptation to fire. Seed dormancy and longevity allow them to appear after fire-free intervals of as long as a century, but they are generally poor colonists. Hopkins and Griffin (1984) supported Kruger's conclusion by pointing out that whole families of plants that are characterized by having wind-dispersed seeds (i.e. Compositae, Poaceae and Orchidaceae) are very poorly represented in Western Australian sand-heaths. These communities are certainly prone to fire. It would be very interesting to examine the dispersal characteristics of the few native representatives of these families. Poor dispersability would be predicted.

In those instances where seed dispersal into burned sites does appear to be responsible for the establishment of particular plant populations, the species involved are often opportunistic weeds that appear in any disturbed area. *Epilobium* species, for example, are widespread roadside weeds as well as 'fire-followers'. Wright and Bailey (p. 184 of 1982) pointed out that many of the weedy, annual forbs (e.g. *Cirsium, Senecio*) that appear after fire in chaparral are also common on adjacent roadsides and may actually invade from there. Bell *et al.* (1984) suggested that several so-called 'fire-ephemeral' species in Western Australian sand-heath also flourish in disturbed areas (i.e. roadsides and recently cleared land) without burning.

For many plant species, the importance of seed dispersal in relation to fire becomes a question of local dispersal rather than long-distance dispersal from the nearest unburned vegetation. For those situations in which long-distance seed dispersal is rare, the seeds that produce the post-fire flush of recruitment must come from within the burned community. Dispersal of seeds may be important here in two ways. Firstly, post-fire dispersal may well differ from dispersal in the absence of fire, thereby producing different patterns of seed dispersion. Post-fire dispersal may result either from reproduction that occurs immediately after fire, or from seeds produced in inter-fire years but stored in the canopy until fire occurs. Secondly, the dispersal of seeds in the years between fires will determine their locations when a fire does occur. This dispersion may affect the survivorship of seeds and the subsequent establishment of seedlings.

A. Post-fire Reproduction and Seed Release

1. Flowering Stimulated by Fire
The appearance of seedlings in high numbers soon after fire may result from post-fire flowering and release of seeds within a site as well as from seeds being dispersed from outside. Fire is thought to be the stimulus for flowering in some plant species (e.g. orchids (Specht, 1981), sundews (Bell *et al.*, 1984), but this effect seems more prevalent among geophytes than perennial shrubs and trees. Gill (1981b) listed Australian species, mostly monocotyledons, that flower best immediately after fire. Grasses do not feature in these Australian examples, but are clearly in this category else-where, such as in the grasslands of North America (Daubenmire, 1968).

Perhaps the lack of trees and shrubs in these examples reflects the fact that few studies have been conducted on these groups. Pyke (1983) showed that there was a peak in flowering in the Proteaceous shrub *Lambertia formosa* two years after fire and Gill and Ingwersen (1976) reported a similar response, in the first year after fire, for *Xanthorrhoea australis*. Several studies suggest that burned populations of some cycads, for example *Macrozamia reidlei* in Western Australia (Baird, 1977; Burbidge and Whelan, 1982; Bell *et al.*, 1984) and *Zamia pumila* in Florida (R. J. Whelan, pers. obs.), flower better than unburned populations. These cycads have large, fleshy fruits, and dispersal as seeds apparently depends upon vertebrates. Synchronous fruiting within a population may increase seed dispersal as long as the dispersal agents are not eliminated or driven away by the fire. However, Ballardie and Whelan (1986) showed that the proportion of the seed crop of individual *Macrozamia communis* plants dispersed further than 2 m was lower in a population displaying substantial, synchronous fruiting than for isolated, fruiting individuals. Much more work remains to be done on the dispersal patterns of seed crops in the immediate post-fire flush of reproduction compared with the sporadic reproduction of later years.

Seeds produced from enhanced post-fire flowering could be dispersed into an environment in which seedlings would experience conditions very different from those in unburned vegetation, and probably more favourable. It would be easy to view such a strategy of massive post-fire flowering and release of seeds as adaptive. However, increased flowering may be caused directly by any of the effects of fire described above, such as increased nutrient availability or decreased competition for light, rather than by an adaptation to release seeds at this particular time. Even for those plant species in which there is no innate dormancy of seeds and post-fire flowering is rapid, seeds may not be released for some months, or even up to a year after fire, by which time there may have been substantial resprouting from rootstocks. *Xanthorrhoea resinosa* and *Blandfordia nobilis* are examples of this response to fire in south-eastern Australian heathlands. It is also a conspicuous characteristic of the small perennial shrubs in longleaf pine (*Pinus palustris*) communities of Florida (Whelan, 1985). There is no information on whether this pattern of flowering and releasing seeds soon after fire determines the effectiveness of seed dispersal. Wind speeds, water flow and animal populations are all likely to be affected by fire, providing a rich area for future study.

The fire-stimulated flowering of woody perennials such as *Lambertia formosa* (Pyke, 1983) and other Australian Proteaceae (R. J. Whelan, pers. obs.) supports the view that enhanced post-fire flowering should not be viewed in itself as an adaptation which puts seeds into a favourable post-fire flowering environment. Despite fire-stimulated flowering seeds of these species are held in woody follicles for many years, until released by the next fire. A substantial proportion of the seeds resulting from peaks of post-fire flowering may in fact become non-viable through insect damage and natural ageing before these seeds can contribute to recruitment after the next fire (e.g. in *Banksia paludosa*, see Fig. 1).

2. Seed Release Stimulated by Fire

Bradyspory, which is the long-term retention of seeds in fruits in the plant canopy (Pate *et al.*, 1984), is commonly viewed as a means of delaying dispersal of seeds until immediately after fire. In this way, seedlings can

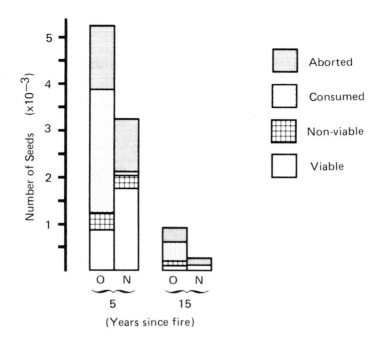

Fig. 1. Seed losses in 1 year old (N) and older (O) confructescences of *Banksia paludosa* at two sites, one 5 years since the last fire, the other 15 years since the last fire. Although there is substantial accumulation of seeds in the 'older' cones, most of the viable seeds released after fire must come from the 1 year old cones. At the 15-year site there were fewer cones in both age classes, reflecting the poor annual flowering and suggesting losses of older cones.

germinate at the same time as resprouts appear, and well before the seedlings of those species, discussed above, which must flower first. Annual flowering produces seeds which are effectively 'deposited' in the plant-stored seed bank until the heat of a fire causes their release from the canopy. This occurs because fire causes desiccation of fruits. In some cases, the heat of the fire is required to break the resin bonds which seal closed the scales of pine cones (Lotan, 1976; Givnish, 1981) or the valves of the follicles in the Proteaceae (Gill, 1976; Wardrop, 1983). The seeds are released after the direct effects of fire have passed but in time to exploit the favourable post-fire conditions. Of course, this will occur only if the fruits on the parent plant protect the seeds from the intense heat during the passage of the fire front. This appears to be the case for many heath and forest species in Australia, North America and South Africa (Gill, 1981b; Rundel, 1982; Kruger and Bigalke, 1984). For example, Siddiqi *et al.* (1976) demonstrated that seeds of three *Banksia* species in south-eastern Australia were killed by temperatures above 100°C. Fire in these same areas, however, produces temperatures in excess of 392°C (Bradstock and Myerscough, 1981). Nevertheless, substantial germination of *Banksia* seeds followed such fires.

Dispersal of seeds from fruits in the canopy after fire may be delayed by a number of mechanisms. After fire, the follicles in *Banksia* species take some time to open sufficiently to release the seeds. Cowling and Lamont (1985a) suggested that for some Western Australian *Banksia* species, several cycles of wet and dry conditions are required before seeds are released. This is because both the valves of the follicle and also the separator between the seeds in the follicle are hygroscopic (see George, 1984 for description). During wet periods, the seeds and separator are jammed inside the follicle (Fig. 2A). The opening and closing of the separator during the wet-dry cycles work it out of the follicle (Fig. 2B), carrying the seeds with it, until the wind is able to catch and disperse them. The requirement of the wet-dry cycle ensures that after summer fires, which are usual in this area, most seeds are not released from the canopy until after the autumn rains start. This interpretation leads to the prediction that fires 'out-of-season' (i.e. in winter or spring in Mediterranean-type climates) would result in poor recruitment. This was exactly the finding reported by Bond *et al* (1984) for *Protea* species in South African heathlands: seedling establishment was poor after winter and spring fires.

3. Wind as a Post-fire Dispersal Agent

Local post-fire dispersal of seeds of bradysporous species is likely to be affected by wind. One direct effect of fire is the removal of foliage, but often the branches and the fruits borne on them are left intact. The buffering effect of a continuous canopy of leaves is removed, facilitating dispersal of seeds by wind. This is probably the case for a variety of plant communities but references to it are mostly only anecdotal or oblique. Zammit (pers. comm., 1984) placed *Banksia ericifolia* seeds into two heathland sites, one beneath *B. ericifolia* plants and the more distant, beneath *B. oblongifolia* plants. None of the seeds placed beneath *B. ericifolia* plants survived to produce seedlings. This location is where those seeds released in the absence of fire would be most likely to land, because of the dense foliage and the fact that the confructescences that bear the seeds are held close to the branches within the canopy. Seeds released after fire are likely to be carried much further, and observations have shown that the winged seeds of this species may be carried 20 m in a strong wind (P. Jordan, pers. comm., 1984). Furthermore, adults are fire-sensitive, so even those seeds which are not dispersed far will have a better chance of establishment.

Other studies provide hints that dispersal of seeds may be better after fire than in an unburned stand. For example, Kruger and Bigalke (p. 107 of 1954) suggested that fire affects wind profiles in South African fynbos;

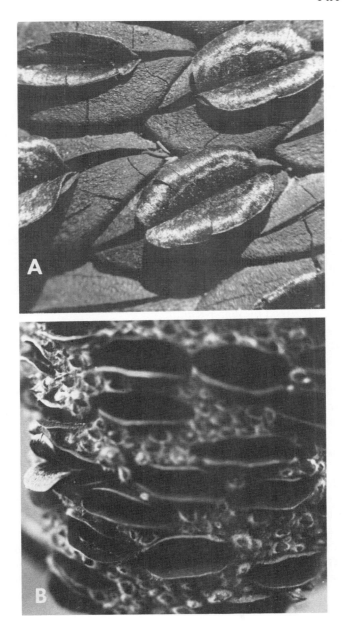

Fig. 2.(A) *Banksia ericifolia* confructescence showing hygroscopic separators, which work the seeds out of the follicles as they expand. The follicles are tightly closed because of high humidity on the day the photo was taken. (B) *Banksia oblongifolia* follicles releasing seeds. The follicles are mostly wide open, and a single separator can be seen almost free of its follicle.

Keeley (1981) stated that Californian chaparral is "not conducive to" wind dispersal; and McMaster and Zedler (1981) considered that for closed-cone pine species, 'cone opening except after fires is sporadic and ineffectual in dispersing seeds'. Some broad biogeographical patterns have also been noted. For example, Clifford and Drake (1981) interpreted the low proportion of wind-dispersed genera in Australian heath and rainforest (8.5%) relative to dry sclerophyll forest (15%) as an evolutionary response to reduced air movements caused by dense, closed canopies of heath and rainforest. For those heath species in which effective dispersal of seeds by wind is important, this requirement is probably satisfied only after fires.

Despite the potential importance of wind as a dispersal agent after fire, few direct measurements of wind speed have been made to compare burned and unburned vegetation. Old (1969) found that wind speed at ground level in a burned prairie in Illinois was 1 m s^{-1} in burned vegetation and 0 m s^{-1} in unburned vegetation. A difference in wind speed occurred between ground level and 2 m above. The implications such differences might have for seed dispersal and subsequent survival and establishment of seedlings remain to be tested for any ecosystem.

4. Post-fire Dispersal by Water Runoff

The removal of above-ground biomass by fire, including both live vegetation and litter, may be expected to result in increased surface runoff after rain (Ahlgren and Ahlgren, 1960; Humphreys and Craig, 1981). Such a result, however, will depend on: (1) the qualities of the soil, which determine infiltration rates; (2) the intensity of the fire, which determines both the amount of biomass destroyed and the rate of recovery of the vegetation; (3) the steepness of the slope; and, (4) the amount and timing of rainfall after fire. The extreme variability of these factors, both among different sites and also within a single site over time, prevents any general conclusion regarding secondary dispersal of seeds by surface water-flow. Nevertheless, small-scale movements of soil and organic matter may be expected to follow rainfall events after fire, even on relatively flat terrain. For example, Adamson et al. (1983) showed that on Hawkesbury sandstone soils near Sydney, both soil and organic matter wash into 'microterraces' which form around small obstructions on the surface of the soil (Fig. 3). I have observed that ash and charred organic matter wash into small depressions in the well-drained sandy soils of both Western Australia and north-central Florida. The importance of this process for seed dispersal, germination and subsequent survival of seedlings deserves much more attention.

Several studies have indicated that water-flow is an important agent of secondary dispersal (e.g. p. 142 of Salisbury, 1942; Bullock, 1976). Miles (1979) reported that seedlings appearing after fire in the *Calluna* dominated heathlands in Britain resulted from seeds that were washed or blown over the burned soil into small cracks. Sims (1951) reported that seedlings of

Fig. 3. (A) A 'microterrace', with walls composed of sand and organic matter. Note the *Acacia* seedlings growing from recently germinated seeds. (B) The build-up of fine organic debris close to the microterrace dam-wall (downhill) contrasts with the washed sand in the foreground.

Fig. 4. A *Hakea* seedling, typical of those concentrated in the dam-wall of a microterrace.

Callitris glauca (Murray pine) in south-eastern Australia were confined to depressions caused by old vehicle tracks. Adamson *et al.* (1983) suggested that seed- or fruit-appendages normally considered to be adaptations for wind-dispersal are equally effective in providing buoyancy and also in ensnaring seeds in small obstructions to water-flow. They also described small, dense clumps of *Acacia* seedlings growing in microterraces (Figs. 3A, 4) It is possible that such microterraces and small depressions form micro-environments that are high in inorganic nutrients, organic matter and perhaps soil moisture. Precise measurements are required to test this.

I mapped all seedlings that germinated after fire in 1500 m² of dry sclerophyll woodland near Perth, Western Australia, and demonstrated that seedlings were highly clumped in small depressions (unpubl. data). This indicates either that seeds were washed into the depressions or that germination was better there. Moreover, survival of seedlings in the clumps was greater than that of isolated seedlings, resulting in an increased measure of clumping over time (Fig. 5). This is contrary to the expectation of increasing evenness that would result from density-dependent thinning (Antonovics and Levin, 1980).

Whatever the demographic consequences of local dispersal, one spatial effect appears to be lateral shifts in boundaries of populations. Bond *et al.* (1984) reported that different dispersability of seeds of various pro-

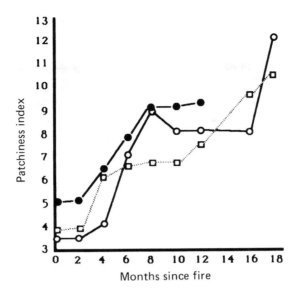

Fig. 5. The change in the degree of clumping of *Stirlingia latifolia* seedlings over time. Seedlings in three 24 x 24 m plots (represented by solid and open circles and squares) were tagged as they germinated, mapped, and regularly scored for mortality. Survivorship was lower among isolated seedlings than among those in clumps.

teaceous species in South African fynbos results in varying proportions of 1 m² sample quadrats supporting 'seedlings but no adults'. Muston (pers. comm., 1985) reported a similar finding, which she interpreted as demonstrating lateral shifts in boundaries of groups of plants without, necessarily, an increase in range (Fig. 6). Do such shifts tend to occur in predictable directions determined by the slope, for water transport of seeds, or by the prevailing post-fire winds of species with wind-dispersed seeds?

B. Soil-stored Seeds

Many of the plant species in fire-prone environments display annual flowering and release of seeds. Nevertheless, germinaton in many of them appears to be substantially greater after fire. Large numbers of herbaceous annuals appear after fire in California chaparral but are absent in unburned sites; these have been called 'pyrophyte endemics' by Hanes (1977). Similar 'fire ephemerals' occur in the Australian heathlands (e.g. Bell *et al.*, 1984) and South African fynbos (Kruger, 1983; Kruger and Bigalke, 1984). These short-lived species undergo their whole growth and reproductive cycle within

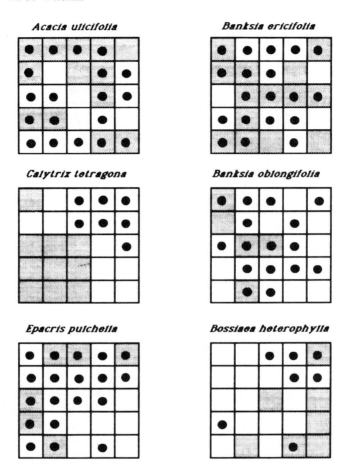

Fig. 6. Changes in boundaries of clumps of woody perennial shrubs in Hawkesbury sandstone soils near Sydney, Australia. Each small square represents one square metre. Stippled squares represent the presence of an adult plant before the fire and filled circles represent those squares in which seedlings became established after fire. *Banksia oblongifolia* and *Bossiaea heterophylla* adults resprout after fire, so seedlings in unstippled squares represent an increased patch size. The other four species are 'obligate seeders' (i.e. adults die in fire), so seedlings in unstippled squares represent a lateral shift in the boundary of a clump. Data of R. Muston.

the post-fire year and leave a long-lived seed bank in the soil. The importance of seed dispersal for these species is not known, but their reappearance after the next fire depends upon the maintenance of seed viability during the interfire years and the protection of seeds from destruction by the heat of the fire.

In addition to this group of plants, many perennial species, some of which tolerate fire as established plants, also show a flush of germination after fires. The importance of seed dispersal for these species includes two factors: (1) the depositing of seeds in sites where they will remain viable and be able to germinate after the next fire; (2) the dispersal of seeds away from the parent plant which will resprout rapidly.

Hard-seededness has been described for many plant species, including legumes (Floyd, 1966; Martin and Cushwa, 1966; Warcup, 1980; Tran and Cavanagh, 1984; Whelan, 1985), Proteaceae and Rutaceae (Kruger and Bigalke, 1984), Anacardiaceae, Compositae, Malvaceae and Poaceae (Gill, 1981b). The impermeability of the seed coat prevents germination until selective damage to the lens or equivalent region of the seed-coats permits imbibition. Fire appears to be the most important factor in overcoming seed-coat impermeability in many communities.

The appearance of seedlings after fire depends largely on the degree to which seeds are subject to a heat treatment during the passage of the fire. Several studies have shown that the density of leguminous seedlings appearing after fire is dependent upon the temperature of the fire (e.g. Christensen and Kimber, 1975; Shea *et al.*, 1979; Christensen, 1980): the lower the temperature of the fire, the lower the density of seedlings. Another factor that affects post-fire germination of hard-seeded species is the depth of seeds in the soil profile. Seeds stored too close to the surface will be incinerated, while those buried too deeply will remain dormant. Ants are responsible for dispersal of seeds of many plant species in Australian forests and heathlands (Chapter 3) and some seeds are stored within ant nests (Shea *et al.*, 1979; Berg, 1981; Majer, 1982). Shea *et al.* (1979) recorded clumps of legume seedlings appearing after wildfire and interpreted the cause of this pattern of dispersion to be ants burying the seeds. Majer (1982) examined this possibility by 'toasting' ant colonies using an infra-red heater and recording the resultant germination. In general, more seeds germinated from nest sites than from bare soil. For some species of plants, germination in nest sites was increased by heating. Dispersal of seeds and subsequent storage by ants has been viewed as an adaptation that permits the survival of the seeds through a fire (Berg, 1981; Buckley, 1982) and which, furthermore, may place seeds in a 'safe site' for establishment after germination. This would result from greater water and nutrient availability provided by ant activity (Berg, 1981; Davidson and Morton, 1981, 1984). Main (1981a) has argued, however, that seeds of species such as these are able to tolerate fire even without such burial. He suggested that if burial by ants has adaptive significance, it is simply that the concealment resulting from burial allows more seeds to remain undetected by seed-eating animals.

Ants are not the only agents of burial of seeds in fire-prone plant communities. Christensen (1980) reported that the woylie (*Bettongia penicillata* — a macropod marsupial) buried seeds in caches 'seemingly at random, in places close to the source of origin.' This behaviour resulted in seedlings of heartleaf (*Gastrolobium bilobum*, Papilionaceae) appearing in small clumps after a high-intensity fire. It is not known whether seeds buried in caches such as these are more likely or less likely to survive than isolated seeds that escape the attention of the woylie.

These studies on dispersal of seeds and the subsequent appearance of seedlings form an interesting line of research which could be pursued much further, particularly as so little is known of the fates of dormant seeds in the between-fire period.

VI. GENERAL THEORIES

Any general theory of seed dispersal in relation to fire needs to account for the variety of patterns described above. This is likely to be a difficult task, given the small amount of information that exists about the general patterns. However, there appear to be regional differences in general community patterns such as: (1) the presence or absence of fire-stimulated annuals; (2) the presence or absence of wind-dispersed ruderals as pioneers in a post-fire 'succession'; (3) the presence or absence of shrubs classified as 'obligate seeders' (adults die in the fire and regeneration of the population is solely from seed); and, (4) the relative abundances of taxa with seed-dormancy broken by fire.

Many species of plants in fire-prone regions have characteristics that are seen as 'fire-adaptive' (Gill, 1981b). These characteristics include: (1) tolerance of heat by established plants (i.e. the ability to resprout after fire from the trunk or from a lignotuber); (2) the protection of seeds with hard seed-coats or by retention in woody fruits; and, (3) dormancy, imposed by the hard seed-coat or the woody fruit, which is removed by fire, thereby concentrating germination immediately after fire. There is a danger in applying the term 'fire adaption' to such characteristics, similar to the danger in the application of the term 'r-selected' to species with certain ruderal characteristics. It implies that fire was the selective force responsible for the evolution of the so-called fire adaptations.

Several authors have attempted to classify species by 'vital attributes' which dictate their recovery after fire (Noble and Slatyer, 1981 and others). These attributes variously include the responses of both established plants and seedlings to a fire, the life-history of the species (e.g. time to first flowering, life span), characteristics of dormancy and release of seeds and competitive ability after establishment. Different fire regimes are seen to favour species with one set of attributes over another and species have been labelled 'invaders', 'evaders', 'resisters' and 'endurers' (Rowe, 1983), depending on the population responses to a fire. These classifications all include long-distance dispersal in at least one category of vital attribute or response to fire.

Two sorts of generalizations are possible about dispersal of seeds in relation to fire. One concerns community-wide patterns, and questions whether the prevalence of a particular type of seed dispersal is related to the characteristics of fire in a given region. The other relates variation in seed dispersal within a species to the characteristics of fire at different sites within the range of the species. In the following discussion, I have listed certain general observations regarding patterns of seed dispersal and have provided a group of hypotheses as an explanation of each observation.

A. Wind-dispersed Pioneers

Observation: Wind-dispersed annuals as pioneers after fire appear to be common in some ecosystems and absent in others.

Hypothesis 1: When fire occurs in ecosystems that have not been prone to fire in the past and in which fire is currently highly unlikely, there are few plant species which possess characteristics allowing them to tolerate the fire (as either established plants or seeds). Thus the pioneer stage of recovery must come from ruderals with long-distance dispersal. The rate of invasion of other species will depend on their powers of dispersal.

Hypothesis 2: In fire-prone regions, past exposure has selected certain adaptations that allow plants to persist, either as seeds or adult plants, through fire. Regeneration by resprouts is rapid, and seedlings of all life-forms (herbs to trees) appear from locally dispersed seeds. Even annuals in these ecosystems have high longevity and poor long-distance dispersal of seeds. Well dispersed weeds are only abundant where there is a source near to the fire and post-fire conditions are appropriate.

The predictability and frequency of fire in the evolutionary past are important because these are the main features that determine the evolution of tolerance to a disturbance. In areas where fire has been predictable, recurring within the evolutionary 'memory' of a species (p. 627 of Harper, 1977), we might expect to see the evolution of adaptations that allow survival of fire and even exploitation of the post-fire environment. Where both predictability of fire and its frequency in relation to the life-spans of plants have been low, we might expect fire to be a catastrophe, causing widespread mortality. The model presented by Noble and Slatyer (1981), based on vital attributes of plants mentioned above, predicts that species with the attribute of 'highly-dispersed propagules' will be favoured when short inter-fire periods (i.e. < generation time) are rare and long interfire periods (i.e. > generation time) are frequent. Keeley (1981) divided the 'options' in reproductive cycles of plants faced with fire into a hierarchy based on production of seeds and resprouting capabilities and considered that a particular fire regime would favour some of these options. For example, frequent fires burning extensive areas would select for species which have local dispersal of long-lived, dormant seeds. At a very general level, the part dispersal of seeds into the site plays in the development of the pioneer plant community after fire may be expected to reflect the fire history in the evolutionary past.

A survey is needed to relate dispersal mechanisms among plant species in a range of communities to current and past patterns of fire. The following is presented as a general hypothesis. Fire appears to be a 'catastrophe' (*sensu* Harper, 1977) in many tropical and temperate broadleaved forests (Bonnicksen and Christensen, 1981), in certain high latitude communities (e.g. tundra and forest-tundra; Auclair, 1983) and in southern US mesic deciduous forests (Christensen, 1981). Heinselman (1981) stated that: 'many northern plants have *highly dispersed propagules* — mostly wind-transported seeds or spores'. In these situations, post-fire regeneration is dominated by a pioneer phase of well dispersed weeds typical of secondary successions described for old fields (Odum, 1971).

This contrasts with the situation in regions in which fire has long been a frequent occurrence (i.e. South African fynbos, Australian eucalypt woodlands and heaths, Californian chaparral). In these areas, not only do many established plants survive fires on the site, but there is also strong post-fire recruitment from seeds dispersed locally either prior to fire or immediately after it. Selection in these regions has favoured pioneers that 'sit-and-wait' for the next disturbance by fire. However, another possibility that needs to be examined is that there is simply no pool of weedy annuals available as colonizers of recently burned sites in these regions.

The availability of weeds to a recently burned site may depend on several factors, including: (1) the proximity and intensity of agriculture or other disturbances (i.e. road verges, railway lines); (2) the history of natural disturbances (landslides, treefalls, river bends) in the region; and (3) the capacity of the weed species to tolerate the post-fire soil conditions. Thus areas that have a long history of agriculture (i.e. Mediterranean) might be expected to have a 'rain' of ruderal weeds falling into any site recently disturbed.

It is likely that floras in regions that have seen long periods of natural disturbance contain taxa that have evolved as colonizing pioneers of such sites (p. 100 of Grime, 1977). For example, Platt (1975) showed that wind-dispersed 'fugitives' were the first colonizers of badger mounds in the North American prairies. The term 'nomad' was introduced to describe those tree species in tropical forests which appear adapted to colonize large gaps caused by windfalls (Keay, 1957, Knight, 1975). Grime (p. 101 of 1977) described two wind-dispersed composites (*Tussilago farfara* and *Petasites hybridus*) which have short-lived seeds that germinate very rapidly under a wide range of conditions. He suggested that these species are specifically adapted for the rapid colonization of bare soil in new river terraces caused by flooding, and in glacial moraines. Such species could be seen to be 'pre-adapted' as colonizers of burned sites.

Even if a pool of weedy species, whatever its origin, is available to a burned site, the conditions at the soil surface are likely to determine whether these species are conspicuous after fires. An obvious regional variable is soil fertility. Clements (1983) showed that weed invasions into bushland on the Hawkesbury sandstone soils of the Sydney region was determined both by the proximity of disturbance and also by the increased fertility to the sites due to increased water and phosphorus runoff. The dry sclerophyll forest sites not adjacent to suburban development, which have low levels of phosphorus, support few weeds even after disturbance.

The hypothesis that the importance of well dispersed seeds in post-fire regeneration is related to the past predictability and frequency of fire may be examined at a local scale, where local climatic conditions determine frequencies of fire. For example, re-establishment of subtropical rainforest on burned sites in south-eastern Australia depends largely on dispersal of seeds into the burned site. The surrounding eucalypt forest community is characterized by species that tolerate fire and also have poor, long-distance seed dispersal. These species also store long-lived, dormant seeds in the soil. Frequent fires will eliminate the rainforest species. Fires of intermediate frequency will permit co-existence of species from the two communities because each fire removes representatives of the rainforest community and stimulates a burst of recruitment of species of the eucalypt forest community. The interfire period sees a re-invasion of rainforest species, dispersed from refugia that were protected from fire (Smith and Guyer, 1983; Erskine, 1984). Absence of fire for a long enough period would result in the depletion of the stored seed bank and a disappearance of eucalypt forest species as dying adults are not replaced. Perhaps lack of fire for such a long time can be seen as a 'catastrophe' for species in this community, because even after the next fire, they will reappear only after invasion, which may be slow due to poor dispersal of seeds.

This pattern of local patchiness of fire-tolerant and fire-intolerant communities, with seed dispersal critical for re-establishment of the intolerant species after fires, may be repeated in many areas. For example, the mesic hardwood 'hammocks' in the coastal plain of Florida are interspersed with fire-prone pine communities of the 'sandhills'. If interfire periods are long, the hardwood species encroach on the pines. Similar interspersion of plant communities with varying degrees of tolerance of fire is seen in the Everglades region of south Florida, and fire influences the spatial pattern of the patches (Wade *et al.*, 1980). The role of seed dispersal in these examples remains unstudied.

B. Patterns of Local Dispersal

Observation: The relative frequencies of different forms of local dispersal of seeds vary between ecosystems. For example, production of post-fire seedlings comes mostly from dormant seeds in some fire-prone ecosystems, and mostly from post-fire flowering in others.

Hypothesis 1: The predictability of timing of naturally-occurring fires in relation to the onset of a rainy season influences the average success of a seed crop that is stimulated to germinate by fire. Fires always likely to be followed by substantial rains might be exploited by elements of the flora as a cue for germination.

Hypothesis 2: The risks of storing seeds between fires outweigh the advantages of increased establishment rates after fire, and interfire recruitment is maximized when flowering is most profuse; i.e. in the first few years after fire.

Hypothesis 3: The characteristics of the flora which determine the response to fire are determined largely by the evolutionary history of the flora. For example, recent introduction of fire as a disturbance in a region previously subjected to other sorts of disturbance (i.e. grazing, drought, landslides) would be expected to result in different patterns of recruitment than if fire had long been the sole source of disturbance.

Many authors have noted regional variation in patterns of germination of seeds in relation to fire. One conspicuous feature of those members of taxa such as legumes that occur in fire-prone ecosystems is 'hard-seededness', which inhibits germination until the seed coat is scarified or shocked in some appropriate way. I have described apparent variation in the prominence of this characteristic among legumes and other shrubs in two fire-prone regions, namely north-central Florida and south-western Australia (Whelan, 1985) and have concluded that much more work remains to be done before reasons for regional differences such as these become obvious. The three hypotheses presented above summarize the conclusions presented by Whelan (1985).

C. Variations within a Species

Observation: Patterns of dispersal of seeds of some species vary over their geographical range. For example, the degree of spontaneous opening in some pine species in North America and in some *Banksia* species in Australia varies among sites.

Hypothesis: Fitness of a particular 'strategy' is determined by trade-offs among several factors. On one hand, fire produces a 'safe site' for establishment of seedlings, so frequent fires provide many opportunities for recruitment. Fires also promote release of seeds from dormancy. Plants that store strongly dormant seeds in the soil or on the canopy should be favoured. On the other hand, frequent appearance of safe sites between fires and a high risk of death of stored seeds between fires would favour spontaneous release of seeds and easy breaking of dormancy.

Once again, more information is needed on regional patterns both of dispersal characteristics of plants and of fire regimes. However, some generalizations have already come to light. For example, Givnish (1981) was able to relate the degree of spontaneous opening of cones within *Pinus rigida* populations in the New Jersey Pine Barrens to the frequency of fires in the region. Cayford and McRae (1983) related the variation in the degree of spontaneous opening of *Pinus banksiana* cones to local fire frequencies. It is assumed that the main recruitment to these pine populations comes after fire, when the litter is removed from the soil surface and a good seed bed is provided. Dispersal of seeds in the interfire period is unlikely to produce seedlings. It is only in the situations of very low fire frequency that recruitment in the sum of interfire years could exceed the episodic, substantial recruitment after fires.

Cowling & Lamont (1985b) proposed a similar explanation for the variation in spontaneous release of seeds from woody follicles in three *Banksia* species in Western Australia. They argued that spontaneous release between fires is favoured in mesic woodlands where site conditions are more conducive to regular recruitment and where fire intensities are unlikely to be high enough to promote dehiscence of fruits. However, recent studies on *Banksia serrata* in south-eastern Australia (Hunt and Whelan, unpubl. data) indicate that spontaneous opening is not consistently related to expected fire frequencies or other environmental factors in four different habitats near Wollongong, New South Wales. In fact, the factor that appeared to be most closely related to spontaneous opening was the degree of damage to follicles by insects prior to seed dispersal. Further studies on the annual losses from stored seed banks would be well worthwhile.

Farrell and Ashton (1978) examined many populations of *Acacia melanoxylon* and discovered differences in the germination characteristics of their seeds. This species is known to have high seed longevity and appears after fire from soil-stored seed that may be more than 50 years old. However, Farrell and Ashton were able to find no relationship between the variation in germination and variation in any of the environmental measurements that were taken. More work is clearly needed to determine which patterns of germination are favoured by particular environmental variables, including fire regime.

VII. SUMMARY

'Seed dispersal' encompasses features such as the timing of seed release, the vectors of seed transport and the fates and dispersion of seeds after dispersal. A study of seed dispersal involves a consideration of the selective factors which have produced current characteristics; in this review, principally fire. Fire is itself very complex, incorporating variation in frequency, intensity, seasonality and extent.

I have attempted to point out that post-fire conditions may produce very favourable sites for germination and subsequent establishment of seedlings. Little is known about how seeds get to recently burned sites, and there appears to be variation between regions in the general patterns. Fire in some areas initiates a typical secondary succession, with pioneers being mostly well dispersed, weedy annuals. Invasion of other species is likely to be determined partially by their dispersability, with wind and animals being the main vectors. To what degree animals are involved in determining which species appear and when is open to question.

In other areas, long-distance dispersal of seeds is insignificant and local dispersal is most important. This can take two forms. Firstly, annual dispersal of dormant seeds between fires produces a dispersion pattern of seeds 'presented to' a fire. Wind, surface water-flow and animals are likely to produce this dispersion, but just how important these three agencies are for survival of seeds is unknown. Secondly, some species of plants hold seeds in the canopy until fire passes, releasing them soon afterwards. Like dormancy, this cues germination to the post-fire year, but it also allows dispersal to occur when foliage is removed and wind speeds through the remaining canopy are likely to be higher. Moreover, post-dispersal movement of seeds is likely to be greater in burned than in unburned vegetation because of greater surface runoff after rains.

One important question that remains in relation to these two patterns of delayed germination is: what is the risk (i.e. annual losses to the seed bank) of storing seeds in the soil or on the plant? The answer might help resolve other questions, for example, what causes regional variation in the frequency of species with dormant seeds and what causes regional variations within a species in the degree of spontaneous fruit opening and hard-seededness?

In the absence of studies that might provide answers to these questions, I have presented a few hypotheses which I hope will stimulate appropriate research.

ACKNOWLEDGEMENTS

I acknowledge discussions with many colleagues, over several years, on the general subjects of responses of floras to fire and seed-dispersal: Ruth Ballardie, Allan Burbidge, Brian Clay, Malcolm Gill, Ross Goldingay, John Harper, Yan Linhart, Bert Main, Ros Muston, Bill Platt, Jan Taylor and Charles Zammit. I am particularly grateful for the constructive comments and other help provided by the following people during the preparation of this chapter: Sue Carthew, Ross Goldingay, Richard and Pat Jordan, Bert and Barbara Main, Ros Muston, Julie Read and Patrick Tap. I also thank Kevin Mills and Ros Muston for permission to use their unpublished data.

Several institutions provided support for the parts of my own research presented here. These were: the University of Western Australia; the University of Florida; Goldhead Branch State Park, Florida; the University of Wollongong; Barren Grounds Nature Reserve, New South Wales; and the National Parks and Wildlife Service of New South Wales. Funding was provided by a Postgraduate Research Studentship (University of Western Australia), an Archie Carr Postdoctoral Fellowship (University of Florida) and the Australian Research Grants Scheme.

REFERENCES

Adamson, D., Selkirk, P.M., and Mitchell, P. (1983). The role of fire and lyre-birds in the landscape of the Sydney Basin. *In* 'Aspects of Australian Sandstone Landscapes' (R. W. Young and G. L. Nanson, eds), pp. 81-93. University of Wollongong, Australia.

Ahlgren, C.E. (1974). Effects of fire on temperate forests: North Central United States. *In* 'Fire and Ecosystems.' (T.T. Kozlowski and C.E. Ahlgren, eds), pp. 195-224. Academic Press, New York.

Ahlgren, I.F., and Ahlgren, C.E. (1960). Ecological effects of forest fires. *Bot. Rev.* **26**, 483-533.

Antonovics, J., and Levin, D.A. (1980). The ecological and genetic consequences of density dependent regulation in plants. *Annu. Rev. Ecol. Syst.* **11**. 411-452.

Ashton, D.H. (1970). The effects of fire on vegetation. *In* 'Second Fire Ecology Symposium'. Monash University, Melbourne, Australia.

Ashton, D.H. (1981). Fire in tall open forests. *In* 'Fire and the Australian Biota' (A.M. Gill, R.H. Groves, and T.R. Noble, eds). pp. 339-366. Australian Academy of Science, Canberra.

Auclair, A.N.D. (1983). The role of fire in lichen-dominated Tundra and Forest-tundra. *In* 'The Role of Fire in Northern Circumpolar Ecosystems.' (W. Wein and D.A. Maclean, eds). pp. 235-256. Wiley, New York.

Baird, A.M. (1977). Regeneration after fire in Kings Park, Perth, W.A. *J. Roy. Soc. W.A.* **60**, 1-22.

Ballardie, R.T. and Whelan R.J. (1986). Masting, seed dispersal and seed predation in the cycad, *Macrozamia communis*. Oecologia *In press.*

Bell, D.T., Hopkins, A.J.M., and Pate, J.S. (1984). Fire in the kwongan. *In* 'Kwongan: Plant Life of the Sandplain' (J.S. Pate and J.S. Beard, eds). pp. 178-204. University of Western Australian Press, Nedlands, Western Australia.

266 R. J. Whelan

Berg, R.Y. (1981). The role of ants in seed dispersal in Australian lowland heathland. *In* 'Heathlands and Related Shrublands of the World. Vol. 9B' (R.L. Specht, ed.). pp. 51-59. Elsevier, Amsterdam.

Bond, W.J. (1984). Fire survival in Cape Proteaceae — influence of fire season and seed predators. Vegetatio 56. 65-74.

Bond, W.J., Vlok, J., and Viviers, M. (1984). Variation in seedling recruitment of Cape Proteaceae after fire. *J. Ecol.* **72** 209-221.

Bonnicksen, T.M., and Christensen, N.L. (1981). Integrating summary. *In* 'Fire Regimes and Ecosystems Properties' (H.A. Mooney et. al., eds), pp. 177-180. United States Department of Agriculture (Gen. Tech. Report WO-26), Washington, D.C.

Booysen, P. de V., and Tainton, N.M. (eds) (1984). 'Ecological Effects of Fire in South African Ecosystems'. Springer-Verlag, Berlin.

Bradstock, R.A., and Myerscough, P.J. (1981). Fire effects on seed release and the emergence and establishment of seedlings of *Banksia ericifolia*. *Aust. J. Bot.* **29**, 521-531.

Buckley, R.C. (1982). Ant-plant interactions: A world review. *In* 'Ant-Plant Interactions in Australia' (R.C. Buckley, ed.). pp.111-141. Junk, The Hague.

Bullock, S.H. (1976). Consequences of limited seed dispersal within simulated annual populations. *Oecologia* **24**, 247-256.

Bullock, S.H., and Primack, R.B. (1977). Comparative experimental study of seed dispersal on animals. *Ecology* **58**, 681-686.

Burbidge, A.H., and Whelan, R.J. (1982) Seed dispersal in the cycad, *Macrozamia riedlei*. *Aust. J. Ecol.* **7**, 63-67.

Cass, A., Savage, M.J. and Wallis, F.M. (1984). The effect of fire on soil and microclimate. *In* 'Ecological Effects of Fire in South African Ecosystems' (P. de V. Booysen, and N.M. Tainton, eds). pp. 311-325. Springer-Verlag, Berlin.

Catling, P.C., and Newsome, A.E. (1981). Responses of the Australian vertebrate fauna to fire: an evolutionary approach. *In* 'Fire and the Australian Biota' (A.M. Gill, R.H. Groves, and I.R. Noble, eds). pp. 273-310. Australian Academy of Science, Canberra.

Cayford, J.H. and McRae, D.J. (1983) The ecological role of fire in Jack Pine forests. *In* 'The Role of Fire in Northern Circumpolar Ecosystems' (W. Wein, and D.A. Maclean, eds), pp. 183-199. Wiley, Chichester.

Christensen, N.L. (1981) Fire regimes in southeastern ecosystems. *In* 'Fire Regimes and Ecosystems Properties' (H.A. Mooney *et. al.*, eds), pp. 112-136. United States Department of Agriculture (Gen. Tech. Report WO-26). Washington, D.C.

Christensen, N.L., and Muller, C.H. (1975). Effects of fire on factors controlling plant growth in *Adenostoma* chaparral. *Ecol. Monogr.* **45**, 29-55.

Christensen, P.E.S. (1980). The biology of *Bettongia penicillata* Gray, 1837 and *Macropus eugenii* (Desmarest, 1817) in relation to fire. Forests Department of Western Australia. Bulletin No. 91, Perth W.A.

Christensen, P.E.S. and Kimber, P.C. (1975). Effect of prescribed burning on the flora and fauna of south-west Australian forests. *Proc. Ecol. Soc. Aust.* **9**, 85-106.

Clements, A. (1983). Suburban development and resultant changes in the vegetation of the bushland of the northern Sydney region. *Aust. J. Ecol.* **8**, 307-319.

Clifford, H.T., and Drake, W.E. (1981). Pollination and dispersal in eastern Australian heathlands. *In* 'Heathlands and Related Shrublands of the World. Vol 9B' (R.L. Specht, ed.), pp. 39-49. Elsevier, Amsterdam.

Cody, M.L. and Mooney, H.A. (1978) Convergence verus nonconvergence in Mediterranean-climate ecosystems. *Annu. Rev. Ecol. Syst.* **9**, 265-321.

Cowling, R.M., and Lamont, B.B. (1985a). Seed release in *Banksia:* the role of wet-dry cycles. *Aust. J. Ecol.* **10**, 169-171.

Cowling, R.M., and Lamont, B.B. (1985b). Variation in serotiny in three *Banksia* species along a climatic gradient. *Aust. J. Ecol.* **10**, 345–350.

Daubenmire, R. (1968). The ecology of fire in grassland. *Adv. Ecol. Res.* **5**, 209-266.

Davidson, D.W. and Morton, S.R. (1981). Competition for dispersal in ant dispersed plants. *Science* **213**, 1259-1261.

Davidson, D.W. and Morton, S.R. (1984). Dispersal adaptations of some *Acacia* species in the Australian arid zone. *Ecology* **65**, 1038-1051.

Erskine, J. (1984). 'The Distributional Ecology of Rainforest in the Illawarra in Relation to Fire'. B.Sc. Honours Thesis, The University of Wollongong Australia.

Farrell, T.P., and Ashton, D.H. (1978). Population studies on *Acacia melanoxylon* R.Br. I: Variation in seed and vegetative characteristics. *Aust. J. Bot.* **26**. 365-379.

Floyd, A.G. (1966). Effect of fire upon weed seeds in the wet sclerophyll forests of Northern New South Wales. *Aust. J. Bot.* **14**, 243-256.

Foster, T. (1976). 'Bushfire: History, Prevention and Control'. Reed, Sydney.

Fox, B.J. (1983). Mammal species and diversity in Australian heathlands: The importance of pyric succession and habitat diversity. *In* 'Mediterranean Type Ecosystems: The Role of Nutrients' (F.J. Kruger, D.T. Mitchell and J.U.M. Jarvis, eds), pp. 473-489. Springer-Verlag, Berlin.

George, A.S. (1984). 'The Banksia Book'. Kangaroo Press and Society for Growing Australian Plants, New South Wales.

Gilbert, J.M. (1959). Forest succession in the Florentine Valley, Tasmania. *Pap. Proc. Roy. Soc. Tas.* **93**, 129-151.

Gill, A.M. (1976). Fire and the opening of *Banksia ornata* F. Muell. follicles. *Aust. J. Bot.* **24**, 329-335.

Gill, A.M. (1981a). Fire adaptive traits of vascular plants. *In* 'Fire Regimes and Ecosystems Properties' (H.A. Mooney, et. al. eds.), pp. 208-230. United States Department of Agriculture (Gen. Tech. Report WO-26). Washington, D.C.

Gill, A.M. (1981b). Adaptive responses of Australian vascular plant species to fires. *In* 'Fire and the Australian Biota' (A.M. Gill, R.H. Groves, and I.R. Noble, eds.), pp. 243-272. Australian Academy of Science, Canberra.

Gill, A.M., and Ingwersen, F. (1976). Growth of *Xanthorrhoea australis* in relation to fire. *J. Appl. Ecol.* **13**, 195-203.

Gill, A.M., Groves, R.H., and Noble, I.R. (eds) (1981). 'Fire and the Australian Biota'. Australian Academy of Science, Canberra.

Givnish, T.J. (1981). Serotiny, geography and fire in the pine barrens of New Jersey. *Evolution* **35**, 101-123.

Gleadow, R.M., and Ashton, D.H. (1981). Invasion by *Pittosporum undulatum* of the forests of Central Victoria. I: Invasion patterns and plant morphology. *Aust. J. Bot.* **29**. 705-720.

Glyphis, J.P., Milton, S.F., and Siegfried, W.R. (1981). Dispersal of *Acacia cyclops* by birds. *Oecologia* **48**, 138-141.

Granger, J.E. (1984). Fire in forest. *In* 'Ecological effects of Fire in South African Ecosystems' (P. de V. Booysen, and N.M. Tainton, eds.), pp. 177-197. Springer-Verlag, Berlin.

Grime, J.P. (1977). 'Plant Strategies and Vegetation Processes'. Wiley, Chichester.

Grubb, P.J. (1977). The maintenance of species-richness in plant communities: the importance of the regeneration niche. *Biol. Rev.* **52**, 107-145.

Hanes, T.L. (1977). California chaparral. *In* 'Terrestrial Vegetation of California' (M.G. Barbour, and J. Major, eds), pp. 417-469. Wiley, New York.

Harper, J.L. (1977). 'The Population Biology of Plants'. Academic Press, London.

Harwood, C.E., and Jackson, W.D. (1975). Atmospheric losses of four plant nutrients during a forest fire. *Aust. For.* **38**, 92-99.

Heinselman, M.L. (1981) Fire intensity and frequency as factors in the distribution and structure of northern ecosystems. *In* 'Fire Regimes and Ecosystems Properties' (H.A. Mooney *et al.*, eds), pp. 7-57. United States Department of Agriculture (Gen. Tech. Report WO-26). Washington, D.C.

Hopkins, A.J.M. and Griffin, E.A. (1984). Floristic patterns. *In* 'Kwongan: Plant Life of the Sandplain' (J.S. Pate, and J.S. Beard, eds), pp. 69-83. University of Western Australia, Nedlands, Western Australia.

Howe, H.F. and Smallwood, J. (1982). Ecology of seed dispersal. *Annu. Rev. Ecol. Syst.* **13**, 201-228.

Humphrey, R.R. (1974). Fire in the desert and desert grasslands of North America. *In* 'Fire and Ecosystems' (T.T. Kozlowski, and C.E. Ahlgren, eds), pp. 366-400. Academic Press, New York.

Humphreys, F.R., and Craig, F.G. (1981). Effects of fire on soil chemical, structural and hydrological properties. *In* 'Fire and the Australian Biota' (A.M. Gill, R.H. Groves, and I.R. Noble, eds), pp. 177-200. Australian Academy of Science, Canberra.

Jarratt, P.H., and Petrie, A.H.K. (1929). The vegetation of the Black's Spur region (2): pyric succession. *J. Ecol.* **17**, 239-281.

Keay, R.W.J. (1957). Wind-dispersed species in a Nigerian forest. *J. Ecol.* **45**, 471-478.

Keeley, J.E. (1981). Reproductive cycles and fire regimes. *In* 'Fire Regimes and Ecosystem Properties' (H.A. Mooney, *et al.*, eds), pp. 231-277. United States Department of Agriculture (Gen. Tech. Report WO-26). Washington D.C.

Kloot, P.M. (1984). The introduced elements of the flora of southern Australia. *J. Biogeog.* **11**, 63-78.

Knight, D.H. (1975). A phytosociological analysis of species-rich tropical forest on Barro Colorado Island, Panama. *Ecol. Monogr.* **45**, 259-284.

Kozlowski, T.T. and Ahlgren, C.E. (eds) (1974). 'Fire and Ecosystems'. Academic Press, New York.

Kruger, F.J. (1983). Plant community diversity and dynamics in relation to fire. *In* 'Mediterranean Type Ecosystems: The Role of Nutrients'. (F.J. Kruger, D.T. Mitchell, and J.U.M. Jarvis, eds), pp. 446-472. Springer-Verlag, Berlin.

Kruger, F.J., and Bigalke, R.C. (1984). *In* 'Ecological Effects of Fire in South African Ecosystems' (P. de V. Booysen, and N.M. Tainton, eds), pp. 67-114. Springer-Verlag, Berlin.

Kruger, F.J., Mitchell, D.T., and Jarvis, J.U.M. (1983). 'Mediterranean Type Ecosystems: The Role of Nutrients'. Springer-Verlag, Berlin.

Lawrence, G.E. (1966). Ecology of vertebrate animals in relation to chaparral fire in the Sierra Nevada foothills. *Ecology* **47**, 278-291.

Leigh, J.H., and Holgate, M.D. (1979). Responses of understorey of forests and woodlands of the Southern Tablelands to grazing and burning. *Aust. J. Ecol.* **4**, 25-45.

Little, S. (1974). Effects of fire on temperate forests: Northeastern United States. *In* 'Fire and Ecosystems'. (T.T. Kozlowski, and C.E. Ahlgren, eds), pp. 225-250. Academic Press, New York.

Lotan, J.E. (1976). Cone serotiny — fire relationships in Lodgepole pine. *Proc. Tall Timb. Fire Ecol. Conf.* **14**, 267-278.

MacLean, D.A., Woodley, S.J., Weber, M.G., and Wein, R.W. (1983). Fire and nutrient cycling. *In* 'The Role of Fire in Northern Circumpolar Ecosystems'. (R.W. Wein, and D.A. MacLean, eds), pp. 111-132. Wiley, Chichester.

Main, A.R. (1981a). Plants as animal food. *In* 'The Biology of Australian Plants'. (J.S. Pate, and A.J. McComb, eds.), pp. 342-361. University of Western Australia Press, Nedlands, Western Australia.

Main, A.R. (1981b). Fire tolerance of heathland animals. *In* 'Heathlands and Related Shrublands of the World. Vol. 9B' (R.L. Specht, ed.), pp. 85-90. Elsevier, Amsterdam.

Majer, J.D. (1982). Ant-plant interactions in the Darling Botanical District of Western Australia. *In* 'Ant-Plant Interactions in Australia' (R.C. Buckley, ed), pp. 45-61. Junk, The Hague.

Martin, R.E., and Cushwa, C.T. (1966). Effects of heat and moisture on leguminous seed. *Proc. Tall Timb. Fire Ecol. Conf.* **5**, 159-175.

McMaster, G.S., and Zedler, P.H. (1981). Delayed seed dispersal in *Pinus torreyana* (Torrey pine). *Oecologia* **51**, 62-66.

Miles, J. (1979). 'Vegetation Dynamics'. Chapman and Hall, London.

Mott, J.J., and Groves, R.H. (1981). Germination strategies. *In* 'The Biology of Australian Plants' (J.S. Pate and A.J. McComb, eds), pp. 307-341. University of Western Australia Press, Nedlands, Western Australia.

Muller, C.H., Hanawalt, R.B., and McPherson, J.K. (1968). Allelopathic control of herb growth in the fire cycle of California chaparral. *Bull. Torrey Bot. Club* **95**. 225-231.

Noble, I.R., and Slatyer, R.O. (1981). Concepts and models of succession in vascular plant communities subject to recurrent fire. *In* 'Fire and the Australian Biota' (A.M. Gill, R.H. Groves, and I.R. Noble, eds.) pp. 311-335. Aust. Acad. Sci., Canberra.

O'Connell, A.M., Grove, T.S., and Dimmock, G.M. (1979). The effects of a high intensity fire on nutrient cycling in jarrah forest. *Aust. J. Ecol.* **4**, 331-337.

Odum, E.P. (1971). 'Fundamentals of Ecology'. Third Edition, Saunders, Philadelphia.

Old, S.M. (1969). Microclimate, fire and plant production in an Illinois prairie. *Ecol. Monogr.* **39**, 355-384.

Pate, J.S., Dixon, K.W., and Orshan, G. (1984). Growth and life form characteristics of the kwongan species. *In* 'Kwongan: Plant Life of the Sandplain' (J.S. Pate, and J.S. Beard, eds), pp. 84-100. University of Western Australia Press, Nedlands, Western Australia.

Platt, W.J. (1975). The colonization and formation of equilibrium plant species associations in badger disturbances in a tall grass prairie. *Ecol. Monogr.* **45**. 285-305.

Purdie, R.W. (1977a). Early stages of regeneration after burning in dry sclerophyll vegetation I: regeneration of the understorey by vegetative means. *Aust. J. Bot.* **25**, 21-34.

Purdie, R.W. (1977b). Early stages of regeneration after burning in dry sclerophyll vegetation II: regeneration by seed germination. *Aust. J. Bot.* **25**, 35-46.

Pyke, G.H. (1983). Relationship between time since the last fire and flowering in *Telopea speciosissima* R.Br. and *Lambertia formosa* Sm. *Aust. J. Bot.* **31**, 293-296.

Raison, R.J. (1979). Modification of the soil environment by vegetation fires, with particular reference to nitrogen transformations: a review. *Plant and Soil* **51**, 73-108.

Raison, R.J. (1980). A review of the role of fire in nutrient cycling in Australian native forests, and of methodology for studying the fire-nutrient interaction. *Aust. J. Ecol.* **5**, 15-21.

Rice, E.L. and Parenti, R.L. (1978). Causes of decreases in productivity in undisturbed tall grass prairie. *Am. J. Bot.* **65**, 1091-1097.

Rowe, J.S. (1983). Concepts of fire effects on plant individuals and species. *In* 'The Role of Fire in Northern Circumpolar Ecosystems' (W. Wein and D.A. Maclean eds.), pp. 135-154. Wiley, Chichester.

Rowe, J.S., and Scotter, G.W. (1973). Fire in the boreal forest. *Quat. Res.* **3**, 446-450.

Rundel, P.W. (1982). Fire as an Ecological Factor. *In* 'Physiological Plant Ecology I. (Encyclopaedia of Plant Physiology. Vol. 12A)' (O.L. Lange, P.S. Noble, C.B. Osmond, and H. Ziegler, eds), pp. 502-538. Springer-Verlag, Berlin.

Salisbury, E.J. (1942). 'Weeds and Aliens'. Collins, London.

Shea, S.R., McCormick, J., and Portlock, C.L. (1979). The effects of fires on the regeneration of leguminous species in the northern jarrah forests of Western Australia. *Aust. J. Ecol.* **4**, 195-206.

Siddiqi, M.Y., Myerscough, P.J., and Carolin, R.C. (1976). Studies in the ecology of coastal heath in N.S.W. IV: seed survival, germination, seedling establishment and early growth in *Banksia serratifolia* Salisb., *B. aspleniifolia* Salisb., and *B. ericifolia* L.F. in relation to fire temperature and nutritional effects. *Aust. J. Ecol.* **1**, 175-183.

Sims, H.J. (1951). The natural regeneration of some trees on sands at Walpeup, Victorian mallee. *Vic. Nat.* **68**, 27-30.

Smith, J.M.B., and Guyer, I.J. (1983). Rainforest-eucalypt interactions and the relevance of the biological nomad concept. *Aust. J. Ecol.* **8**, 55-62.

Specht, R.L. (1981). Responses to fires in heathlands and related shrublands. *In* 'Fire and the Australian Biota'. (A.M. Gill, R.H. Groves, and I.R. Noble, eds), pp. 394-415. Australian Academy of Science, Canberra.

Springett, J.A. (1976). The effects of prescribed burning on the soil fauna and on litter decomposition in Western Australian forests. *Aust. J. Ecol.* **1**, 83-87.

Sweeney, J.R. (1956). Responses of vegetation to fire: a study of the herbaceous vegetation following chaparral fires. *Univ. Calif. Publ. Bot.* **28**, 143-150.

Sweeney, J.R. (1968). Ecology of some "fire-type" vegetations in northern California, *Proc. Tall Timb. Fire Ecol. Conf.* **8**, 111-125.

Tap, P.M. and Whelan, R.J. (1984). The effect of fire on populations of heathland invertebrates. *In* 'Proceedings of the Fourth International Conference on Mediterranean Ecosystems'. pp. 147-148. Perth, Western Australia.

Tothill, J.C. (1969). Soil temperatures and seed burial in relation to the performance of *Heteropogon contortus* and *Themeda australis* in burnt native woodland pastures in eastern Australia. *Aust. J. Bot.* **17**, 269-275.

Tothill, J.C. (1977) Seed germination studies with *Heteropogon contortus*. *Aust. J. Ecol.* **2**, 477-484.

Tran, V.N. and Cavanagh, A.K. (1984) Structural aspects of dormancy. *In* 'Seed Physiology'. (D.R. Murray, ed.) pp. 1-44. Academic Press, Sydney.

Van Cleve, K., and Viereck, L.A. (1981). Forest succession in relation to nutrient cycling in the boreal forest of Alaska. *In* 'Forest Succession: Concepts and Applications.' (D.C. West, H.H. Shugart, and D.B. Botkin, eds), pp. 185-211. Springer-Verlag, New York.

Viereck, L.A. (1973). Wildfire in the Taiga of Alaska. *Quat. Res.* **3**, 471-474.

Vines, R.G. (1974). Air movements above large bush fires. *Proc. Tall Timb. Fire Ecol. Conf.* **13**, 295-301.

Vines, R.G. (1981). Physics and chemistry of rural fires. *In* 'Fire and the Australian Biota'. (A.M. Gill, R.H. Groves, and I.R. Noble, eds.), pp. 129-149. Australian Academy of Science, Canberra.

Viro, P.J. (1974). Effects of forest fire on soil. *In* 'Fire and Ecosystems' (T.T. Kozlowski, and C.E. Ahlgren, eds), pp. 7-46. Academic Press, New York.

Vitousek, P.M., and Reiners, W.A. (1975). Ecosystem succession and nutrient retention: A hypothesis. *BioScience* **25**. 376-381.

Vogl, R.J. (1974). Effects of fire on grasslands. *In* 'Fire and Ecosystems' (T.T. Kozlowski, and C.E. Ahlgren, eds), pp. 139-194. Academic Press, New York.

Wade, D., Ewel, J.J., and Hofsetter, R. (1980). 'Fire in South Florida Ecosystems' General Technical Report SE17. USDA Forest Service (Asheville, North Carolina).

Warcup, J.H. (1980). Effect of heat treatment of forest soil on germination of buried seed. *Aust. J. Bot.* **28**, 567-571.

Wardrop, A.B. (1983). The opening mechanism of follicles of some species of *Banksia. Aust. J. Bot.* **31**, 485-500.

Wein, R.W., and MacLean, D.A. (eds) (1983). 'The Role of Fire in Northern Circumpolar Ecosystems'. Wiley, New York.

Whelan, R.J. (1977). 'The Influence of Insect Grazers on the Establishment of Post-fire Plant Populations'. Ph.D. Thesis, University of Western Australia, Perth, Australia.

Whelan, R.J.(1985). Patterns of recruitment of plant populations after fire in Western Australia and Florida. *In* 'Are Australian Ecosystems Different?' *Proc. Ecol. Soc. Aust.* **14**, 169-178.

Whelan, R.J., Langedyk, W., and Pashby, A.S. (1980). The effects of wildfire on arthropod populations in Jarrah-*Banksia* woodland. *W.A. Nat.* **14**, 214-220.

Whelan, R.J., and Main, A.R. (1979). Insect grazing and post-fire plant succession in southwest Australian woodland. *Aust. J. Ecol.* **4**, 387-398.

Wilson, R.E., and Rice, E.L. (1968). Allelopathy as expressed by *Helianthus annus* and its role in old field succession. *Bull. Torrey Bot. Club.* **95**, 432-448.

Woodmansee, R.G., and Wallach, L.S. (1981). Effects of fire regimes on biogeochemical cycles. *In* 'Fire Regimes and Ecosystems Properties' (H.A. Mooney *et al.*, eds.), pp. 379-400. United States Department of Agriculture (Gen. Tech. Report WO-26). Washington, D.C.

Wright, H.A., and Bailey, A.W. (1982). 'Fire Ecology: United States and Southern Canada'. Wiley, New York.

Zimmer, W.J. (1940). Plant invasions in the mallee. *Vic. Nat.* **56**, 143-147.

CHAPTER 7

Evolution of Seed Dispersal Syndromes According to the Fossil Record

B. H. TIFFNEY

I.	Introduction	274
II.	Materials and Methods	
	A. Inferences from Modern Ecology	275
	B. The Fossil Record	278
III.	The Fossil Record — Data	
	A. Dispersal in Pre-seed Communities	281
	B. Earliest seeds: Latest Devonian-Mississippian	282
	C. Pennsylvanian	283
	D. Permian	285
	E. Triassic	285
	F. Jurassic	286
	G. Cretaceous	287
	H. Tertiary	288
IV.	Discussion	
	A. Devonian-Mississippian	289
	B. Pennsylvanian	290
	C. Permian into Triassic	293
	D. Triassic-Early Cretaceous	295
	E. Cretaceous and the Angiosperms	295
	F. Tertiary	296
V.	Summary	
	A. History of Dispersal	298
	B. Dispersal Ecology	299
	Acknowledgement	
	Text References	
	Data References	

273

SEED DISPERSAL
ISBN   0 12 511900 3

I. INTRODUCTION

Dispersal plays a central role in the biology of extant seed plants. Howe and Smallwood (1982) indicate three central ecological 'goals' for dispersal. First, it provides escape from excessive mortality of seeds or seedlings caused by competition or consumption in the vicinity of the parent plant. The reality of this mortality, initially a theoretical concept, has been underscored by Howe and Schupp (1985). Second, dispersal permits the colonization of ecologically favorable sites that are scattered in space and time. Third, dispersal may permit directed movement of seeds to specific microsites appropriate to the ecology of the parent plant. These 'goals' are not independent, and all may simultaneously influence the selective pressures that favor dispersal. Dispersal also has energetic costs involving: (1) the metabolic price of the tissues necessary to effect dispersal; and, (2) the loss inherent in dispersal-related seed consumption. Different dispersal mechanisms involve different cost-benefit solutions and influence the effects of selective pressures.

Dispersal may be achieved in two broad ways — abiotic or biotic, the latter involving the intercession of an animal. Each has its own morphological and energetic correlates. Abiotic dispersal can achieve the first and second 'goals' of dispersal mentioned above. Biotic dispersal can answer all three 'goals', and is almost specifically necessary for the 'goal' of directed dispersal; animals become the plant's 'legs'.

Biotic dispersal involves the appearance of what are popularly viewed as 'co-evolved' syndromes of characters linking dispersal agents and the propagules they disperse. The concept of co-evolution has grown over the past decade, until it has become a popular metaphor for the complexity and interdependence of natural systems. In its most rigid formulation, it implies a long history of reciprocal interaction of the two partners, each evolutionary step of one being met by a response from the other — a constant ping-pong match of shifting selective forces. Recent authors, including Futuyma and Slatkin (1983) have suggested that the true situation is far more loose-knit; while there indeed are morphological adaptations of dispersal agent to its fruit or seed and *vice versa*, these are the result of a general interaction — the 'diffuse co-evolution' of Janzen (1980) or 'coarse co-evolution' of Herrera (1985). Herrera (1985) gives a particularly elegant summary of this perspective, indicating that, while the pre-existing view of tightly-linked disperser-plant interactions is probably incorrect, the new perspective of looser interactions is no less biologically important or fascinating, and dispersal remains an integral and important part of seed plant biology.

In spite of the scientific excitement and ecological importance of dispersal in the present day, its historical roots have rarely been examined. However, the tacit assumption that, as long as vertebrates and plants co-existed on the earth's surface, dispersal mechanisms have been qualitatively and quantitatively similar to those of the present day, is unwarranted. Van der Pijl (1982) provided a brief review of suspected dispersal relationships of extinct gymnosperms and vertebrates. Weishampel (1984a) examined Mesozoic vertebrates and plants and concluded that 'dispersal occurred' in much the same manner as in the present, although both animal and plant communities were very different from present ones. I examined dispersal and size characteristics of Cretaceous and Tertiary angiosperm propagules (Tiffney, 1984) and concluded that the two periods were dominated by quite different patterns. This led to an examination of Mesozoic and Cenozoic vertebrate-plant interactions (Wing and Tiffney, in press) and the further conclusion that the structure of herbivore-plant interactions in the Cretaceous was quite different than in the Tertiary or the present.

This chapter incorporates data from Tiffney (1984) and Wing and Tiffney (in press), together with a survey of dispersal commencing with the origin of the seed habit. It presents a broad overview of seed dispersal through time, and of necessity omits detailed consideration of many specific questions.

II. MATERIALS AND METHODS

Two data sources contribute to the understanding of past dispersal biology. In uniformitarian style, the first is modern information on the ecological correlates of fruit-seed morphology. The second involves data on seed and fruit morphology in the fossil record, and more limited data on what herbivores actually ate in the past. In ensuing discussion, I use the term 'diaspore' to indicate the unit that is transported, which may be an ovule, a seed or a fruit containing seed(s).

A. Inferences from Modern Ecology

1. Dispersal Syndromes
Neontologists have recognized suites of morphological characteristics of diaspores which are broadly, although not absolutely, associated with particular dispersal modes or agents. These have been summarized by van

der Pijl (1982). If recognizable in the fossil record, these provide an indication of dispersal mode. The most important of these syndromes include:

a. Anemochory — Wind-borne diaspores may be extraordinarily small (e.g. orchids); possess wings (e.g. *Pinus* among gymnosperms, members of the Bignoniaceae of the angiosperms); possess plumes of hairs (e.g. *Asclepias* of the Asclepiadaceae, many members of Compositae) or other means of increasing the surface area of a light diaspore. For examples, see Chapter 1.

b. Hydrochory — Water-borne diaspores generally lack specific morphological adaptations, save in some cases spongy or hollow tissue that permits flotation (e.g. *Cocos, Barringtonia*). Many seeds will float for at least some length of time. For discussion, see Chapter 2.

c. Autochory — Some plants may disperse diaspores actively, often involving changes in tension of the fruit or receptacle wall resulting from water loss, e.g. *Hura crepitans* of the New World tropics may throw its seeds many meters when the woody fruit breaks. Other autochorous plants may cast seeds from pods which vibrate in the wind (e.g. *Papaver*).

d. Epizoochory — Transport of diaspores on the outer surface of animals usually involves a sticky substance, or the presence of hooks or stiff hairs on the surface of the diaspore (e.g. Fig. 1 of Chapter 2).

e. Myrmecochory — Ant dispersal usually involves the presence of an oily or fat-rich elaiosome which is consumed by the ant (Chapter 3).

f. Fish dispersal — The importance of this mode has been indicated by the work of Goulding (1980), but it is not associated with any specific diaspore morphology. I suspect that fish dispersal is facultative, and is important only in unusual, geomorphologically-controlled environments (e.g. the modern Amazon river Basin, Chapter 2, II, B).

g. Saurochory — Reptile dispersal is associated with diaspores that possess brightly-colored and often strongly-scented flesh, which are borne near the ground or dropped to the ground at maturity.

h. Ornithochory — Bird dispersal involves diaspores clearly displayed on the plant, often brightly-colored (commonly red) or bedecked with contrasting colors, without smell, and with a hard inner protective portion (Tables II, IV of Chapter 4).

i. Mammal dispersal — Mammals vary widely in their sizes and abilities, and characteristics of mammal-dispersed diaspores are equally varied. In general, mammal-dispersed diaspores may or may not be brightly-colored, often have distinct odor, and possess a hard protective wall about the kernel. Sizes and methods of display on parent plants are also variable (Tables III, IV of Chapter 4; Chapter 5).

2. Size and Ecology of Seeds
The size of a seed generally reflects the size of the nutritive reserve contained

within. In gymnosperms, this nutritive tissue is provided by the megagametophyte and is mainly formed before fertilization of the egg. In many angiosperms the nutritive tissue is endosperm, formed by the process of double fertilization (for discussion, see Murray, 1984). Salisbury (1942; see also Harper *et al.*, 1970) demonstrated that the amount of nutritive tissue available to the germinating seedling influenced its success in different environments. Small seeds are associated with plants that germinate and grow in light-rich environments; the seedling is provided with enough energy by the seed reserves to grow its initial leaves, with which it quickly becomes self-supporting. By contrast, large seeds are associated with plants that grow in closed, light-poor communities. Here the reserves permit the seedling to establish a large leaf area before it becomes self-supporting. The amount of nutritive reserve does not dictate mature plant size, as quite large trees may have small seeds which germinate and grow in light-rich environments.

This relationship of seed size and light environment dictates a general association between seed size and the successional status of the plant that bears the seed. Small-seeded plants are often early-successional, while large-seeded plants are often late-successional. Plants with seeds of intermediate size occupy an intermediate zone of succession, the question of size and successional status being a relative one. Salisbury (1942) demonstrated this relationship in temperate English communities, and subsequent work (e.g. Foster and Janson, 1985) indicates that the same relationship holds in tropical communities. Some variation does occur in seed size depending on the ecological situation of individual plants (Baker, 1972), and the absolute sizes of seeds are larger in the tropics than in the temperate zone (Levin, 1974).

3. Size and Dispersal

Small seeds of early successional plants may be abiotically or biotically dispersed. Small herbs often have wind-borne diaspores, but a large number of temperate early-successional shrubs and trees are bird-dispersed (Sorenson, 1981; McDonnell and Stiles, 1983), and granivorous rodents (Chapter 5) are a significant factor in the movement of small, dry diaspores (e.g. grasses). Small seeds may even become aggregated in a large fruit (e.g. *Ficus*) which becomes the diaspore, often biotically-dispersed. While a few large diaspores move abiotically, particularly by water (e.g. *Cocos, Barringtonia*), the majority of large seeds and diaspores are generally biotically-dispersed. Furthermore, circumstantial evidence (Janzen and Martin, 1983; Tiffney, 1984) suggests that in the absence of developed characteristics for abiotic dispersal, plants with large diaspores cannot maintain their populations without biotic dispersal. In short, large diaspores of later-successional plants generally require the intercession of a biotic dispersal agent.

B. The Fossil Record

1. Morphology of Fossils

a. Data available — Certain morphological characteristics of dispersal syndromes are fossilizable, others are not. The wings and hairs of airborne diaspores, the spongy coats of water-borne ones, the hooks of epizoochorous ones and occasionally the flesh of endozoochorous diaspores may all be preserved. The sticky substances that may have glued diaspores to animals and the elaiosomes that attracted ants are unlikely to be preserved; likewise the colors and odors of fleshy diaspores are ephemeral. Autochorous structures are unlikely to be clear in the fossil record. Abiotic dispersal is thus fairly easy to discern; the full suite of characteristics of biotic dispersal is unlikely to be found. Still, given the assumption that plants do not evolve and maintain strictly functionless and energetically-expensive structures, the simple presence of an external flesh generally implies the participation of an animal. In many periods of earth history, the diversity of animals and the ranges of their sizes are sufficiently limited that one can reasonably guess at the nature and importance of dispersal activities.

b. Size measurements — Seed size is presented here in cubic millimeters (c.f. Table 1, Fig. 1). This measurement appears to be the best way of comparing fossil diaspores, which are often available for study only as published descriptions, and which are frequently preserved by minerals of differing specific gravities. Volume appears to be a good indicator of weight (Tiffney, 1984), to judge from modern diaspores. Diaspores are generally spherical or ovoid. An estimate of volume based on length x width x breadth assumes a body constrained by right angles, and thus over-estimates the actual volume. However, this error is consistent across the data presented here.

c. Sources — Data on the morphological characteristics, sizes of seeds and sizes of diaspores of Devonian through Cretaceous gymnosperms were assembled from the literature (see 'Data References'). Those for angiosperms come from Tiffney (1984). The angiosperm data were collected by flora — thus each summary reflects a roughly co-existent community of plants in time and space. Published seed floras generally do not exist for pre-angiosperm communities, and I assembled data from reports of individual seeds and summarized it by period. This has the serendipitous advantage of avoiding potential bias of sampling a single, ecologically-anomalous, community. These data are not numerous before the rise of the angiosperms. However, I believe that the trends which I infer from these data are real; while further data would alter numerical conclusions, the broad patterns would remain (Table 1; Fig. 1).

Table I. Calculated data on diaspore size, by period.

	Dev	Miss	Penn	Perm	Triassic		Jurassic		Cretaceous		Tertiary	
					General	Grnland.	General	Yorkshre	Gymno	Angio	Pe	Ne
N	4	9	7	16	12	30	9	19	12	203	485	578
$\bar{X}$*	7.7	67 (12.6)	19,712	877	2,140	188	1,950	304	205	1.6	925	671
Standard deviation	4.6	164 (7.2)	47,042	2,972	5,486	402	5,230	756	621	n.a.	n.a.	n.a.
Maximum size*	12.6	500	317,600	12,000	19,360	1,600	15,870	3,050	2,180	55	61,000	25,000
Minimum size*	2.0	0.16	7.0	1.0	12.0	0.64	8.0	1.0	18	0.02	0.07	0.04

* In mm^3

Geologic abbreviations as in Figure 1. Under Triassic and Jurassic the columns are separated into one for summed data from isolated reports and one from one major flora. Cretaceous data are split into gymnosperm and angiosperm columns. In Mississippian column unbracketed values are for all data (N = 9); bracketed values are for same data minus *Salpingostoma dasu*. Standard deviation values are not given for angiosperms in the Cretaceous, nor for the angiosperm-dominated floras of the Paleogene and Neogene, as the figures given for average size are a summation of values for many separate floras.

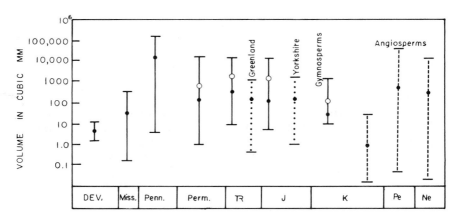

Fig. 1. Graphic representation of diaspore size over time (see also Table I).

Vertical axis, volume on a logarithmic scale; horizontal axis, time: Dev = Devonian, Miss = Mississippian, Penn = Pennsylvanian, Perm = Permian, TR = Triassic, J = Jurassic, K = Cretaceous, Pe = Paleogene (Tertiary), Ne = Neogene (Tertiary). Upper bar = maximum diaspore volume in flora; lower bar = minimum diaspore volume. Central dot = mean diaspore volume.

The data for the Triassic and Jurassic are split. In each case the left-hand line represents a summation of isolated diaspore sizes while the right-hand line (dotted) represents the diaspore sizes in one large flora (see text for details).

The data for the Cretaceous are split into gymnosperms (left-hand line) and angiosperms (right-hand line).

Data for the Permian, Triassic and Jurassic isolated diaspores and the Cretaceous gymnospersms are presented with two mean values. Open circles represent means calculated on all diaspores present. Closed (lower) dots are the same data, minus the largest diaspore in each case. This provides graphic illustration of the strong bias exerted on average seed size by one outlying value in these small samples.

Geological time scale. Dates after Harland *et al* (1982). The Mississippian and Pennsylvanian of North American usage correspond to the Lower and Upper Carboniferous of European usage.

d. Biases — Several potential biases should be recognized at the outset. Seed size may be altered by compression, erosion or shrinkage. Clearly eroded or deformed specimens were not included in this study, but otherwise this bias must be assumed to average out over time. It is difficult to judge the maturity of many fossil seeds which lack living homologs. While I have attempted to include only well developed diaspores, some estimates may include immature, and thus anomalously small representatives. Certain time periods (Permian, Triassic) are notorious for the small numbers of known fossils, apparently reflecting both the availability of preserving environments and the actual existing diversity. While these periods are poorly represented in the sample, I can only presume that the available data provide a representative sample of contemporary diaspore morphologies.

e. Ecological biases — The application of uniformitarian principles dictates that we infer past dispersal biology from the concordance of past and present morphology (Section II, A, *1*). However, we must examine the possibility that the morphological responses of plants to dispersal agents have altered over time, and that morphological structures of the past are not functionally analogous to those of the present. Diaspore morphology is largely shaped by two external forces, dispersal and consumption. There is no reason to assume that the ecological 'goals' of dispersal (c.f. Section I) have changed over time. Similarly, there is no reason to assume that the forms ('quality') of seed consumption (Smith, 1975) have changed greatly over time, although the pressure ('quantity') of seed consumption may have varied with changes in herbivore types and diversities. I conclude that the continuity of certain morphological characters (wings, flesh) over the history of land plant diaspore production indicates a continuity of similar dispersal relationships.

2. Evidence for Seed Consumption by Animals

Evidence for biotic dispersal is also provided by three aspects of the potential dispersal agents themselves.

a. Dental morphology — It is possible to distinguish carnivores from omnivores and herbivores on the basis of tooth structure; to a lesser degree of certainty, one can use this method to separate omnivores and generalized herbivores. Certain mammalian groups have developed specialized herbivorous teeth, and may, for example, be recognized as seed specialists (frugivores) or leaf specialists (folivores). By and large, the primary data of importance to the present discussion are the times of appearance of herbivory among early tetrapods. More specialized herbivorous dentitions appear to be a feature of mammalian evolution in the Tertiary.

b. Stomach contents — In rare cases, animals with intact stomach contents have been preserved. These contents cannot be taken as indicative of the average diet of the organism, but only to show what was within range of its diet.

c. Coprolites — Similarly, fossil fecal material provides evidence of the diet of extinct animals. Again, such fossils are rare, and indicate only a meal, not necessarily the average diet.

   The animal-based data are scanty and often inferential, but may illuminate and perhaps constrain conclusions derived from plants alone.

III. THE FOSSIL RECORD — DATA

A. Dispersal in Pre-seed Communities

The earliest vascular plants possessed pteridophytic, homosporous reproduction, in which the gametophytes were free-living and presumably bisexual (Taylor, 1981; Tiffney, 1981). The central obstacle faced by these plants was the heritage of their aqueous origin; the desiccating terrestrial environment curtailed success of sexual reproduction involving free-swimming sperm. Thus it seems likely that asexual, vegetative reproduction occupied a dominant place in the life history of these organisms (Tiffney and Niklas, 1985).

The dominance of pteridophytes in the Devonian and Carboniferous, and the diversity of the ferns from the Mesozoic to the present day, bespeak the eventual successful response of early homosporous land plants to the challenge of the terrestrial environment. However, this was only an intermediate step in the evolution of reproduction in the terrestrial sphere. Just as seed reserves support the embryo, the storage products within spores provide support for the initial growth of the free-living gametophyte. Larger gametophytes would be more successful in bearing gametes and, in the case of the egg, provisioning it with reserves (see Tiffney, 1981 for an extended discussion). This logic suggests the evolution of larger spores through the Devonian, which is exactly what is observed (Chaloner and Sheerin, 1981). Eventually the gametic responsibilities became split, small (micro-) spores giving rise to antheridiate gametophytes, and large (mega-) spores giving rise to archegoniate gametophytes.

Over time, the number of megaspores per sporangium became reduced until only one remained. At some point this spore was no longer dispersed to the environment, but was retained in the megasporangium, the microspore being borne to it by the wind. In this we see the origin of 'pollination'. However, it is likely that in the earliest cases, once pollination could occur, the micro- and megaspores were shed together to disperse as a joint unit, perhaps similar to the joint dispersal of associated *Selaginella* micro- and megaspores in the present day.

Also over time, the associated branches around the megasporangium came to invest it, presumably first to aid in the trapping of microspores, but ultimately to form a protective cover — the integument (see Niklas, 1981, 1983, for a summary). Through this process, the first seed, an 'integumented megasporangium' evolved. However, certain 'habits' of its spore-based ancestry lingered on for tens of millions of years. It is probable that pollination was effected while the unfertilized seed (technically an ovule) was still on the parent plant, but fertilization may not have occurred until after the ovule was dispersed, similar to the case in *Selaginella* spp. or *Ginkgo biloba*.

B. Earliest Seeds: Latest Devonian—Mississippian

The earliest seeds appeared in the Late Devonian, and by the Early Mississippian seeds were fairly abundant. Limited stratigraphic resolution and the simplicity of biological characters preclude recognition of clear ancestor-descendant relationships; it is possible that the origin of the seed was polyphyletic.

These early seeds were tiny, averaging about 10 mm^3 with a maximum of roughly 20 mm^3. Seeds of the Devonian appear to be generally smaller (average 7.7 mm^3) than those of the Mississippian (average 15 mm^3), but this pattern cannot be statistically supported. The majority of these seeds were borne in cupules, enclosing structures which held one to 48 radially-symmetrical ovules. The cupule presumably mimicked the effect of the integumentary lobes, causing the deflection of microspores in the wind stream and thereby aiding pollination. These cupule-borne seeds lacked strongly developed seed-coats, but while the cupule may have afforded some protection, evidence suggests that many of these seeds were often shed from their cupules on maturity. A second and less common morphology was represented by *Spermolithus* (Chaloner *et al.*, 1977) of the Upper Devonian and *Lyrasperma* (Long, 1960) of the Lower Mississippian. Here the seed was bilaterally-symmetrical and possessed two rudimentary lateral wings.

C. Pennsylvanian

By the Middle Pennsylvanian, a wide variety of seed plants and seed types were present. The popular image of the Pennsylvanian is of lowland coal swamps, but these are only part of the vegetational story. Co-eval communities existed in mesic lowlands adjacent to the coal swamps, and in drier upland areas. Furthermore, the classic swamps of the textbooks were restricted to the equatorial Euramerican region; to the north-east were the cooler and drier environs of the Angaran flora, and to the south were the beginnings of the seasonal, occasionally cold, environments of the Gondwana flora. The ecologies of seeds in these environments need to be treated separately.

1. Euramerican Swamps
These communities were dominated by arborescent pteridophytes, pteridosperms, and cordaites (extinct gymnosperms related to conifers). The genus *Cordaites* encompassed a wide range of ecologies (Raymond and Phillips, 1983). Seeds of the *Mitrospermum* morphology were flattened, winged, with a volume of roughly 122 mm^3. Seeds of the genera *Cordaicarpon* and *Nucellangium* were ovoid, with a well-developed fleshy sarcotesta and a sclerotic, often ornamented, sclerotesta. These were much larger, ranging from perhaps 350 mm^3 to 12 000 mm^3.

The swamps were also inhabited by many taxa of seed ferns, the most numerous and well known of which belonged to the Lyginopteridaceae and the Medullosaceae. The lyginopterid seed ferns tended to possess smallish seeds (average 35 mm^3; maximum 106 mm^3) which were borne singly in a cupule, although *Gnetopsis elliptica* possessed two to four ovules per cupule. The cupular structure is very similar to that of the Devonian, and may reflect a phylogenetic link. However, as with the Devonian seeds, it is not clear how often lyginopterid seeds were shed from the cupule and how often dispersed in it. The seeds do possess a hard sclerotesta, and the cupule could have served as a fleshy attractant.

The medullosans were represented by many genera and species of seeds. The two most important are *Trigonocarpus* and *Pachytesta*, which probably represent different preservational states of the same seed (c.f. Taylor, 1981). Both possessed thick sclerotestas and often well developed external sarcotestas. The average volume of *Trigonocarpus* was 21 398 mm^3 with a maximum of 160 000 mm^3; the average for *Pachytesta* was 47 870 mm^3 with a maximum of 317 625 mm^3. These largest sizes approached the volume of the largest range of extant angiosperm seeds. Other taxa of seeds associated with the Medullosaceae exhibited similar structures and sizes. A third family of seed ferns, the Callistophytaceae, (Rothwell, 1981) is represented by two species of small seeds (20 mm^3) with a sarcotesta and a sclerotesta of average proportions.

2. Euramerican Uplands

Frederiksen (1972) has convincingly argued that the coal swamps of the Pennsylvanian were not steaming jungles, but rather subtropical in nature, and that the uplands surrounding them were cool and seasonal. This upland environment was apparently too dry for the majority of Pennsylvanian plants, and hosted (by contrast to the lowlands) a low diversity of conifers, perhaps with a scattering of advanced ferns and cycadophyte ancestors. I am unable to locate any convincing records of Pennsylvanian conifer seeds, but by analogy to the present and to conifers of the Permian and Mesozoic, it seems likely that they were small and abiotically-dispersed.

3. Angaran Floras

Meyen (1982) summarizes the Angaran floras of the later Pennsylvanian and Early Permian which existed in the cooler latitudes to the north-east of the Euramerican swamps. The gymnosperm component of these floras was dominated by cordaitalean taxa which uniformly possessed winged seeds, with the wing either circumferential (*Bardocarpus*) or asymmetric (*Sylvella*). The average volume of these seeds was 132 mm^3 with a range from 34 to 200 mm^3. Seed ferns comprised a minor portion of the flora; medullosan seed ferns were rare, and were represented by only one seed type similar to *Holcospermum*. Other seed fern groups appeared in the area in the Permian (Peltaspermaceae, Cardiolepidaceae), but left a limited seed record; *Cardiolepis* possessed a multi-seeded capsule, but it is unclear whether the seeds were dispersed in the capsule or separately, since separate dispersed seeds (*Nucicarpum*) are reported.

D. Permian

1. Gondwanan Floras

Like the Angaran floras, but to a greater degree, the Gondwanan floras of the Southern Hemisphere Paleozoic cross the Pennsylvanian-Permian boundary, and could be considered appropriately under either heading. Gondwanan gymnosperms were dominated largely by members of the Glossopteridales: arborescent, deciduous seed ferns. Glossopterid seeds were small by contrast to northern counterparts, and occasionally possessed wings (e.g. Raynor and Coventry, 1985). The largest reported seed (Kovács-Endródy, 1974) had a volume of 560 mm^3, but the remainder were much smaller, averaging 19 mm^3. They were most frequently borne on multi-seeded receptacles associated with the leaves of *Glossopteris* and associated genera, but apparently were shed at maturity. The only possible exception might be the structure reported by Gould and Delevoryas (1977), in which the small (about 1 mm^3) seeds were borne inside an inrolled and perhaps fleshy leaf-like structure. It is not clear whether this structure was mature.

2. Northern Hemisphere Floras

Seed data are sparse for the Permian of the Northern Hemisphere. The patterns that develop include: (1) the demise of the medullosan seed ferns with their large seeds; (2) the appearance of new seed fern groups (Cardiolepidaceae, Peltaspermaceae) with much smaller seeds in Angara (Section III, C, 3); (3) the reduction and ultimate disappearance of the cordaites group; (4) the appearance of ancestral ginkgos; and, (5) the rise in importance of conifers. Discounting the Early Permian medullosan remnants, representative seeds of this time range between 30 and 200 mm^3. The conifers and remnant cordaites possess adaptations for wind dispersal, while the dispersal types in the remainder are ambiguous.

E. Triassic

Data are again sparse, but may be presented from two sources: a general survey of isolated occurrences; and a summary of the diaspore remains in one major flora.

1. Isolated Sources

Sphenobaiera, a ginkgophyte, possessed an apparently fleshy diaspore of intermediate size (1100 mm^3). The seed fern *Dicroidium* apparently possessed diaspores that ranged up to 3000 mm^3, while the seed fern fructification *Umkomasia* had one to two smallish (50 mm^3) seeds borne in a fleshy cupule; the nature of the diaspore is not clear. The remainder of the isolated seeds which were described in the literature were of unknown affinity and ranged from 12 to 700 mm^3, with one specimen of 19 360 mm^3, although I am not convinced that it represents plant material.

2. The Scoresby Sound Flora

Harris (1932a, 1932b, 1935) described the fossil flora of the Late Triassic of Scoresby Sound, Greenland, including 30 morphologies of seeds. This provides the best sample of Triassic diaspore structures, although it may be biased by its derivation from a single vegetational unit. Excluding the three largest seeds (ranging from 1000 to 1600 mm^3) the average size was 63 mm^3, the smallest being approximately 1 mm^3 (represented by 7 seeds) and the largest 500 mm^3.

The seed ferns were represented by *Lepidopteris* (112 mm^3) and *Caytonia* (about 1 mm^3), the latter possessing several seeds borne in a fleshy 'berry'. The cycad *Nilssonia incisoserrata* had a seed of about 1000 mm^3, and presumably possessed a fleshy layer around the sclerotesta. The cycadeoids had seeds from 1 to 140 mm^3 which were apparently wind dispersed. These small seeds were borne many to a 1.5 to 2.5 cm diameter receptacle, but apparently were shed on maturity (Section IV, D). The conifers (n = 8) were small seeded (average 19 mm^3) and three possessed wings. The ginkgoalean seeds were much larger (average 525 mm^3, n = 4) and appeared to possess an external fleshy layer. Unknowns ranged from 1 to 1600 mm^3 in volume, some possessing flesh, the remainder of unclear dispersal mode.

F. Jurassic

As for the Triassic, the data come from isolated reports and one flora.

1. Isolated Sources

Carpolithes conicus (Seward, 1917), attributed to the cycads, was the largest seed of this period (15 900 mm^3). *Problematospermum ovale* is attributed to the Cycadeoids (Krassilov, 1973). It was small (18 mm^3) and possessed an apical tuft of hairs strongly suggestive of wind dispersal. The ginkgoalean seed *Burejospermum* was large (1000 mm^3) and by analogy to modern *Ginkgo biloba* may have possessed an external flesh. The conifer *Araucaria mirabilis* (Taylor, 1981) had a large, unwinged seed (470 mm³) while *Para-raucaria patagonica* (Stockey, 1977) was smaller (110 mm³) and winged. The seed of *Pentoxylon* was unspecialized and about 8 mm³.

2. The Yorkshire Jurassic Flora

Harris (1964, 1969, 1979) and Harris *et al.* (1974) summarized this rich flora, reporting 19 seed types. *Caytonia* (seed ferns) was represented by three species with an average seed size of 1.6 mm³; in each case the seeds were borne in a fleshy structure and Harris (1964) reported coprolites containing *Caytonia* seeds. However, individual seeds separated from their receptacle are frequently found. Cycadeoids were represented by two species of small seeds (average volume 5.2 mm³). Hill (1976) reported small coprolites from this deposit containing cuticular remains of *Ptilophyllum* and *Wil-liamsonia*, the latter a genus for cycadeoidalean receptacles. However, the coprolites did not contain any actual seeds (Section IV,D). The cycad seed *Beania* was large (3000 mm³ — although it is not clearly a mature specimen) and presumably possessed an external flesh. Ginkgos were represented by two species of presumably fleshy seeds of 116 and 300 mm³ volume. *Lep-tostrobus*, the ovulate reproductive organ of *Czekanowskia*, was a valvate capsule that released a few small (8 mm³) seeds on maturity. The six species of conifers averaged 9.5 mm³ (maximum 43 mm³; all but one of the others less than 3 mm³), and none with wings.

G. Cretaceous

The Caytoniales, cycadeoids, ginkgos and conifers all carried on into the Cretaceous with seed characteristics similar to those of their earlier repre-sentatives. The former two groups became extinct in this period, while the ginkgos continued without great change and the conifers witnessed the beginnings of the radiation of the Pinaceae, the dominant modern group. Kräusel (1922) reports a specimen of the herbivorous dinosaur *Trachodon* containing fragments of conifer foliage, cones and seeds, indicating a diet that included seeds (see Krassilov, 1981 for details).

The major botanical significance of the Cretaceous, however, lies in the origin and rise of the angiosperms (c.f. Doyle and Hickey, 1976; Hickey and Doyle, 1977; Tiffney, 1981). The patterns of dispersal biology evident in the history of the angiosperms have been reviewed by Tiffney (1984), and are briefly summarized here.

The earliest angiosperms possessed small, abiotically-dispersed seeds. Where larger, clearly angiospermous structures have been found, they were follicles or capsules containing small seeds. This pattern generally was maintained through the Cretaceous-Tertiary boundary, although there are isolated reports that potentially larger seeds and diaspores occurred in the later Cretaceous. Dispersal adaptations were generally unspecialized (e.g. small size), or may have involved winged fruits or seeds (e.g. the fruits described by Krassilov, 1982).

Animal interaction with the newly-appeared angiosperms is reflected in the spread of ornithopod dinosaur herbivores. Bakker (1978) notes that these animals tended to feed at lower levels than their sauropod predecessors. However, no specific information exists to indicate a direct link between these new herbivores and angiosperm diaspore morphology.

H. Tertiary

Tertiary dispersal biologies and diaspore morphologies indicate circumstances much like those of the present day. The cycads were present, possessed large seeds with fleshy coverings, but became increasingly restricted to a few tropical or near-tropical sites. *Ginkgo* maintained its seed morphology, but became increasingly restricted in geography, and was nearly extinct in the Pleistocene (c.f. Tralau, 1968). Amongst the conifers, the Pinaceae, with small, winged seeds radiated in the Northern Hemisphere, while the Podocarpaceae, with small to intermediate sized seeds covered with a fleshy structure of debated morphological origin, apparently radiated in the Southern Hemisphere.

A major radiation and modernization of angiosperms began in the Late Cretaceous; by the Early Tertiary, many modern families and genera were present. Analogy to modern taxa, together with morphological information from the fossil fruits and seeds, indicates that much of the diverse array of dispersal syndromes that characterize the group in the modern day evolved during the Late Cretaceous and Early Tertiary. These newly-appeared seeds include many that were as large (50 000-100 000 mm^3) as those seen in the Pennsylvanian seed ferns, although others were as small as those that were initially characteristic of the angiosperms in the Early Cretaceous.

Systematic and morphological diversity during the later Tertiary indicates little change in the kinds of dispersal syndromes of the angiosperms, although the spread of grasslands (Tiffney, 1981) in the middle and late Tertiary may have led to the increasing importance of certain syndromes. Smaller seed sizes became more common in temperate floras as the result of the increasing importance of herbs, but large seeds remained. Almost no evidence exists for the middle and late Tertiary record of tropical forests, and I am forced to surmise that dispersal syndromes and seed sizes in these communities were much as they are today, given the presence in the fossil pollen record of so many modern genera.

IV. DISCUSSION

A. Devonian-Mississippian

The earliest seeds were quite small. This is not surprising in light of their derivation from megaspores. However, relative to the megaspores of coeval plants, these seeds represented an advance as they encompassed 'dispersable' megagametophytes with great initial energy reserves.

The origin of the seed presumably involved the reduction and coalesence (both energetic and morphological) of several megasporangia (c.f. Andrews, 1963; Tiffney, 1981; Niklas, 1981) and their associated stalks. The aggregation of the stalks (telomes) initially fused to form the integument of the seed. A subsequent cycle of telomes formed a second investing structure, the cupule, which apparently contributed to the efficacy of pollination. The fleshy appearance of the cupule suggests a function in dispersal, but I believe that, at this stage of seed evolution, the structure served no purpose beyond pollination. (In later stages, the cupule may have provided a 'preadaptation' [or 'exaptation of Gould and Vrba, 1982] leading to the development of fleshy seeds.) This interpretation is supported by the frequent description of seeds shed from their cupules. These seeds lacked obvious morphological adaptations for dispersal, and may simply have floated, or been blown by wind. Specimens of the Mississippian *Gnetopsis hispida* possessed hairs which might have aided in flotation (Gensel and Skog, 1977) or in aerial dispersal (Niklas, 1983).

Among cupulate early seeds, there seems to have been a trend away from the production of many seeds towards fewer large seeds per cupule, culminating in the single-seeded cupules of the lyginopteridalean seed ferns of the Pennsylvanian. This is again logical in light of the trend towards reduction in number of megasporangia that gave rise to the earliest seeds. Continued selection for energetic support of the young embryo would have resulted in reduction of the number of ovules per cupule.

The presence or absence of dormancy in these earliest seeds is conjectural, but I suspect that they lacked dormancy. Being an encased megagametophyte, the ovule could be expected to commence growing as soon as fertilized. If fertilized before dispersal, it might have grown too large for efficient dispersal, as well as commencing its growth in the absence of the soil and adequate water. Unless the microspore was held in stasis between pollination and fertilization, a specialization that one would expect to require considerable time to evolve, it seems likely that fertilization followed pollination in a matter of hours or days, and that the subsequent seed lacked a real dormancy period.

The exceptions to the cupulate structure of early seeds are the two early platyspermic genera, *Spermolithus* and *Lyrasperma* (Section III.B). The origin of the wings of these seeds is unclear, but they offer the first evidence of possible morphological specialization to wind dispersal.

B. Pennsylvanian

1. Major Lineages

By the Pennsylvanian, three major groups of seed plants were entrenched, each with its distinctive seed morphology.

a. Seed ferns — The lyginopterids were characterized by small seeds (average 35 mm³) borne singly in cupules, the exception being *Gnetopsis elliptica* of the Upper Pennsylvanian, which had two to four seeds per cupule. The predominance of isolated seeds in the fossil record suggests that the ovules or seeds were generally shed from the cupule at maturity, possibly post-pollination. In light of this, while the cupule may have aided pollination and protected the developing ovule, it does not appear to have played a role in dispersal. The individual seeds show no particular dispersal adaptations. They often possessed a thin sarcotesta and a weakly-developed sclerotesta. Because they are small in size, one assumes they may have been wind and water dispersed. The seeds of the Callistophytaceae were similar in structure and dimensions to those of the Lyginopteridaceae, although they were not borne in cupules.

The medullosan pteridosperms, by contrast, possessed large to huge seeds which were not borne in a cupule. Out of 51 species of *Trigonocarpus* and *Pachytesta* (probably an overestimate due to duplication of species in the two genera), the smallest seed was 18 mm³; the next smallest was 80 mm³, and the averages for the two genera lay at 19 600 and 47 900 mm³. Only the two smallest medullosan seeds entered the size range of the lyginopterid seeds. The largest medullosan seeds (160 000 and 317 000 mm³) were not exceeded in size until the appearance of large angiosperm seeds in the Tertiary. Medullosan seeds were almost uniformly characterized by thick sclerotestas and well developed fleshy sarcotestas.

b. Cordaites — The cordaitaleans split into two groups in terms of ecology and seed morphology (Raymond and Phillips, 1983). One group of Euramerican species possessed a mangrove ecology, growing in salt water. Members of this group possessed seeds of the *Nucellangium* and *Cardiocarpus* type, which were fairly large (1000 to 12 000 mm³), and possessed well developed sarcotestas underlain by a thick sclerotesta. In *C. spinatus*, the surface of the sclerotesta was dramatically spinose. A second group of Euramerican cordaites was apparently restricted to fresh-water sites. This group included seeds of the genus *Mitrospermum*, which were much smaller (average volume of 120 mm³), and possessed wide lateral wings. While these wings appear well adapted for wind dispersal, A. Raymond (pers. comm. 1984) has suggested that they would work equally well as floats to aid in water dispersal. The Permo-Carboniferous members of the group characteristic of the Angaran region all possessed winged seeds, some with circumferential wings and some with asymmetric wings like those of extant Pinaceae. At an average of 132 mm³, they were somewhat larger than the seeds of *Mitrospermum*, but not nearly as big as the seeds of the mangrove-ecology species.

c. *Conifers* — The conifers first appeared in the Pennsylvanian (Scott and Chaloner, 1983), but apparently stayed in the mesic uplands outside of the swamps in which the other Pennsylvanian seed plants grew. From the size of their cones and from the morphology of their successors, we can infer small, abiotically-dispersed seeds.

2. Interpretations

Morphology suggests wind dispersal for seeds of fresh-water cordaitaleans and perhaps the conifers, but provides no distinct clues as to the dispersal of the lyginopterid or callistophycean seed ferns. However, the mangrove-like cordaites and the medullosans possessed distinctively thick sclerotestas overlain by fleshy sarcotestas. What is the significance of these features?

a. *Seed consumption by insects.* — Carpenter and Burnham (1985) detail the diversity of the Carboniferous entomofauna. Baxendale (1979) reports insect coprolites consisting of the remains of *Nucellangium*, and Scott and Taylor (1983) depict a *Trigonocarpus* seed with a putative insect hole in its sclerotesta. Clearly the utilization of seeds by insects could enhance selection for mechanical protection of ovules and seeds, but this does not explain the size of these seeds, nor their possession of a sarcotesta.

b. *Floating dispersal.* — Is it possible that these seeds dispersed by floating on water? This seems unlikely, as there is no evidence that they possessed air chambers or air-trapping fibrous coverings. The exception is *Aethotesta* of the Medullosaceae, which Krassilov (1975) reports as possessing air chambers. It is possible that the sarcotesta or storage tissue of these seeds was oil-rich and thus less dense than water, but this is not testable; also, such nutrient-rich oil is energetically-expensive, and would degrade rapidly in an aqueous environment. Finally, while the mangrove-like cordaites grew in an environment amenable to water transport, the medullosans tended to occupy higher, more mesic sites in the coal swamp (DiMichele and Phillips, 1985), and it seems improbable, although not impossible, that they would have evolved water dispersal. In short, while some of these seeds probably were dispersed by water, the observed morphological and ecological characteristics do not agree with this conclusion.

c. *Fish dispersal.* — Goulding (1980) has described the importance of fish dispersal in the modern Amazon, and others have queried its possible significance in Carboniferous swamps (Gensel and Skog, 1977; W. Chaloner, pers. comm. 1981). This would explain the hard sclerotesta and fleshy sarcotesta of medullosans and mangrove cordaites. It fits better with the ecological situation of the cordaites, as it is hard to envision fish consistently depositing seeds in the more mesic sites favored by medullosans.

d. Tetrapod dispersal. — A classic assumption made by many researchers is that vertebrate herbivores did not use terrestrial plant resources before the Permian. For this reason, most paleobotanists have ignored the possibility of vertebrate dispersal in the Pennsylvanian. However, vertebrate paleontologists (Panchen, 1973; Olson, 1976) have described herbivorous reptiles from the Upper Pennsylvanian, particularly those belonging to the Edaphosauridae and Caseidae, and suggest that they possessed herbivorous forebears. While they were possibly semi-aquatic, their habitat and largish size (*Edaphosaurus* stretched to almost 3 meters; Romer, 1966) suggest that they, and their relatives, would have served as efficient dispersal agents. Modern and many fossil reptilian herbivores tend to gulp food whole, the notable exception being the ornithopod herbivores of the Cretaceous, which possessed specialized grinding teeth (Weishampel, 1984a, b). In this light, the seed would require little morphological specialization beyond an attractive flesh and a resistant sclerotesta, both of which the medullosans possessed.

This putative vertebrate dispersal of large medullosan seeds is supported by circumstantial biogeographic evidence suggesting parallelism of ranges of early tetrapods and medullosans. Medullosan seeds are known primarily from the Euramerican coal swamps; only one report (Meyen, 1982) is made of similar seeds from the Angaran region. Panchen (1973) observes that early tetrapods appear to have been restricted in occurrence to the Euramerican swamps until the very end of the Middle Pennsylvanian, an observation explainable by paleogeographic considerations (Johnson, 1980). In the Late Pennsylvanian, tetrapod faunas appear to have expanded slightly to eastern Europe and western North America; in a paleogeographic context, this represents a gradual spread from the paleoequator. This pattern of poleward expansion carried on through the Permian. If reptiles dispersed medullosan seeds, and if they did not expand into the Angaran phytogeographic unit until the Permian, this could explain the very low frequency of occurrence of medullosan seeds in Angaran floras.

Modern reptile-dispersed seeds are characterized by being borne low to the ground or dropped on the ground, and possessing a brightly-colored and often odoriferous external flesh. It is tempting to speculate that these characteristics may have had their roots in the biology of the seed ferns, whose seeds possessed a fleshy outer layer and appear to have been shed following pollination.

3. Dormancy and Dispersal

A classic observation of students of Paleozoic seed fern seeds is the consistent absence of fossil embryos. The seeds (technically ovules) are found occupied by gametophytes bearing eggs, or are empty. Preservational bias is an inappropriate solution to this puzzle, particularly in light of the large number of specimens examined. It is possible that some of these fossils represent ovules shed before pollination in the mode of 'self-thinning'. However, I find the most logical explanation to be the hypothesis that Paleozoic seeds were like those of extant *Ginkgo biloba* (see Favre-Duchartre, 1958); following pollination, the ovule was shed and fertilization occurred at some subsequent time (Arnold, 1948). Presumably ovules which had failed to be fertilized had a higher chance of being preserved than those that had been fertilized, the latter promptly germinating and producing plants.

Rothwell (1982) expanded this interpretation, arguing that dispersal after pollination but before fertilization was necessitated by the lack of dormancy mechanisms in these seeds; dispersal was initiated immediately following pollination in order to optimize its efficacy. This was apparently true of cordaitalean seeds (Rothwell, 1982) and seed ferns, and would be a logical consquence of the lack of dormancy in their ancestral heterosporous pteridophytes.

The origin of dormancy mechanisms in seeds is not clear. Post-Pennsylvanian seed fern seeds are not preserved in a manner that permits recognition of the embryo, and cordaites became extinct in the Permian. Miller and Brown (1973) describe a Permian voltzialean conifer-cone that bore seeds with embryos. This suggests that fertilization preceded dispersal in this plant. This characteristic may have existed in earlier conifers, but we lack fossil evidence. Miller and Brown suggest that this dormancy mechanism may have developed in response to the cooling and drying of Permian climates, although it would have been equally useful in upland Pennsylvanian environments.

C. Permian into Triassic

Starting in the Late Pennsylvanian, and continuing through the Triassic, the movements of continental plates caused the spread of dry and seasonal continental climates from the poles to the equator (Schwarzbach, 1963; Bambach et al., 1980). In the lowland Euramerican swamps of the Pennsylvanian this change commenced with a Late Pennsylvanian drying period that led to the reduction in importance of arborescent lycopods and seed ferns, and the expansion of tree ferns (Phillips et al., 1974; Pfefferkorn and Thomson, 1982). Continuing climatic 'deterioration' resulted in the extinction of the lyginopterid seed ferns in the Early Permian and the medullosan seed ferns and cordaites by the Middle Permian. The conifers suffered some extinction, but the mesic heritage of the group permitted them to radiate into the drier and more seasonal environments (c.f. Frederiksen, 1972). The increasing importance of water retention is mirrored in vegetative features; sclerophylly becomes common in conifers and in the earliest cycadophytes. The latter first appear during this interval, presumably derived from seed fern ancestors. These floristic changes were paralleled in the changing seed biology. Dormancy mechanisms may have become more prevalent (c.f. Section IV, B, *3*). Potentially water-dispersed seeds disappear. The large seeds of the Medullosaceae disappear, presumably because of the extinction of their parent sporophytes rather than the simple loss of dispersal agents, as large herbivores radiated during this time (see below). The seed fern seeds were replaced by the smaller, but still sizable (50 to perhaps 1000 mm^3) seeds of the cycads and early ginkgo relatives. Air-borne conifer seeds are presumed to have become more common, although actual fossils with wings are not found until the Triassic.

The development of these ecological circumstances was presaged by the seed biology of the Permo-Carboniferous floras of Gondwana, dominated by *Glossopteris*. Here seeds were consistently smaller and apparently abiotically dispersed. This presumably reflected the cool, seasonal climates, but also may have been influenced by the late arrival of herbivorous tetrapods as members of the ecosystem (c.f. Panchen, 1973; Johnson, 1980).

The changing climates of the Pennsylvanian and Permian also affected the tetrapod community. Rohdendorf (1970) and Olson (1976) both note the dependence of many early terrestrial tetrapods on insects as a food source. The drying climate of the Late Pennsylvanian and Permian caused the extinction of many insect groups. Olson (1976, as an indirect observation) and Niklas (in press) have suggested that the reduction of the insect resource led to the gradual diversification of first, omnivorous, and then herbivorous reptiles. Many of these were derived from large ancestors (3 m or more). Further, plant material offers far less concentrated nutrition than flesh, and the newly-evolving herbivores required larger volumes of food. As a result, the effect of herbivory on the plant world became more pronounced during this period.

The largest of these herbivores were probably not selective, but ate all parts of the plant. This is curcumstantially borne out by the mixture of twigs, leaves and seeds found in the gut of a Permian rhyncocephalian (Weigelt, 1930) and a Cretaceous hadrosaur (Krausel, 1922). This absence of selective herbivory would mitigate against the widespread occurrence of specializations for biotic seed dispersal (Janzen, 1984), although some adaptations must have been present (e.g. Schweitzer's (1968) report of thick-walled gymnosperm seeds in a Late Permian coprolite).

D. Triassic—Early Cretaceous

The changes of the Permo-Triassic set the stage for most of the rest of the Mesozoic. Conifers by and large possessed small, generally winged seeds, although *Araucaria mirabilis* possessed a large (almost 500 mm^3), wingless seed. The cycads and ginkgos became more common. Seeds of cycads appear rarely in the record, but measured between 300 and 15 000 mm^3. *Ginkgo*-related seeds were more common and ranged from perhaps 70 mm^3 (immature) to roughly 1000 mm^3. To judge from extant relatives, both groups were dominated by long-lived, slow-growing plants, and were dispersed by animals, presumably small dinosaurs.

The cycadeoids (Bennetitales of some authors) possessed small seeds, perhaps with hairs, and were presumably wind-dispersed. This fits their postulated early-successional status (Retallack and Dilcher, 1981). Weishampel (1984a) intimated that perhaps the multi-seeded receptacles of the cycadeoids were the diaspores of the group; this interpretation could be supported by Hill's (1976) observation of the cuticle of a cycadeoidean receptacle in a coprolite. However, Harris (1969) has stated that fossil evidence demonstrates that cycadeoidean receptacles fell apart on maturity. This, and the morphology of the seeds, suggests that cycadeoids were normally wind dispersed. A similar problem involves interpretation of the dispersal of the seeds of the Mesozoic seed fern *Caytonia*, which bore several tiny seeds in an inrolled, fleshy receptacle. Harris (1964) described several Jurassic coprolites laden with receptacles and seeds of *Caytonia*, suggesting animal dispersal. However, both he and Reymanówna (1973) have described large quantities of dispersed *Caytonia* seeds, and Reymanówna has suggested that the seeds were normally released from the cupule at maturity.

The time of immense herbivores, starting in the Triassic (e.g., Galton, 1985) continued through the Mesozoic. Wing and Tiffney (in press) suggest that this dominance of large herbivores maintained plant communities in a constant state of successional flux, selecting for early successional ecologies and against specific animal-plant interactions like those that characterize the present day. Thus, while animal dispersal of seeds occurred, it was not of central importance to the vegetation of the time; it appears to have been facultative or even fortuitous in nature, rather than obligate.

E. Cretaceous and the Angiosperms

The appearance of the angiosperms had little initial effect on dispersal strategies. Their rapid life cycle, weedy growth, developmental plasticity and apparent early co-evolutionary interaction with insect pollinators (Hickey and Doyle, 1977; Doyle 1978; Tiffney, 1981; Crepet, 1984) led to their rapid diversification and their numerical dominance of world floras by the middle of the Upper Cretaceous. However, their seeds were small, and generally possessed abiotic dispersal morphologies until the early Tertiary. While some developed definite wings for wind dispersal (e.g. Krassilov, 1982), the majority retained a general morphology, and the small size indicative of generalized abiotic dispersal (e.g. Friis, 1984; Tiffney, 1984). It is possible that some multi-seeded fruits may have been consumed and dispersed by animals. Similarly, there are a few large structures in the Late Cretaceous which appear to be both angiospermous, and the true diaspore of the plant. However, these structures are distinctly in the minority.

Wing and Tiffney (in press) postulate that this pattern resulted from the entry of the angiosperms into an ecosystem dominated by large herbivores, which as noted above, would maintain the ecosystem in a constant state of flux and select for early successional, 'r-strategy' type reproductive traits.

The effect of angiosperm evolution on the gymnosperms is debated. Krassilov (1978) claims that gymnosperms showed no signs of extinction until the Cretaceous-Tertiary boundary. Vachrameev (1981) claims in contrast that gymnosperms underwent a gradual decline from the Middle Cretaceous on, save in a few isolated refugia. Overall, diversity data (Niklas *et al.*, 1985) seem to support the latter view. This suggests that while angiosperms shared general dispersal characteristics with co-eval gymnosperms, the other characteristics of the former group ultimately gave them a competitive edge. Certainly the latest Cretaceous saw an increased pace of angiosperm evolution leading to the appearance of modern genera and families (c.f. Muller, 1970, 1981), together with the appearance of larger diaspores.

F. Tertiary

With the demise of the large herbivorous dinosaurs, angiosperms, birds and mammals all radiated, and modern dispersal syndromes (as inferred from modern counterparts and from the morphology of the fossils) became established for the first time. Smaller birds, terrestrial mammals and ultimately bats commenced using seeds and fruits as an important element of their diet, thereby selecting morphological traits which enhanced the relationship (surficial flesh, thick testas, colors, odors, hooks). The diaspore characteristics associated with saurochory were apparently not inviting to the newly radiating dispersal agents, and became a 'relict' syndrome, although occasionally mammals will disperse cycads or ginkgos (Burbidge and Whelan, 1982; van de Pijl, 1982; Chapter 2. III, A).

The strength of the selective pressures for biotic dispersal may be judged by the fact that at least two families can be demonstrated to have switched from abiotic to predominantly biotic dispersal across the Cretaceous-Tertiary boundary (Tiffney, in press). Relatives of the Juglandaceae were abiotically-dispersed in the Cretaceous. The early Tertiary witnessed a development of increasingly larger wind-borne fruits in the family, presumably in response to the general increase in seed size of the early Tertiary (Tiffney, 1984). However, by the later Tertiary, the family was dominated by the large, animal (largely mammal) dispersed fruits of *Carya* and *Juglans*. A similar case, although far less well documented by fossil evidence, can be argued for a switch from abiotic to biotic dispersal in the Fagaceae. As a whole, this pattern suggests the selective advantages of biotic dispersal, and it is a measure of the unique nature of Cretaceous ecosystems that it was not until these systems began to crumble that the diversity of modern dispersal relationships was established.

During this period, surviving gymnosperms appear either to have become slowly reduced in diversity, relying on original dispersal mechanisms, or to have adopted specific solutions for competing with the angiosperms. The cycads (Chapter 2, III, A) and *Ginkgo* became relict groups, with little active dispersal (Tiffney, 1984), and many conifer families became reduced in diversity, but maintained successful populations in specific environments. The notable exceptions to this pattern are the Pinaceae and the Podocarpaceae, the two most diverse extant gymnosperm families.

The Pinaceae capitalized upon broad environmental tolerances combined with efficient abiotic dispersal of winged seeds. Several taxa within the group secondarily became associated with mammal, especially rodent (Chapter 5), and bird dispersal agents. The lack of morphological specialization to attract these biotic agents is interesting on two levels. First, it demonstrates that caution must be used in inferring dispersal mode from diaspore morphology alone. Second, the failure of these gymnosperms to show any morphological specializations to biotic dispersal adds circumstantial evidence to the idea that the success of the angiosperms is partially based upon their developmental plasticity and ability to evolve new structures (c.f. ideas in Doyle, 1978 or Tiffney, 1981). The Podocarpaceae adopted the 'if you can't beat them, join them' strategy, and evolved flesh-covered seeds attractive to mammals and birds.

The middle and late Tertiary radiation of cool-temperate annual and biennial herbs not only placed renewed emphasis on abiotic dispersal in the angiosperms (e.g. the wind dispersed pappus of some Compositae), but also gave rise to further biotic dispersal adaptations such as the hooks of *Bidens* or *Desmodium*. However, to judge by modern systematic affinities more than by actual fossil evidence, most morphological dispersal syndromes had evolved by the later part of the early Tertiary. Aspects of the middle and late Tertiary were simply variations on an established theme.

V. SUMMARY

A. History of Dispersal

The structure of the earliest seeds was dictated by their developmental origin and the pressures for efficient wind pollination. As a result, with the exception of two winged platyspermic forms, the earliest seeds lacked specific dispersal adaptations. There is little evidence at present to suggest antecedent evolution of the dispersal structures seen in the Pennsylvanian. Adaptations for dispersal probably developed through the Mississippian, drawing on 'exaptations' (*sensu* Gould and Vrba, 1982) provided by earlier, non-dispersal related, morphologies.

Pennsylvanian plants included forms with well developed abiotic wind dispersal (wings) and perhaps water dispersal, although clear morphological adaptations to the latter are not seen. Biotic dispersal began in the Pennsylvanian, perhaps with the consumption of seed fern or cordaitalean seeds by fish or amphibians. Strong circumstantial evidence suggests that, by Late Pennsylvanian, terrestrial reptiles were consuming and dispersing at least the seeds of some seed ferns.

The climatic changes of the Permian and Triassic usher in alterations in both plant and animal communities. I think it unlikely that changes in animal dispersers caused the extinction of the plants they dispersed; it seems more probable that changing climates caused the extinction of the sporophyte, forcing the herbivores to seek other food sources.

In response to the changing climates of this time, plants become more drought-tolerant, and abiotic dispersal becomes more common (e.g. conifers). In some lineages of limited diversity, simple animal-dispersed seeds (ginkgos, cycads) evolved. Among tetrapod communities terrestrial herbivores became increasingly important, many being of quite large size and presumably requiring high caloric intake. As a result, herbivores began to physically dictate the structure of plant communities by the pressure of their feeding. The dominant large herbivores consumed vegetable material without regard to its composition. Seeds were consumed with other material, and little selection existed for the establishment of complex dispersal relationships. As a result, while animal dispersal of seeds occurred, it never became a dominant or specialized situation; potential 'fruits' like those of *Caytonia* did not spur radiations, and dispersal morphology remained generalized. This situation extended through the Mesozoic until the demise of the dinosaurs.

Only with the latest Cretaceous-early Tertiary co-radiation of developmentally-plastic angiosperms and a host of small mammals and birds did biotic dispersal activity surpass the apparent previous peak of success of the Pennsylvanian.

B. Dispersal Ecology

1. Ovules, Seeds and Diaspores

Multi-seeded diaspores pose interesting questions concerning seedling competition and gene flow in present ecological considerations. While several fossil gymnosperms possessed multi-seeded receptacles, in no case is it clear that the receptacle consistently served as a diaspore, resulting in the simultaneous dispersal of many seeds. Possible cases include the Permian seed fern *Glossopteris* and perhaps the Triassic seed fern *Umkomasia*. The fleshy receptacles of *Caytonia* (Triassic-Cretaceous) were apparently consumed by herbivores, and may have been dispersed, although we have no idea of their viability. However, there is no clear proof that this was a consistent feature of the group. It appears that the dispersal of multi-seeded diaspores is primarily a trait of angiosperms.

A similar ecological question of the modern day involves the number of ovules present as contrasted to the number of seeds produced. The fossil record cannot presently provide information on this question.

2. A Disperser does not Dictate the Syndrome

Herrera (1985) and Howe (Chapter 4, VI, B) have recently suggested that dispersal syndromes are not specific to a single disperser species, but are maintained in a lineage of plants by a succession of different dispersal agents. The fossil record strongly supports this contention. On a geological time scale, I suspect that the survival of particular dispersal syndromes depended far more on the survival of the sporophyte plant than on the survival of the particular dispersal agent. In this light, the classic story of *Calvaria major* and the dodo (*Raphus cucullatus*) (Temple, 1977), in which it was suggested that the extinction of the dispersal agent nearly caused the extinction of the plant, is more a measure of the isolation of the ecosystem in which the dodo was the sole large herbivore, than it is an argument for the specificity of dispersal relationships.

3. Dispersal and Diversity

While it is incorrect to suggest that dispersal is a primary determinant of terrestrial plant diversity, there is a clear correlation between the diversity of dispersal mechanisms present and overall diversity in the history of land plants. Niklas *et al.* (1980, 1985) have traced changes in the diversity of the terrestrial flora. Diversity increased from the Devonian through the Carboniferous, and then maintained a relative plateau through to the appearance of the angiosperms. Angiosperm diversity increased in two pulses, the first in the middle Cretaceous and the second in the late Cretaceous and Tertiary, corresponding to the widening distribution of modern dispersal agents. Tiffney (in press) has limited evidence for greater species diversity in extant angiosperm families dominated by biotic dispersal than in those dominated by abiotic dispersal.

4. The Continuity of Morphological Solutions

This overview of dispersal modes suggests that the basic mechanical structures of dispersal have remained similar from the Pennsylvanian to the present day. Thus wings and hair have long been important in permitting wind dispersal, and flesh has long been the primary attractant of biotic dispersal agents. However, the relative proportions of various dispersal modes in any given flora were influenced by the co-eval animal communities. I do not discern any dispersal modes that have become 'extinct', but it must be admitted that they may have been present, though unrecognizable in the fossil record.

5. The Present is not the Past

The predominance and importance of biotic dispersal in modern communities (Howe and Smallwood, 1982; Janson, 1983) leads one to infer that, given the existence of herbivores in past communities, sophisticated dispersal syndromes must have been present (c.f. Weishampel, 1984a). I believe that this is not the case, and that the present relative complexity of dispersal syndromes is a feature of the last 70 million years (approximately). Prior to that point, differing herbivore communities dictated a more general dispersal relationship (see Wing and Tiffney, in press).

ACKNOWLEDGEMENTS

I thank Karl J. Niklas (Cornell University) and Scott L. Wing (Smithsonian Institution) for their review of this manuscript, and Robin M. Gowen for rendering the figures. The research was supported by NSF grant BSR 83-06002.

TEXT REFERENCES

Andrews, H. N. (1963). *Science* **142**, 925–931.
Arnold, C. A. (1948). *Bot. Rev.* **14**, 450–472.
Baker, H. G. (1972). *Ecology* **53**, 997–1010.
Bakker, R. T. (1978). *Nature* **274**, 661–663.
Bambach, R. K., Scotese, C. R., and Ziegler, A. M. (1980). *Am. Sci.* **68**, 26–38.
Baxendale, R. W. (1979). *Palaeontol.* **22**, 537–548.
Burbidge, A. H., and Whelan, R. J. (1982). *Aust. J. Ecol.* **7**, 63–67.
Carpenter, F. M., and Burnham, L. (1985). *Annu. Rev. Earth Planet. Sci.* **13**, 297–314.
Chaloner, W. G., and Sheering, A. (1981). *In* 'Evolution Today. (Proc. Second Intl. Cong. Syst. Evol. Biol.)' (G. G. E. Schudder and J. L. Reveal, eds), pp. 93–100. Hunt Institute, Carnegie-Mellon University, Pittsburgh.
Chaloner, W. G., Hill, A. J., and Lacey, W. S. (1977). *Nature* **265**, 233–235.
Crepet, W. L. (1984). *Ann. Missouri Bot. Gard.* **71**, 607–630.
DiMichele, W. A., and Phillips, T. L. (1985). *In* 'Geological Factors and the Evolution of Plants' (B. H. Tiffney, ed.), pp. 223–256. Yale University Press, New Haven.
Doyle, J. A. (1978). *Annu. Rev. Ecol. Syst.* **9**, 365–392.
Doyle, J. A., and Hickey, L. J. (1976). *In* 'Origin and Early Evolution of Angiosperms' (C. B. Beck, ed.), pp. 139–206. Columbia University Press, New York.
Favre-Duchartre, M. (1958). *Phytomorphol.* **8**, 377–390.
Foster, A. A., and Janson, C. H. (1985). *Ecology* **66**, 773–780.
Frederiksen, N. O. (1972). *Geoscience and Man* **4**, 17–28.
Friis, E. M. (1984). *Ann. Missouri Bot. Gard.* **71**, 403–418.
Futuyma, D., and Slatkin, M. (1983). 'Coevolution'. Sinauer Associates, Sunderland, Massachusetts.
Galton, P. M. (1985). *Lethaia* **18**, 105–123.
Gensel, P. G., and Skog, J. E. (1977). *Brittonia* **29**, 332–351.
Gould, R. E., and Delevoryas, T. (1977). *Alcheringa* **1**, 387–399.
Gould, S. J., and Vrba, E. S. (1982). *Paleobiology* **8**, 4–15.
Goulding, M. (1980). 'The Fishes and the Forest'. University of California Press, Berkeley.
Harland, W. B., Cox, A. V., Llewellyn, P. G., Pickton, C. A. G., Smith, A. G., and Walter, R. (1982). 'A Geologic Time Scale' Cambridge University Press, Cambridge.
Harper, J. L., Lovell, P. H., and Moore, K. G. (1970). *Annu. Rev. Ecol. Syst.* **1**, 327–356.
Harris, T. M. (1932a). *Meddel. Grønland* **85**(3), 1–114.
Harris, T. M. (1932b). *Meddel. Grønland* **85**(5), 1–133.
Harris, T. M. (1935). *Meddel. Grønland* **112**(1)), 1–176.
Harris, T. M. (1964). 'The Yorkshire Jurassic Flora. II. Caytoniales, Cycadales & Pteridosperms'. British Museum (Natural History), London.
Harris, T. M. (1969). 'The Yorkshire Jurassic Flora. III Bennettitales'. British Museum (Natural History), London.
Harris, T. M. (1979). 'The Yorkshire Jurassic Flora. V Coniferales'. British Museum (Natural History), London.
Harris, T. M., Millington, W., and Miller, J. (1974). 'The Yorkshire Jurassic Flora. IV 1. Ginkgoales. 2. Czekanowskiales'. British Museum (Natural History), London.
Herrara, C. M. (1985). *Oikos* **44**, 132–141.

Hickey, L. J., and Doyle, J. A. (1977). *Bot. Rev.* **43**, 3–104.
Hill, C. R. (1976). *Bull. Brit. Mus. (Nat. Hist.) Geol.* **27**, 289–294.
Howe, H. F., and Smallwood, J. (1982). *Annu. Rev. Ecol. Syst.* **13**, 201–228.
Howe, H. F., and Schupp, E. W. (1985). *Ecology* **66**, 781–791.
Janson, C. H. (1983). *Science* **219**, 187–189.
Janzen, D. H. (1980). *Evolution* **23**, 1–27.
Janzen, D. H. (1984). *Am. Nat.* **123**, 338–353.
Janzen, D. H., and Martin, P. H. (1982). *Science* **215**, 19–27.
Johnson, G. A. L. (1980). *Syst. Assoc. Spec. Pap.* **15**, 39–54.
Kovács-Endródy, E. (1974). *Palaeontol. Afr.* **17**, 11–14.
Krassilov, V. A. (1973). *Geophytol* **3**, 1–4.
Krassilov, V. A. (1975). 'Paleoecology of Terrestrial Plants. Basic Principles and Techniques'. Wiley-Halsted, New York.
Krassilov, V. A. (1978). *Palaeontol.* **21**, 893–905.
Krassilov, V. A. (1981). *Palaeogeog. Palaeoclimatol., Palaeoecol.* **34**, 207–224.
Krassilov, V. A. (1982). *Palaeontographica, Abt. B., Paläophytol.* **181**, 1–43.
Kräusel, R. (1922). *Paläontol. z.* **4**, 80.
Levin, D. A. (1974). *Am. Nat.* **108**, 193–206.
Long, A. G. (1960). *Trans. Roy. Soc. Edin.* **64**, 261–280.
McDonnell, M. J., and Stiles, E. W. (1983). *Oecologia* **56**, 109–116.
Meyen, S. V. (1982). 'The Carboniferous and Permian Floras of Angaraland (A Synthesis)'. International Publishers, Lucknow.
Miller, C. N., Jr., and Brown, J. T. (1973). *Am. J. Bot.* **60**, 561–569.
Muller, J. (1970). *Biol. Rev. Biol. Proc. Cambridge Philos. Soc.* **45**, 417–450.
Muller, J. (1981). *Bot. Rev.* **47**, 1–142.
Murray, D. R. (1984). *In* 'Seed Physiology. Vol. 1. Development' (D. R. Murray, ed.), pp. 1–40. Academic Press, Sydney.
Niklas, K. J. (1981). *Am. J. Bot.* **68**, 635–650.
Niklas, K. J. (1983). *Evolution* **37**, 968–986.
Niklas, K. J. (in press). Dahlem Conf. Publication, June, 1985.
Niklas, K. J., Tiffney, B. H., and Knoll, A. H. (1980). *In* 'Evolutionary Biology, Vol. 12' (M. K. Hecht, W. C. Steere and B. Wallace, eds), pp. 1–89. Plenum, New York.
Niklas, K. J., Tiffney, B. H., and Knoll, A. H. (1985). *In* 'Phanerozoic Diversity Patterns: Profiles in Macroevolution' (J. W. Valentine, ed.), pp. 97–128. Princeton University Press, Princeton.
Olson, E. C. (1976). *In* 'Morphology and Biology of Reptiles' (A. d'A. Bellairs and C. B. Cox, eds), pp. 1–30. Linn. Soc. Symp. Ser. 3, Academic Press, London.
Panchen, A. L. (1973). *In* 'Atlas of Palaeobiogeography' (A. Hallam, ed.), pp. 117–125. Elsevier, Amsterdam.
Pfefferkorn, H. W., and Thomson, M. C. (1982). *Geology* **10**, 641–644.
Phillips, T. L., Peppers, R. A., Avcin, M. J., and Laughnan, P. F. (1974). *Science* **187**, 1367–1369.
Raymond, A., and Phillips, T. L. (1983). *In* 'Tasks for Vegetation Science, Vol. 8' (H. J. Teas, ed.), pp. 19–30. Junk, The Hague.
Rayner, R. J., and Coventry, M. K. (1985). *South Afr. J. Sci.* **81**, 21–32.
Retallack, G., and Dilcher, D. L. (1981). *In* 'Paleobotany, Paleoecology and Evolution, Vol. II' (K. J. Niklas, ed.), pp. 27–78. Praeger, New York.
Reymanowna, M. (1973). *Acta Palaeobot.* **14**, 75–87.
Rohdendorf, B. B. (1970). *Paleontol. Zurn.* **1970**(1), 10–18.
Romer, A. S. (1966). 'Vertebrate Paleontology' (Third ed.). University of Chicago Press, Chicago.

Rothwell, G. W. (1981). *Rev. Palaeobot. Palynol.* **32**, 103–121.
Rothwell, G. W. (1982). *Am. J. Bot.* **69**, 239–247.
Salisbury, E. J. (1942). 'The Reproductive Capacity of Plants. Studies in Quantitative Biology'. G. Bell & Sons, London.
Schwarzbach, M. (1963). 'Climates of the Past. An Introduction to Paleoclimatology'. D. van Nostrand, London.
Schweitzer, H. -J. (1968). *Naturwiss. Rundschau* (Stuttgart). **21**, 93–102.
Scott, A. C., and Chaloner, W. G. (1983). *Proc. Roy. Soc. London B.* **220**, 163–182.
Scott, A. C., and Taylor, T. N. (1983). *Bot. Rev.* **49**, 259–307.
Seward, A. C. (1917). 'Fossil Plants, Volume III. Pteridospermeae, Cycadofilices, Cordaitales, Cycadophyta'. Cambridge University Press, Cambrige.
Smith, C. C. (1975). *In* 'Coevolution of Animals and Plants' (L. E. Gilbert and P. H. Raven, eds), pp. 53–77. University of Texas Press, Austin.
Sorensen, A. E. (1981). *Oecologia* **50**, 242–249.
Stockey, R. A. (1977). *Am. J. Bot.* **64**, 733–744.
Taylor, T. N. (1981). 'Paleobotany. An Introduction to Fossil Plant Biology'. McGraw-Hill, New York.
Temple, S. (1977). *Science* **197**, 885–886.
Tiffney, B. H. (1981). *In* 'Paleobotany, Paleoecology and Evolution, Vol. II'. (K. J. Niklas, ed.), pp. 193–230. Praeger, New York.
Tiffney, B. H. (1984). *Ann. Missouri Bot. Gard.* **71**, 551–576.
Tiffney, B. H. (in press). *Ann. Missouri Bot. Gard..*
Tiffney, B. H., and Niklas, K. J. (1985). *In* 'Population Biology and Evolution of Clonal Organisms' (L. W. Buss, J. B. C. Jackson and R. Cook, eds), pp. 35–66. Yale University Press, New Haven.
Tralau, H. (1968). *Lethaia* **1**, 63–101.
Vachrameev, V. A. (1981). *Paleontol Zurn.* **1981**(2), 3–14.
van der Pijl, L. (1982). 'Principles of Dispersal in Higher Plants'. (Third edition). Springer-Verlag, Berlin.
Weigelt, J. (1930). *Leopoldiana* **6**, 269–280.
Weishampel, D. B. (1984a). *Jahrb. Geol. Paläontol.* **167**, 224–250.
Weishampel, D. B. (1984b). *Advances Anatomy, Embryol. Cell Biol.* **87**, 1–110.
Wing, S. L., and Tiffney, B. H. (in press). *In* 'Angiosperm Origins and Biological Consequences' (E. M. Friis, W. G. Chaloner and P. R. Crane, eds), pp. xx–xx. Cambridge University Press, Cambridge.

DATA REFERENCES

Devonian-Mississippian
Andrews, H. N. (1963). *Science* **142**, 925–931.
Chaloner, W. G., Hill, A. J., and Lacey, W. S. (1977). *Nature* **265**, 233–235.
Gensel, P. G., and Skog, J. E. (1977). *Brittonia* **29**, 332–351.
Gillespie, W. H., Rothwell, G. W., and Scheckler, S. E. (1981). *Nature* **293**, 462–464.
Long, A. G. (1960a). *Trans. Roy. Soc. Edin.* **64**, 201–215.
Long, A. G. (1960b). *Trans. Roy. Soc. Edin.* **64**, 261–280.
Matten, L. C., and Lacey, W. S. (1981). *In* 'Geobotany II' (R. C. Romans, ed.), pp. 221–234. Plenum, New York.
Matten, L. C., Lacey, W. S., and Lucas, R. C. (1980). *J. Linn. Soc. Biol.* **81**, 249–273.

Pettitt, J. M., and Beck, C. B. (1968). *Contrib. Mus. Paleontol. Univ. Michigan.* **2**, 139–154.
Walton, J. (1949). *Trans. Roy. Soc. Edin.* **61**, 719–728.

Pennsylvanian
Baxter, R. W. (1971). *Phytomorphol.* **21**, 108–121.
Brogniart, A. (1881). 'Recherches sur les Graines Fossiles Silicifiées'. Imprimerie Nationale, Paris.
Gastaldo, R. A., and L. C. Matten. (1978). *Am. J. Bot.* **65**, 882–890.
Grove, G. G., and Rothwell, G. W. (1980). *Am. J. Bot.* **67**, 1051–1058.
Hoskins, J. H., and Cross, A. T. (1946a). *Am. Midland Naturalist* **36**, 207–250.
Hoskins, J. H., and Cross, A. T. (1946b). *Am. Midland Naturalist* **36**, 331–361.
Leisman, G. A. (1964). *Am. J. Bot.* **51**,1069–1075.
Mamay, S. H. (1954). U. S. Geol. Surv. Prof. Pap. 254-D, 81–95.
Mapes, G., and Rothwell, G. W. (1984). *Palaeontol.* **27**, 69–94.
Meyen, S. V. (1982a). *Paleontol. Zurn.* **1982**(2), 109–120.
Meyen, S. V. (1982b). 'The Carboniferous and Permian Floras of Angaraland (A Synthesis)'. International Publishers, Lucknow.
Rothwell, G. W. (1971a). *Am. J. Bot.* **58**, 706–715.
Rothwell, G. W. (1971b). *Palaeontographica, Abt. B., Paläophytol.* **131**, 167–178.
Rothwell, G. W. (1981). *Rev. Palaeobot. Palynol.* **32**, 103–121.
Rothwell, G. W. (1982). *Am. J. Bot.* **69**, 239–247.
Rothwell, G. W., and Eggert, D. A. (1970). *Bot. Gaz.* **131**, 359–366.
Rothwell, G. W., Taylor, T. N., and Clarkson, C. (1979). *J. Paleontol.* **53**, 49–54.
Smoot, E. L., and Taylor, T. N. (1983). *Rev. Palaeobot. Palynol.* **40**, 165–176.
Stidd, B. M., and Cosentino, K. (1976). *Bot. Gaz.* **137**, 242–249.
Stubblefield, S. P., and Rothwell, G. W. (1980). *J. Paleontol.* **54**, 1012–1016.
Stubblefield, S. P., Rothwell, G. W., and Taylor, T. N. (1984). *Can. J. Bot.* **62**, 96–101.
Taylor, T. N. (1965). *Palaeontographica, Abt. B. Paläophytol.* **117**, 1–46.
Taylor, T. N. (1966). *Am. J. Bot.* **53**, 185–192.
Taylor, T. N. (1981). 'Paleobotany: An Introduction to Fossil Plant Biology'. McGraw-Hill, New York.
Taylor, T. N., and Delevoryas, T. (1964). *Am. J. Bot.* **51**, 189–195.
Taylor, T. N., and Leisman, G. A. (1963). *Am. J. Bot.* **50**, 574–580.
Taylor, T. N., and Millay, M. A. (1981). *Rev. Palaeobot. Palynol.* **32**, 27–62.

Permian
Cichan, M. A., and Taylor, T. N. (1981). *Rev. Palaeobot. Palynol.* **34**, 359–367.
Florin, R. (1949). *Acta Horti Bergiani* **15**, 79–109.
Goman'kov, A. V., and Meyen, S. V. (1979). *Paleontol. Zurn.* **1979**(2), 124–138.
Gould, R. E., and Delevroyas, T. (1977). *Alcheringa* **1**, 387–399.
Kovács-Endrödy, E. (1974). *Palaeontol. Afr.* **17**, 11–14.
Mamay, S. H., and Watt, A. D. (1971). U. S. Geol. Surv. Prof. Pap. 750-C, C48–C51.
Meyen, S. V. (1982a). *Paleontol. Zurn.* **1982**(2), 109–120.
Meyen, S. V. (1982b). 'The Carboniferous and Permian Floras of Angaraland (A Synthesis)' International Publishers, Lucknow.
Miller, C. N. Jr., and Brown, J. T. (1973). *Am. J. Bot.* **60**, 561–569.
Rayner, R. J., and Coventry, M. K. (1985). *South Afr. J. Sci.* **81**, 21–32.
Schopf, J. M. (1976). *Rev. Palaeobot. Palynol.* **21**, 25–64.
Surange, K. R., and Singh, P. (1952). *Curr. Sci.* **21**, 40–41.
Taylor, T. N. (1981). 'Paleobotany: An Introduction to Fossil Plant Biology'. McGraw-Hill, New York.

Triassic

Anderson, J. M., and Anderson, H. M. (1984). *Palaeontol. Afr.* **25**, 39–59.

Bock, W. (1969). 'The American Triassic Flora and Global Distribution'. Geological Center, North Wales, Pennsylvania.

Daugherty, L. H. (1941). *Publ. Carnegie Inst. Wash.* **526**, 1–108.

Delevoryas, T., and Hope, R. C. (1981). *Am. J. Bot.* **68**, 1003–1007.

Harris, T. M. (1932a). *Meddel. Grønland* **85**(3), 1–114.

Harris, T. M. (1932b). *Meddel. Grønland* **85**(5), 1–133.

Harris, T. M. (1935). *Meddel. Grønland* **112**(1), 1–176.

Retallack, G. J. (1983). *J. Roy. Soc. N.Z.* **13**, 129–154.

Taylor, T. N. (1981). 'Paleobotany: An Introduction to Fossil Plant Biology'. McGraw-Hill, New York.

Jurassic

Harris, T. M. (1964). 'The Yorkshire Jurassic Flora. II. Caytoniales, Cycadales & Pteridosperms'. British Museum (Natural History), London.

Harris, T. M. (1969). 'The Yorkshire Jurassic Flora. III. Bennettitales'. British Museum (Natural History), London.

Harris, T. M. (1979). 'The Yorkshire Jurassic Flora. V. Coniferales'. British Museum (Natural History), London.

Harris, T. M., Millington, W., and Miller, J. (1974). 'The Yorkshire Jurassic Flora. IV. 1. Ginkgoales. 2. Czekanowskiales'. British Museum (Natural History), London.

Krassilov, V. A. (1969). *Palaeobotanist* **18**, 12–19.

Krassilov, V. A. (1973a). *Geophytol.* **3**, 1–4.

Krassilov, V. A. (1973b). *Lethaia* **6**, 163–178.

Krassilov, V. A. (1977). *Rev. Palaeobot. Palynol.* **24**, 155–178.

Reymanówna, M. (1973). *Acta Palaeobot.* **14**, 45–87.

Seward, A. C. (1917). 'Fossil Plants, Volume III. Pteridospermeae, Cycadofilices, Cordaitales, Cycadophyta'. Cambridge Univ. Press, Cambridge.

Sharma, B. D., and Bohra, D. R. (1977). *Geophytol.* **7**, 107–112.

Stockey, R. A. (1977). *Am. J. Bot.* **64**, 733–744.

Taylor, T. N. (1981). 'Paleobotany: An Introduction to Fossil Plant Biology'. McGraw-Hill, New York.

Cretaceous

Krassilov, V. A. (1982). *Palaeontographica, Abt. B., Paläophytol.* **181**, 1–43.

Krassilov, V. A., and Bugdaeva, E. V. (1982). *Rev. Palaeobot. Palynol.* **36**, 279–295.

LaPasha, C. A., and Miller, C. N. Jr. (1981). *Am. J. Bot.* **68**, 1374–1382.

Miller, C. N. Jr., and Robison, C. R. (1975). *J. Paleontol.* **49**, 138–150.

Seward, A. C. (1917). 'Fossil Plants, Volume III. Pteridospermeae, Cycadofilices, Cordaitales, Cycadophyta'. Cambridge University Press, Cambridge.

Tiffney, B. H. (1984). *Ann. Missouri Bot. Gard.* **71**, 551–576.

Wieland, G. R. (1906). 'American Fossil Cycads, Vol. I'. Carnegie Institute, Washington, D.C.

Wieland, G. R. (1916). 'American Fossil Cycads, Vol. II'. Carnegie Institute, Washington, D.C.

Tertiary

Tiffney, B. H. (1984). *Ann. Missouri Bot. Gard.* **71**, 551–576.

SPECIES INDEX

A

Abies
  balsamea, 202
  lasiocarpa, 209
Abrus precatorius, 62, 63
Acacia, 63–66, 87–121, 146, 253, 254
  acinacea, 92
  acuminata, 92
  adunca, 92
  alleniana, 92
  aneura, 92, 98
  aspera, 92
  aulacocarpa, 51, 92
  auriculiformis, 92, 105, 107
  baileyana, 92, 98
  bakeri, 88
  beckleri, 92
  bidwillii, 92, 97, 107, 108, 110
  binervata, 92
  brachystachya, 92
  buxifolia, 92
  calamifolia, 92
  cardiophylla, 92
  collinsii, 116
  complanata, 92
  conferta, 92
  confusa, 64, 65
  cornigera, 116
  cowleana, 89, 92
  craspedocarpa, 92, 95, 107
  cultriformis, 92
  cyclops, 89, 90, 91, 93, 97, 98, 102, 104,
      105, 107, 113, 115, 142, 150, 245
  cyperophylla, 93, 105
  dawsonii, 93
  dealbata, 93, 98
  deanei, subsp. *paucijuga*, 93
  decora, 93
  decurrens, 93, 105
  doratoxylon, 93
  drummondii, 93
  elata, 89, 93, 105
  excelsa, 93
  extensa, 93, 98
  falcata, 89, 93
  falciformis, 64, 93
  farnesiana, 64, 93, 98, 101, 108, 110, 116
  filicifolia, 93
  fimbriata, 93
  floribunda, 93
  hakeoides, 93
  havlandii, 93, 97
  hemiteles, 93
  heterophylla, 64, 65
  implexa, 93, 102
  insulae-iacobi, 66
  iteaphylla, 93
  ixiophylla, 93
  juncifolia, 93
  kauaiensis, 64, 65
  koa, 64, 65
  koaia, 64, 65
  kybeanensis, 93
  lanigera, 93
  lasiocalyx, 93
  latericola, 93
  laurifolia, 64
  leprosa, 93
  ligulata, 89, 94, 99, 105, 106, 107, 113
  linifolia, 94, 99
  longifolia, 94, 102
  macracantha, 66
  maidenii, 94, 99, 105
  mangium, 64, 65, 94, 105
  mathuataensis, 64
  mayana, 116
  mearnsii, 94, 99
  melanoxylon, 64, 89, 91, 94, 99, 105, 107,
      115, 263
  microcarpa, 94
  montana, 94
  mucronata, var. *mucronata*, 89, 94
  myrtifolia, 89, 91, 94, 99, 105, 106, 111
  neriifolia, 94
  nilotica, 116
  obliquinervia, 94

307

obtusata, 94
oxycedrus, 94, 102
pendula, 94
penninervis, 94, 105
podalyriaefolia, 64, 94
polybotrya, 94
pravissima, 94
prominens, 94, 105
pulchella, 94, 99
pycnantha, 91, 95, 99, 105, 106
quornensis, 95
richei, 64
rorudiana, 66
salicina, 51, 89, 95, 100, 102, 105, 107
saligna, 95, 100
schinoides, 95
semirigida, 95
siculiformis, 95, 107
simplicifolia, 64
sophorae, 95, 100, 102, 105
spectabilis, 95
sphaerocephala, 116
spirorbis, 64
stenophylla, 51
stricta, 95, 105
suaveolens, 95, 100, 105
terminalis, 89, 90, 95, 100, 104, 105, 110
tetragonophylla, 89, 95, 100, 105, 107,
    113, 115
tricoptera, 110
ulicifolia, 95, 100, 107, 110, 256
uncinata, 95
venulosa, 95
verniciflua, 95
vestita, 95
victoriae, 110
viscidula, 95
willardiana, 64
Acanthagenys rufogularis, 99, 100
Acanthiza uropygialis, 106
Acanthus ilicifolius, 73
Acer
    pseudoplatanus, 28
    rubrum, 211
Aceros, 129
Adenostoma fasiculatum, 220
Adrisa, 113
Aegialitis annulata, 73
Aegiceras corniculatum, 73, 74

Aepyceros, 135, 153
Aesculus hippocastanum, 214, 219
Aethotesta, 291
Aglaia, 166
agouti, *see Agouti paca, Dasyprocta
    punctata*
Agouti paca, 166
Ailurus, 134
Alectryon, 174
Alouatta, 133, 145, 174
    palliata, 145, 157
Amazona, 141
    autumnalis, 128
Anacardiaceae, 257
Andira, 153
Annona, 173
Annonaceae, 173
Anthochaera carunculata, 98
Aotus, 133
Aphaenogaster longiceps, 99, 100
Apocynaceae, 125, 137, 157
Apodemus sylvaticus, 207, 225
Apodytes, 157
aracaris, *see Pteroglossus torquatus*
Araliaceae, 112, 148, 159, 174
Araucaria mirabilis, 287, 295
Arctitis, 132
Arctogalidia, 134
Ariteus, 132
Artamus cinereus, 99
Artibeus, 132, 154, 174
    jamaicensis, 143, 144, 149
    lituratus, 143
Asclepiadaceae, 276
Asclepias, 276
Astrocaryum, 146
Ateles, 132, 145
    geoffroyi, 145, 154
Atriplex, 51
    inflata, 55
    semibaccata, 112
    spongiosa, 55
Aulacorhynchus, 129
Avena, 205
    fatua, 200, 224
Avicennia,
    germinans, 74
    marina, 72, 73, 74
Azadirachta indica, 144, 146

B

baboon, *see Papio*
Baionella taxisperma, 154
balsam fir, *see Abies balsamea*
Banksia, 249, 250, 263
    ericifolia, 250, 251, 256
    oblongifolia, 250, 251, 256
    paludosa, 248, 249
    serrata, 263
bank vole, *see Clethrionomys gapperi*
Bardocarpus, 284
Barringtonia, 276, 277
    racemosa, 73
Baryphthengus martii, 171, 180
Bassaricyon, 132
bat, *see Amazona, Artibeus, Carollia,*
    Glossophaga, Uroderma, Vampyressa,
    Vampyrodes, Vampyrops
Beania, 287
Beilschmiedia, 142
Bertholletia excelsa, 51
Bettongia penicillata, 257
Betula, 244
Bidens, 297
Bignoniaceae, 276
birch, *see Betula*
bird of paradise, *see Paradisaea*
Birgus latro, 79
Bison, 135
bitterbrush, *see Purschia tridentata*
blackcap, *see Sylvia atricapilla*
black oak, *see Quercus velutina*
black walnut, *see Juglans nigra*
Blandfordia nobilis, 248
Bobea, 174
Bombacaceae, 168
Bombycilla, 130
Bos, 135
    taurus, 53, 98, 110, 116, 146, 210, 215
Bossiaea
    foliosa, 112
    heterophylla, 256
Brachyteles, 132
Brazil nut, *see Bertholletia excelsa*
Bromus
    diandrus, 224
    tectorum, 214
Brosimum, 153
Brotogeris, 128

Bruguierra,
    cylindrica, 73
    exaristata, 73
    gymnorhiza, 73, 74
    parviflora, 73
    sexangula, 73
Bryonia, 159
Bucerotidae, 129
Burejospermum, 286
bur oak, *see Quercus macrocarpa*
burrow weed, *see Haplopappus tenuisectus*
Bursera graveolens, 165
Burseraceae, 141, 145, 165
Bycanistes, 129

C

cabbage palm, *see Livistona australis*
Cactaceae, 170
Caesalpinia bonduc, 62, 63
Caesalpiniaceae, 62, 63, 64
Calla palustris, 55
Callistophytaceae, 284, 290
Callithrix, 133
Callitris glauca, 254
Calluna, 252
Caluromys, 133
Calvaria major, 142, 299
Calystegia soldanella, 80
Calytrix tetragona, 256
camel, *see Camelus*
Camelus, 135
Camponotus, 99
Camptostemon schultzii, 73
Canavalia maritima, 63, 80
Canis, 134
    latrans, 215
Capra hircus, 98, 110, 116
Caprifoliaceae, 159
carao, *see Cassia grandis*
Cardiocarpus, 290
    spinatus, 290
Cardiolepidaceae, 285
Cardiolepis, 285
Carex pendula, 28
Carollia, 132, 154, 174
    perspicillata, 144, 161
Carpodectes, 130
Carpolithes conicus, 286

Carya, 297
    glabra, 212, 213, 214, 219
    ovata, 202, 212
Casearia corymbosa, 148, 156, 161, 180
Caseidae, 292
Cassia grandis, 146
cassowary, *see Casuarius casuarius*
Castanospermum australe, 63
Casuarius, 129
    casuarius, 142
Catharus, 130
cattle, *see Bos taurus*
Caytonia, 286, 287, 295, 298, 299
Ceanothus greggii, 220
Cebuella, 133
Cebus, 133, 145
    capucinus, 145
Cecropia, 168
Celastraceae, 159
Centaurea nigra, 36
Centranthus ruber, 36
Cephalophus, 135
Ceratogymna, 129
Cercatetus, 133
Circopithecus, 133
Cereus giganteus, 170
Ceriops
    decandra, 73
    tagal, 73, 74
Chamaenerion, see Epilobium
chamise, *see Adenostoma fasiculatum*
cheatgrass, *see Bromus tectorum*
Chelaner, 99
Chenopodiaceae, 112
cherry, *see Prunus*
chestnut oak, *see Quercus prinus*
chestnut-rumped thornbill, *see Acanthiza uropygialis*
chicken, *see Gallus gallus*
Chimpansee, 133
chimpanzee, *see Chimpansee, Pan*
Chisocheton, 148
Chlorophora, 144
Chrysocyon, 134
Cirsium, 246
Cissa, 131
Clark's nutcracker, *see Nucifraga columbiana*
Clethrionomys
    gapperi, 202, 207, 210, 213, 225
    glareolus, 207

Clusia rosea, 175
Clusiaceae, 175
coco-de-mer, *see Lodoicea seychellarum*
coconut, *see Cocos nucifera*
Cocos, 276, 277
    nucifera, 60, 61, 79
Colobus, 145
Compositae, 36, 38, 75, 244, 246, 257, 261, 276
Conotrachelus, 145, 165, 166, 171
Conyza
    angustifolia, 75
    indica, 75
Coprosma hirtella, 112
Cordaicarpon, 283
Cordaites, 283
Corvus, 131
    bennetti, 99
    coronoides, 99
Cotinga, 130
coyote, *see Canis latrans*
Cracticus torquatus, 99
Crax, 129
creosote bush, *see Larrea tridentata*
Crossarchus, 134
Croton hircinus, 175
Cryptocarya, 174
cuckoo, *see Eudynamis scolopacea*
Cucurbitaceae, 159
cycad, *see Beania, Carpolithes, Cycas, Macrozamia, Nilsonnia, Zamia*
Cycas
    circinalis, 55–60, 75, 80
    revoluta, 55, 56, 57
    rumphii, 55–57, 60
    thouarsii, 55–57, 60
Cynometra iripa, 73
Cynopterus, 132
Cypripedium parviflorum, 55
Czekanowskia, 287

D

Dasyprocta punctata, 166, 210, 219
Datura meteloides, 200, 202
Daviesia mimosoides, 112
deer, *see Odocoilus*
Dendrocitta, 131
Dendrolagus, 133
Dendrophthoe, 169
Desmodium, 297
    umbellatum, 63

Dicaeum, 130
Diceros, 134
Dichroidium, 286
Didelphis, 174
Dilleniaceae, 112
Dillwynia floribunda, 112
dinosaur, *see* Subject Index
Dioclea reflexa, 63
Dioscoreaceae, 159
Diospyros
    ferrea, 73
    mespiliformis, 146
Diphyllodes, 130
Dipodomys, 207, 218
    heermanni, 209, 223, 224
    ingens, 208
    merriami, 203, 208, 210, 215, 218, 221
    panamintinus, 208
    spectabilis, 208
    venustus, 208, 209
Dipterocarpaceae, 168
Dipteryx panamensis, 143, 153, 165
Distoechurus, 133
Dobsonia, 132
dodo, *see Raphus cucullatus*
Dolichandrone spathacea, 73
Dolichoderus, 99
Dorcopsulus, 133
Douglas fir, *see Pseudotsuga menziesii*
Douglas squirrel, *see Tamiasciurus douglasi*
Dromaius, 130
    novae-hollandiae, 142
Ducula, 129
Dumatella, 131

E

Eastern chipmunk *see Tamias striatus*
Ebenaceae, 146
Edaphosauridae, 292
Edaphosaurus, 292
Edwardsia
    chilensis, see Sophora macrocarpa
    macnabiana, see Sophora microphylla
    microphylla, see Sophora microphylla
Eira, 134
Elaenia, 131
Elaeocarpaceae, 174, 176
Elaeocarpus, 174
    reticulatus, 112
elephant, *see Elephas, Loxodonta*
Elephas, 134

emu, *see Dromaius novae-hollandiae*
Enchylaena tomentosa, 112
Entada gigas, 63
Enterolobium cyclocarpum, 146, 200
Epilobium, 246
    angustifolium, 36, 244
Epimachus, 130
Epomophorus gambianus, 132, 144
Epomops, 132
Equus, 134
    cabalus, 53, 98, 110, 116, 139, 146
Erator, 130
Erithacus, 130
Erodium, 209
    botrys, 224
    cicutarium, 203, 209
Erythrina
    indica, see Erythrina variegata
    sandwicensis, 77
    variegata, 63, 77
Eucalyptus, 239, 240, 243, 244, 245, 252, 261
Eudynamis scolopacea, 148
Eugenia, 173
Euonymus, 159
Euphonia gouldi, 147, 149, 159
Euphorbiaceae, 146, 175
Eutamias ruficaudus, 207
Excoecaria agallocha, 73
Exocarpos cupressiformis, 112

F

Fagaceae, 297
Ficus, 51, 143, 144, 149, 153, 154, 180
    insipida, 143, 176
    yoponensis, 143
fig, *see Ficus*
filaree, *see Erodium cicutarium*
Flacourtiaceae, 148, 180
flowerpecker, *see Dicaeum*
flycatcher, *see Megarhynchus, Myiozetetes, Myiarchus, Pitangus, Tityra*
Fraxinus pennsylvanica, 45
fox squirrel, *see Sciurus niger*

G

Gallus gallus, 81
Gangamopteris, 53
Gastrolobium bilobum, 258
Gazella, 116, 153
Geranium bicknellii, 23
giant kangaroo rat, *see Dipodomys ingens*

Ginkgo, 288, 295, 297
    biloba, 282, 286, 293
Glossophaga, 133, 154, 174
    soricina, 144
Glossopteris, 53, 285, 294, 299
Gnetopsis
    elliptica, 284, 290
    hispida, 289
goat, *see Capra hircus*
gomphothere, 173
Gorilla, 133·
Gossypium
    barbadense var. *darwinii*, 78
    davidsonii, 77
    klotschianum, 77
    sandvicense, 76
gray squirrel, *see Sciurus carolinensis*
grey currawong, *see Strepera versicolor*
ground parrot, *see Pezoporus wallicus*
ground squirrel, *see Spermophilus beechiyi*
guan, *see Penelope purpurascens*
guanacaste, *see Enterolobium cyclocarpum*
Guapira fragrans, 175
Gustavia superba, 145, 154, 168
Gutierrezia sarothrae, 244
Gymnoderus, 130
Gymnorhina
    dorsalis, 98
    tibicen, 99

H

hadrosaur, 294
Hakea, 254
Haloragis prostrata, 81
Haplopappus tenuisectus, 244
Hawaiian cotton, *see Gossypium sandvicense*
Hedera, 159
Helianthus annuus, 202
Helogale, 134
hemlock, *see Tsuga canadensis*
Heritiera littoralis, 73
Heteromys desmarestianus, 207
Hibbertia exutiaces, 112
Hibiscus tiliaceus, 73, 74
hickory, *see Carya ovata*
Hieracium, 244
Holcospermum, 284
Hordeum
    leporinum, 224
hornbill, *see* Bucerotidae

horse, *see Equus cabalus*
horse chestnut, *see Aesculus hippocastanum*
howler monkey, *see Alouatta*
Hura crepitans, 276
Hybanthus prunifolius, 156
Hyemoschus, 135
Hylobates, 133, 145
Hymenanthera dentata, 112
Hypsignathus, 132

I

Icacinaceae, 157
Icterus, 116
Indian rice grass, *see Oryzopsis hymenoides*
Ipomoea pes-caprae, 80
Iriartea, 146
Iridomyrmex, 98, 100
    purpureus, 99

J

Jessenia, 146
jojoba, *see Simmondsia chinensis*
Juglandaceae, 297
Juglans, 297
    nigra, 201, 202, 211
juniper, *see Juniperus communis*
Juniperus, 149
    communis, 170
    virginiana, 170

K

kangaroo rat, *see Dipodomys*
Kennedia prostrata, 112

L

Lagothrix, 132
Lambertia formosa, 247
Larrea tridentata, 204, 220
Lauraceae, 142, 174
Lecythidaceae, 145, 153, 154, 168
Lecythis, 153
Legatus, 131
Lemur, 133
Lepidopteris, 286
Leptonycteris, 132
Leptostrobus, 287
Lepturus repens, 80
Lepus europaeus, 222
Liliaceae, 161

Lindackeria, 157
Lipangus, 130
Liriodendron tulipifera, 44
Livistona australis, 245
lodgepole pine, *see Pinus contorta*
Lodoicea seychellarum, 62
Lolium multiflorum, 224
Lonicera, 159
Lophorina, 130
loulu palm, *see Pritchardia*
Loxodonta, 134, 153
   africanus, 116, 124
Lumnitzera
   littorea, 73
   racemosa, 73
Lycoperdon perlatum, 22, 23
Lyginopteridaceae, 284, 290
Lyrasperma, 283, 290

M
Macaca, 133, 145
Macropus, 133
Macrozamia, 124
   communis, 248
   riedlei, 60, 124, 247
Madoqua, 153
Malvaceae, 257
Manacus, 130, 174
   candei, 147
manakin, *see Manacus, Pipra*
Manorina flavigula, 99
Manucodia, 130
Mayriella, 109
Mazama, 135
Medullosaceae, 284, 291, 294
Megarhynchus, 131
   pitangua, 141
Melanerpes, 131
Melastomataceae, 147, 157, 175
Meliaceae, 144, 146, 148, 157
Meliphaga
   chrysops, 99, 100
   virescens, 98, 100
Melophorus, 98, 101, 109
Menyanthes trifoliata, 55
Merriam's kangaroo rat, *see Dipodomys merriami*
Miconia laevigata, 147, 157, 175
Microcebus, 133

Microtus
   californicus, 200, 224
   pennsylvanicus, 202
millet, *see Panicum miliaceum*
Mimosaceae, 62, 63, 146
Mimus, 131
Mionectes, 131
Mirabilis hirsuta, 127, 164
mistle thrush, *see Turdus viscivorus*
mistletoe, *see Dendrophthoe, Viscum*
Mitella, 22, 23
Mitrospermum, 283, 290, 291
monkey, *see Alouatta, Ateles, Cebus, Colobus, Hylobates, Macaca, Pongo, Presbytis*
Monomorium, 109
Monophyllus, 133
Moraceae, 51, 143, 144, 150, 153, 168, 176
Morus nigra, 150
motmot, *see Baryphthengus martii*
mourning dove, *see Zenaida macroura*
Mucuna sloanei, 63
Muntiacus, 135
Muntingia calabura, 128, 144, 154, 156, 176, 180
Murray pine, *see Callitris glauca*
Mus musculus, 200, 224
Myadestes townsendi, 149
Myiarchus nuttingi, 141
Myiozetetes, 131
   granadensis, 141
   similis, 141
Myriodynastes, 131
Myristicaceae, 127, 149, 165, 170, 176
Myrsinaceae, 175
Myrtaceae, 173

N
Nandinia, 132
Nasua, 134
Nauclea latifolia, 146
Nilssonia incisoserrata, 286
Nitidulidae, 166
Nitraria billardieri, 142
Noogoora burr, *see Xanthium occidentale*
Nucellangium, 283, 290, 291
Nucicarpum, 285
Nucifraga columbiana, 207, 211, 228
Nuttall's violet, *see Viola nuttalli*
Nyctaginaceae, 127, 175

Nycticebus, 133
Nypa fruticans, 73, 74

O
Ochroma, 168
Odocoileus, 215
   virginiana, 166
oilbirds, *see Steatornis*
opossum, *see Didelphis*
orchid, 246, 247, 276
   see also Cypripedium parviflorum
oriole, *see Icterus, Oriolus*
Oriolus, 130
Oryzopsis hymenoides, 202, 208, 214, 218, 225
Osbornia octodonta,
Ovis aries, 53

P
Pachytesta, 284, 290
Paguma, 134
Pan, 133
Panicum miliaceum, 200
Papaver, 276
Papilionaceae, 62, 63, 112, 143, 153, 165, 169, 258
Papio, 133
   anubis, 145
Paradisaea, 130
   rudolphi, 148
Paradoxurus, 134
Pararaucaria patagonica, 287
Parkinsonia aculeata, 63
Parotia, 130
Parus, 159
Passer domesticus, 229, 230
peccary, *see Tayassu tajacu*
Peltaspermaceae, 285
Pemphis acidula, 73, 74
Penelope, 120
   purpurascens, 171
Pentoxylon, 54, 287
Perodicticus, 133
Perognathus, 207, 208
   amplus, 207
   baileyi, 202
   formosus, 208
   longimembris, 208

Peromyscus, 203
   leucopus, 202, 204, 207, 210, 213
   maniculatus, 207
Petasites hybridus, 261
Pezoporus wallicus, 241
Phainopepla nitens, 141, 142
Phalanger, 133
Pharomacrus, 129
   mocinno, 142
Pheidole, 98, 109
Phillipia evansii, 244
Phloeoceastes, 131
Phyllostomus, 133
Picea glauca, 210, 244
pig, *see Sus scrofa*
pigeon, *see Reinwardtoena reinwardsti*
pignut hickory, *see Carya glabra*
pin cherry, *see Prunus pennsylvanicus*
Pinaceae, 287, 297
Pinus, 263, 276
   albicaulis, 207, 208
   banksiana, 263
   contorta, 201, 227
   edulis, 207, 211
   palustris, 248
   ponderosa, 214
   rigida, 263
   strobus, 210, 213
pinyon pine, *see Pinus edulis*
Piper, 144, 153
   amalago, 161
Piperaceae, 144, 153
Pipra, 130, 174
   mentalis, 147
Pitangus sulphuratus, 141
Pittosporum undulatum, 245
Platypodium elegans, 165, 169
Poaceae, 77, 246, 257
pocket gopher, *see Thomomys bottae*
pocket mice, *see Heteromys, Perognathus*
Podocarpaceae, 288, 297
Polyscias sambucifolia, 112
Pongo, 133, 145
Potos, 132, 152
Presbytis, 145
Pritchardia, 62, 76
   lowreyana, 62
   martii, 62
   remota, 61, 62

Problematospermum ovale, 286
Procnias, 130
Procyon, 134
Prosopis juliflora, 63, 210, 215, 218, 220
Protea, 250
    repens, 243
Proteaceae, 243, 248, 249, 255, 257
Protium, 141
Prunus, 159
    mahaleb, 159, 163
    pensylvanica, 168
    serotina, 151
Pseudocheirus, 133
Pseudotsuga menziesii, 207
Pterissocephalus, 130
Pteroglossus, 128, 129
    torquatus, 180
Pteropus, 132
Ptilionopus, 129
Ptilophyllum, 287
Pulsatilla vulgaris, 36
Pultenaea daphnoides, 112
Purschia tridentata, 212, 214

Q
Quercus, 201, 202, 214, 219, 220
    agrifolia, 219
    alba, 201, 212, 213, 218, 219
    douglasii, 219
    lobata, 219
    macrocarpa, 201
    petraea, 205, 207, 219, 225
    prinus, 219
    rubra, 204, 205, 218
    shumardii, 201
    velutina, 210, 219
Querula, 130
quetzal, *see Pharomacrus mocinno*

R
rabbit, *see Lepus, Sylvilagus*
Ramphastos, 128, 129, 173, 180
    sulfuratus, 141, 156, 171
    swainsonii, 141, 148, 156, 171
Ranunculaceae, 36
Rapanea guianensis, 175
Raphus cucullatus, 142, 299
red backed vole, *see Clethrionomys gapperi*
red maple, *see Acer rubrum*

red oak, *see Quercus rubra*
red squirrel, *see Tamiasciurus hudsonicus*
Reinwardtoena reinwardsti, 148
red tailed chipmunk, *see Eutamias*
    ruficaudus
Rhinoceros, 116, 134
Rhizophora
    apiculata, 73
    X lamarckii, 73
    mangle, 72, 74
    mucronata, 73
    stylosa, 73
Rhytidoponera, 98, 100, 109, 110
    inornata, 99
    metallica, 99, 100, 110
    violacea, 98
Rhytipterna, 130
Rosa, 159
Rosaceae, 151, 159, 168, 174
Rubiaceae, 112, 146, 159, 161, 173, 174
Rubus, 159, 174
    ulmifolius, 161
Rupicola, 130
Rutaceae, 157, 257

S
sacred datura, *see Datura meteloides*
saguaro, *see Cereus giganteus*
Saguinus, 133
Salix
    bebbiana, 244
    scouleriana, 244
Salsola paulsenii, 202
Saltator, 142
saltbush, *see Atriplex*
Sambucus, 159
Santalaceae, 112
Sapindaceae, 174
Sapotaceae, 142, 154
Scheelea, 153
    rostrata, 229
Schefflera
    chaetorrhachis, 148
    pachystyla, 148
Sciurus, 219
    carolinensis, 201, 202, 208, 212, 213, 214, 218
    niger, 201, 202, 208, 210, 211, 212
Scyphiphora hydrophyllacea, 73

Securingia virosa, 146
Selaginella, 282
Senecio, 75, 246
sequoia, *see Sequoiadendron giganteum*
Sequoiadendron giganteum, 209, 219
Sesarma, 79
sessile oak, *see Quercus petraea*
sheep, *see Ovis aries*
Shorea
    leprosula, 168
    maxweliana, 168
shumard oak, *see Quercus shumardii*
Silene vulgaris, 28
silky flycatcher, *see Phainopepla nitens*
Simmondsia chinensis, 202
Smilax aspera, 161
snakeweed, *see Gutierrezia sarothrae*
solitaire, *see Myadestes townsendi*
Sonchus palustris, 36
Sonneratia
    alba, 73, 74
    caseolaris, 73
    X gulngai, 73
    lanceolata, 73
Sophora, 66–72
    howinsula, 67, 69
    macrocarpa, 67, 69–72
    masafuerana, 67, 69
    microphylla, 66–72, 78
    prostrata, 66, 67
    tetraptera, 66, 67
    tomentosa, 63, 66
    toromiro, 67, 69
Sowerbaea juncea, 242
sparrow, *see Passer domesticus*
Spermolithus, 283, 290
Spermophilus beechiyi, 223, 224
Sphaerobolus stellatus, 23
Sphenobaiera, 286
spider monkey, *see Ateles*
spiny pocket mice, *see Heteromys*
    desmarestianus
Steatornis, 129
    caripensis, 128, 141, 166, 174
Stemmadenia donnell-smithiii, 125, 137, 157
Stirlingia latifolia, 255
Strepera versicolor, 98, 113
Stylosanthes sympodialis, 63
subalpine fir, *see Abies lasiocarpa*
sunflower, *see Helianthus annuus*
Sus scrofa, 81

swamp wallaby, *see Wallabia bicolor*
Sylvella, 284
Sylvia, 130
    atricapilla, 158, 159, 163
Sylvilagus, 222

T
Tamias striatus, 202, 209, 210
Tamiasciurus
    douglasi, 208, 209, 219
    hudsonicus, 199, 201, 207, 208, 209, 210,
        227, 228
Tamus, 159
tanager, *see Euphonia gouldi*
Tangara, 130
tapir, *see Tapirus*
Tapirus, 134
    bairdii, 146, 166
Tayassu, 128, 146, 174
    tajacu, 166, 215
Tetragastris panamensis, 125, 138, 145, 151,
    156, 157, 160, 161, 168, 180
Tetraplasandra, 174
Tetratheca pilosa, 112
Thomomys bottae, 223
Tijuca, 130
tits, *see Parus*
Tityra semifasciata, 130, 141, 180
thrush, *see Turdus*
Tockus, 131
toucan, *see Ramphastos*
Trachodon, 287
Tragopogon porrifolius, 36
Tragulus, 135
Trema, 157, 168
Tremandraceae, 112
Trichilia tuberculata, 157
Trichoglossus chlorolepidotus, 99
Trichosurus, 133
Trigonocarpus, 284, 290, 291
Trogon, 129
    massena, 171, 180
Tsuga
    canadensis, 209
    heterophylla, 205
Turdus, 130, 141, 159, 162, 170
    merula, 159
    viscivorus, 149
Tussilago farfara, 28, 261
Tyranniscus, 131

U

Ulmaceae, 157, 168
Umkomasia, 286, 299
Urera, 147
Urocyon, 134
Uroderma, 174
Ursus, 134
Urticaceae, 147

V

Valerianaceae, 36
Vampyressa, 174
Vampyrodes, 174
Vampyrops, 174
velvet mesquite, *see Prosopis juliflora*
Vepris, 157
Vermivora, 131
Viola
  hirta, 169
  nuttallii, 228, 229
  odorata, 169
Violaceae, 112
Vireo, 130, 131
  flavoviridis, 141, 180
  olivaceous, 141
Virola
  sebifera, 149, 156, 157, 176, 180
  surinamensis, 125, 127, 138, 144, 145,
    154, 156, 157, 160, 161, 162, 165, 166,
    169, 170, 171, 176
Viscum, 141, 169
Viverra, 134
Viverricula, 134
Vulpes, 134

W

Wallabia bicolor, 241, 242
waxwing, *see* Bombycillidae
western hemlock, *see Tsuga heterophylla*
whitebark pine, *see Pinus albicaulis*
white footed mice, *see Peromyscus leucopus*
white oak, *see Quercus alba*
white pine, *see Pinus strobus*
white spruce, *see Picea glauca*
wild oat, *see Avena fatua*
wiliwili, *see Erythrina sandwicensis*
Williamsonia, 287
woodmice, *see Apodemus sylvaticus*
woylie, *see Bettongia penicillata*

X

Xanthium
  occidentale, 52, 53, 54
  strumarium, 52
Xanthorrhoea
  australis, 247
  resinosa, 241, 248
Xipholena, 130
Xylocarpus
  australasicus, 73
  granatum, 73

Z

Zamia pumila, 247
Zarhynchus, 131
Zenaida macroura, 241
Zosterops, 131
  lateralis, 98
Zygophyllaceae, 142

A

adaptation
  definition, 126
aerodynamic
  classification of diaspores, 2–13, 24–46
  drag, 4–8, 10, 11, 12, 14, 25, 26, 27, 28,
    31, 36, 38, 39
  lift, 4–8, 10, 11, 14, 25, 27, 33, 39, 42, 45
Africa, 144, 145, 146, 153, 157, 239, 249, 250,
  255, 260
air cavities
  in seeds and fruits, 52, 53, 55, 56, 57, 59,
    254, 291
air currents, 65, 238
albedo, 239, 240
Aldabra, 61, 78
allelopathy, 239
Amazon River basin, 51, 276, 292
Andean flora, 77, 78
Angara
  floras, 283, 284, 292
  seeds, 284
angiosperms
  early seed characters, 278, 279, 280, 296
  see also evolution, of angiosperms
ants, see seed dispersal, biotic, by ants
ants' nests
  as seed germination sites, 113, 169, 257
aril, 126, 127, 142, 204
  colour, 89, 91, 92–97, 101, 102, 125
  functional definition, 88
  nutrient content, 90, 91–97, 103–105, 125,
    138, 156, 160
  reward ratio, 90, 92–95, 102, 104, 161
  shape, 89, 92–97, 102
  size, 89, 92–97, 102
Australia, 52, 53, 63, 64, 65, 66, 142, 240,
  241, 244, 247, 248, 249, 252, 255, 261
  climate changes, 115
autorotation, 7, 42

B

birds
  behaviour of, 101, 128, 138–143, 146–149,
    228, 241
  flightless, 129, 130, 142, 299
  frugivory of, 128, 138–143, 146–152, 158,
    159, 172, 242, 245
  migration, 115, 158, 159, 173
  Tertiary radiation, 115, 296, 299
  vision
    bicoloured displays, 97, 101, 102, 107,
      125, 150, 151
    spectrum, 101, 151
  see also seed dispersal, by birds
Brazil, 51, 62, 175
British Isles, 50, 62, 78, 79
buoyant density
  of seeds in water, 53, 62, 66, 67

C

caches, 205
  types, 206–210
  recovery of, 212–214
  seedling establishment from, 218–220
canopy, 250, 252
  light-gaps, 156, 164, 168, 169, 171, 175,
    176, 219, 220, 240
  seed retention times, 91, 106, 107, 248
Carboniferous, 123, 282, 300
  see also Mississippian, Pennsylvanian
chaparral, 252, 255, 260
Chatham Island, 67
Chile, 67–72, 77
coal swamp floras, 283, 284, 291, 292
co-evolution, 51, 172, 295
  diffuse, 110, 179–181, 274
colonization hypothesis, 167–169, 175
competition, 151, 173, 206, 240
  see also seedling, competition

conifers, *see* evolution, of conifers
continental drift, *see* plate tectonics
coprolites, 281, 287, 291, 294, 295
cordaites, 283, 284, 285, 290, 293
Costa Rica, 143, 147, 148, 157, 158, 161, 162, 168, 176, 229
Cretaceous
    angiosperms, 123, 275, 279, 280, 287, 288, 295, 296, 300
    gymnosperms, 123, 279, 280, 287, 296
    herbivores, 123, 124, 288, 292
Cretaceous-Tertiary boundary, 288, 297
cupules, 283, 284, 286, 289, 290
cycadeoids, 286, 287
cycads, 55–60, 75, 80, 123, 124, 243, 247, 286, 288, 296

D
Darwin, Charles, 76, 77, 78
dental morphology, 51, 191, 281, 292
Devonian
    plant diversity, 279, 282, 284, 300
    plant evolution, 282, 283, 289
diaspores
    definition, 88, 275
    multi-seeded, 229, 277, 295
dinosaurs, herbivory of, 287, 288
    see also seed dispersal, by dinosaurs
dioecious species
    on islands, 80
directed dispersal, 113, 116, 169–172, 228, 243, 257, 274
disturbance, of vegetation, 81, 168, 175–177, 239, 240, 259, 260, 262
dormancy, of seeds, 63, 67, 156, 167, 168, 218, 246, 248, 257, 260, 261, 262, 264, 289, 293, 294

E
Easter Island, 69
echolocation, 144
elaiosome, *see* aril
Eocene, 191
ephemerals
    desert, 51
    fire, 244, 245, 246, 247, 255
equations of motion, 13–17, 19, 24–31, 37
escape, of seeds, 124, 163–167, 168, 169, 171, 229, 274
eucalypt forests, 239, 240, 243, 244, 245, 252, 261

Europe, 150, 159
evolution
    of *Acacia*, 105, 113–118
    of angiosperms, 69, 76, 81, 123, 155, 287, 288, 295–298
    of conifers, 284, 285, 286, 287, 288, 291, 294, 295, 297
    of cycads, 56, 60, 123, 284, 286, 288
    of gymnosperms, 287, 288, 297
    of Juglandaceae, 297
    of mammals, 281
    of reptiles, 292, 295
    of rodents, 191
    see also co-evolution
exaptations, 289
extinctions
    of herbivores, 142, 153, 158, 275, 299
    of insects, 294
    of plant species, 275, 287, 288, 293, 294, 296, 297, 298, 299

F
faeces, seeds from, 110, 124, 142, 143, 144, 145, 146, 153, 169, 170, 171, 200, 215, 242, 281
fate path, for seeds and seedlings, 193–220, 222, 223, 224, 226, 227, 230
fatty acids
    of aril lipids, 102–105
    of fruits, 104, 105
    of seed storage lipids, 103
fire, and seed dispersal, 116, 227, 228, 237–271
fireweeds, 244, 245, 247
fitness, 124, 136, 160, 163, 215–218, 226, 227, 230, 263
flavonoids
    embryo, 69, 72
    leaf, 67
    seedcoat, 67, 69, 72
floating fruits, largest, 62
floating seeds, earliest reports, 50
flowering
    post-fire, 247, 248, 262
    seasonal, 156, 157, 162, 173, 223, 249, 255
foraging theory, 128–149, 159, 192–208
    see also profitability
forests, 65, 124, 126, 155, 157, 158, 161, 172, 219, 288
    see also eucalypt forests, rainforests
fossil record, 273–305

frugivore, 129–135
    definition, 127
    obligate, 128, 141, 142, 158
    opportunistic, 128, 141, 158, 159
    reproduction, 174
frugivory, evolution of, 153
fruit, 126, 127
    adaptations
        bat, 143, 152, 153, 154
        bird, 106, 150–152, 154
        mammal, 152, 153, 154
        megafaunal, 153, 158
    colour, 150, 151, 152, 153, 154
    development, 155, 158
    energy value, 138, 155
    nutrient content, 138, 141, 151, 152, 153, 156, 157, 158, 159, 160
    odour, 152, 160
    profitability, 90, 137–140
funicle, swollen, *see* aril
fynbos, South African, 243, 250, 255, 260

G
Galapagos Islands
    age, 78
    flora, 62, 66, 74, 77, 78, 165
Gondwanaland, 53, 54
    dislocation, 54
    floras, 53, 54, 283, 285, 294
gymnosperms
    evolution of, 56, 57, 60, 287, 288, 294, 297
    seed characters, 55–61, 227, 228, 279, 282, 283, 285

H
habitat islands, 50, 174, 175
hadrosaurs, herbivory of, 294
Hawaiian Islands
    age, 78
    flora, 61, 64, 76, 77, 78
helicopter rotors, 43
herbivory
    of birds, 123–182, 299
    of insects, 145, 165, 166, 171
    of mammals, 123–182, 299
    of reptiles, 294, 298
Hooker, J. D., 77

I
insects
    in avian diets, 109, 140, 141, 142, 241
    pollination by, 181, 295
    seed consumption by, 229, 291
    see also ants
integuments, 55
    evolution, 56, 282, 283, 289
    impermeability, 55, 56, 62, 67, 257, 283
    nutrients of, 60
    pigments of, 60
    texture, 56, 108, 229
    thickness, 55, 62, 63, 66, 67, 68, 108, 257, 283
    see also sarcotesta, sclerotesta
island
    biogeography, 49, 50, 80, 81, 124, 174, 181, 182
    emergence, 54
    submergence, 54
    vegetation, 53–81, 174, 176
    see also habitat islands

J
Jurassic, floras, 279, 286
    see also Yorkshire (Jurassic) flora

K
Kermadec Islands, 67
Krakatau
    eruption, 50, 74
    expeditions, 75
    revegetation, 50, 61, 62, 75, 80

L
land-bridges, 54, 173
Lord Howe Island, 62, 69, 74
lycopods, arborescent, 283, 293

M
Madagascar, 54, 57, 61
mangroves
    definition, 72
    dispersal, 72–74
    distribution, 74, 77
    early forms, 290
Mascarene Islands, 64, 65
mast-seeding, 206, 208, 211, 214, 222, 248
Mesozoic, 282, 284
    herbivory, 275, 298, 299
microterrace, 252, 253, 254

Miocene, 115
Mississippian, 279, 282, 283
    seed dispersal, 289, 290, 298
mutualism, 143
mycorrhizae, 219

N

New Caledonia, 54, 64
New Guinea, 62, 148, 151
New Hebrides, 74, 81
New Zealand, 53, 54, 65, 66, 238
non-protein amino acids, 63, 64, 65
nutrients, release by fire, 239, 243

O

ocean currents, 50, 53, 54, 57, 60, 61, 62, 63,
    65, 67, 74–79

P

Paleocene, flora, 279, 280
Panama, 143, 145, 157, 171, 173, 176, 229
pappus, 36, 38, 238, 244
parachutes, 12, 25, 36
Pennsylvanian
    conifers, 284
    dispersal syndromes, 289, 290, 292, 298,
        300
    floras, 279, 280, 283, 284, 288, 292, 293
    see also coal swamps
Permian
    environments, 293
    extinctions, 293, 294
    floras, 279, 280, 284, 285, 292, 294
phyllode, 63, 64, 65
pioneer species, 61, 75, 77, 260
    see also fireweeds
plate tectonics, 53, 54, 115, 293
Pleistocene, 65, 80, 288
pollen grains
    aerial movement, 11, 27–32
    age of fossil grains, 65, 288
    from *Acacia*, 65
pollination, 80, 181
    origin of, 282, 283, 293, 295, 298
population ecology, 126, 162–181, 221–231,
    258–264
porous bundles, 10, 13
predator avoidance, 143, 149, 161
preservational bias, 278, 280, 286, 293
profitability, 90, 136, 137–140, 158, 159, 161,
    162, 227
    definition, 136, 137

Q

Queensland, 52, 65

R

raindrops, 50, 51
rainfall, 51, 115, 156, 223, 229, 230, 252, 262
rainforests, 62, 149, 157, 168, 171, 245, 252,
    261
regeneration, post-fire, 259
Reynolds number, 4, 7, 9, 10, 19, 20, 25–29,
    33, 34, 39
rodents
    adaptation, 191, 220
    caching behaviour, 192, 193, 197–215,
        218, 219, 220, 221, 225
    diets, 200–202, 220
    evolution, 191
    olfaction, 193, 198, 212
    population density, 205, 220, 221, 222
r-strategy, 258, 296

S

samara, 16, 42
sampling, 160, 181, 182, 278
sarcotesta, 55, 56, 57, 283, 284, 286, 290, 291,
    292
sclerotesta, 55, 283, 284, 286, 290, 291, 292
seed anatomy, 52, 55–60
seed coats, *see* integuments
seed consumption, 145, 149, 161, 171, 203,
    206, 220, 223, 229, 274, 287, 291
seed dispersal
    abiotic
        by water, 49–121, 240, 248, 252–255,
            276, 289, 291
        by wind, 19–22, 36, 37, 75, 76, 77,
            106, 110, 164, 168, 238, 244, 245,
            246, 250, 252, 254, 255, 258, 259,
            260, 261, 264, 276, 285, 289, 290,
            291, 295, 297
    biotic
        by ants, 90, 97, 98–100, 104, 106, 116,
            164, 169, 222, 229, 276
        by arboreal primates, 145, 152, 154,
            157
        by bats, 60, 143, 144, 152, 153, 158,
            175, 296
        by birds, 65, 76, 77, 78, 90, 91, 97, 98–
            100, 106, 123–189, 245, 276, 296

by dinosaurs, 124, 292
by fish, 51, 276, 291, 292
by terrestrial mammals, 53, 116, 117,
    123–189, 215, 245, 276, 296
by reptiles, 123, 276, 292, 296, 298
by rodents, 146, 191–231
definition, 88, 127, 264
distance, 36, 37, 38, 49, 50, 124, 163–167,
    210, 211, 238, 247, 250, 259
dynamic release, 19–23, 276
multiple pathways, 53, 54, 180, 228, 229
passive release, 18, 19, 53
syndromes, 88, 89–91, 150–155, 273–300
    ant-, 91, 101, 102, 103, 117
    bird-, 91, 101, 102, 103, 117, 150–152
    megafaunal-, 116, 117, 153, 154
seed fern, 53, 54, 283, 284, 285, 286, 287,
    288, 289, 290, 291, 292, 293
seedling
    clumping 163, 170, 214, 215, 254, 255,
        257, 258
    competition, 153, 164, 165, 177, 217, 223
    consumption, 76, 78, 79, 164, 171, 220,
        241
    emergence, 213, 223, 228, 239, 247, 248,
        249, 252, 254, 257
    establishment, 78, 164, 165, 170, 213, 214,
        215–220
seed longevity, 63, 64, 168, 246, 256, 259
seed nutrients, 69, 71, 90, 92–97, 201, 276,
    277, 291
seed shadows, 113, 117, 163, 164, 176–179,
    230
seed storage
    in plants, 238, 248, 249, 263, 264
    in soil, 116, 175, 193, 205, 222, 238, 246,
        255, 256, 257, 261, 263, 264
Seychelle Islands, 54, 62
soil temperature, 78
speciation, 69
spherical shape, 10, 11, 16, 26
Stokes, 4, 11, 19, 26, 27, 29, 32

succession
    light-gap, 175–177
    post-fire, 116, 244, 260, 264
    seed size, 277
Sumatra, 75

T
Tertiary
    angiosperms, 275, 279, 280, 288, 296, 297,
        298, 300
    herbivore radiation, 191, 296
    see also Paleocene, Eocene, Miocene
trajectory analysis, 10, 13, 24–46
trajectory number, Tn, 29
Triassic
    floras, 279, 280, 285, 286, 295
Tristan da Cunha, 79

V
viscosity, 4, 27
volcanic activity
    origin of islands, 50, 53, 54, 74

W
water run-off, 51, 240, 248, 252–255, 261,
    264
wind
    dispersal range prediction, 39, 46
    speed, 19, 33, 248, 252
        fluctuations, 21, 22, 29, 30, 36, 42, 44,
            46, 106
        relative, 14
winged fruits or seeds
    origins, 284, 285, 286, 287, 288, 289, 290,
        291, 296, 297
    shapes, 6, 7, 10–13, 39, 42
    trajectories, 12, 39–46
    see also samara; seed dispersal, abiotic,
        by wind

Y
Yorkshire (Jurassic) flora, 279, 287

6 7 8 9 0 1 2 3 4 5
A B C D E F G H I J

DATE DUE

DEMCO 38-297